石油和化工行业"十四五"规划教材

新形态教材

英汉对照
分子生物学导论

第三版

An Introduction to Molecular Biology
with Chinese Translation

（美）西尔维恩·W. 勒潘（Sylvain W. Lapan）

王 勇　齐向辉

编著

化学工业出版社

·北京·

内容简介

本教材前两版广受欢迎,被许多高校用作本科和研究生教材,新版入选石油和化工行业"十四五"规划教材。教材内容涵盖分子生物学的中心问题,即 DNA 复制、转录、翻译和重组。从该领域的基本内容(生物大分子的结构与功能)开始,系统、完整地阐述了上述主题,使读者能贯通现代知识体系,从而深入理解这些过程是如何进行调控并创造出我们称之为"生命"的动态系统的。新版新增以新冠病毒核酸检测、亲子鉴定为实例介绍分子生物学科学知识在社会生活中的应用,以及几种基因编辑系统的工作原理及组成元件。

教材使用简洁的语言和详尽的插图来强化双语教学。书中通过嵌入二维码的方式配备了英语课文、单词表音频录音以及彩图,由此进一步强化双语教学目标。二维码内容还包含一些重要的实验技术双语内容,可以帮助学生将课堂所学分子生物学知识与课外学习活动联系起来。课程授课老师还可通过化工教育网(www.cipedu.com.cn)注册下载课件等教学资源。

本教材为分子生物学双语教学而编写,可以用作高校生物类专业本科教材,也适合自学。

图书在版编目(CIP)数据

英汉对照分子生物学导论/(美)西尔维恩・W.勒潘(Sylvain W. Lapan),王勇,齐向辉编著. —3 版. —北京:化学工业出版社,2024.4
石油和化工行业"十四五"规划教材
ISBN 978-7-122-44659-6

Ⅰ.①英… Ⅱ.①西…②王…③齐… Ⅲ.①分子生物学-高等学校-教材-英、汉 Ⅳ.①Q7

中国国家版本馆 CIP 数据核字(2024)第 046398 号

责任编辑:傅四周 赵玉清   装帧设计:王晓宇
责任校对:田睿涵

出版发行:化学工业出版社(北京市东城区青年湖南街 13 号 邮政编码 100011)
印 装:高教社(天津)印务有限公司
787mm×1092mm 1/16 印张 20 字数 558 千字 2024 年 7 月北京第 3 版第 1 次印刷

购书咨询:010-64518888   售后服务:010-64518899
网   址:http://www.cip.com.cn

凡购买本书,如有缺损质量问题,本社销售中心负责调换。

定 价:59.00元   版权所有 违者必究

Sylvain W. Lapan, Ph. D.

Sylvain W. Lapan earned a B. A. degree in Biochemistry from Columbia University in 2005 and obtained a Ph. D. degree in Biology from the Massachusetts Institute of Technology in 2012. From August to December of 2005, he taught as a lecturer in Molecular Biology at the School of Food and Biological Engineering, Jiangsu University, China at the graduate and undergraduate levels. The "courseware for bilingual teaching of Molecular Biology" with his contribution was awarded the second level prize of excellent multimedia for higher education by Jiangsu Province. The "design and fulfillment of bilingual teaching for Molecular Biology" with his participation was awarded the first level prize of teaching achievement for higher education by Jiangsu University. From 2011 to 2019, he worked as a research fellow at Harvard Medical School, USA. The application of developmental genetics, molecular biology, and transcriptomics to the study of eye development and regeneration in diverse animals has been his primary research focus. He has authored 16 papers in periodicals including *Cell*, *PLoS Genetics*, *Nature Neuroscience*, and *Journal of Nanjing University*.

西尔维恩·W. 勒潘，博士

西尔维恩·W. 勒潘，2005年毕业于哥伦比亚大学，获生物化学学士学位；2012年毕业于麻省理工学院，获生物学博士学位。2005年8～12月，任教于中国江苏大学食品与生物工程学院，讲授硕士研究生与本科生"分子生物学"双语课程。参与制作的"分子生物学双语教学课件"获江苏省高等学校优秀多媒体教学课件二等奖，参与实施的"分子生物学双语课程建设与实践"获江苏大学高等教育教学成果一等奖。2011年至2019年在美国哈佛大学医学院任研究员，主要从事不同动物眼器官发育与再生的遗传学、分子生物学与转录组学研究。已在 *Cell*、*PLoS Genetics*、*Nature Neuroscience*、《南京大学学报》等期刊上发表论文16篇。

Yong Wang, Ph.D.

Yong Wang graduated from the Department of Biology, Fudan University, China and acquired B. Sc. degree in 1985. His graduate studies were conducted at the Institute of Life Sciences, Jiangsu University, China, and he earned a Ph.D. degree in 2008. He worked as an assistant researcher at Institute of Sericulture, Chinese Academy of Agricultural Sciences from 1985 to 1997, and as an English translator at Sonoco Hongwen Paper Co. Ltd. (Shanghai) from 1997 to 2001. Since 2001, he has been working at the School of Food and Biological Engineering, Jiangsu University as an associate professor, lecturing bilingual courses of Molecular Biology for graduates and undergraduates and of Bioinformatics for undergraduates. He was entitled "the Advanced Lecturer in Undergraduate Education" and "the Most Welcomed Ten Lecturers" by Jiangsu University. His "courseware for bilingual teaching of Molecular Biology" was awarded the second level prize of excellent multimedia for higher education by Jiangsu Province. His "design and fulfillment of bilingual teaching for Molecular Biology" was awarded the first level prize of teaching achievement for higher education by Jiangsu University. He has conducted and participated in 16 national and provincial research programs, and published 42 papers in scientific progress and teaching improvement.

王勇,博士

王勇,1985年毕业于中国复旦大学生物系,获理学学士学位;2008年毕业于江苏大学生命科学研究院,获工学博士学位。1985—1997年在中国农业科学院蚕业研究所工作,任助理研究员;1997—2001年就职于中美合作上海实宏纸业有限公司,任英语翻译;2001年至今任江苏大学食品与生物工程学院副教授,讲授硕士研究生与本科生"分子生物学"双语课程、本科生"生物信息学"双语课程。获得江苏大学"本科教学先进个人""最受学生欢迎十佳教师"荣誉称号。制作的"分子生物学双语教学课件"获江苏省优秀多媒体教学课件二等奖,"分子生物学双语课程建设与实践"获江苏大学高等教育教学成果一等奖。主持、参与省部级科研项目16个,发表科研与教改论文共42篇。

Xianghui Qi, Professor, Doctoral Supervisor

Xianghui Qi graduated from Central South University of Forestry and Technology entitled with B. Sc. degree in 1998, and from Guangxi University entitled with Ph.D. degree in 2006. Since 2007, he has worked at the School of Food and Biological Engineering, Jiangsu University, acting as Leader of Microbial Biosynthesis and Intelligent Manufacturing team, High-level talent of Jiangsu Province, and "Minjiang Scholar" Distinguished Chair Professor of Fujian Province. He teaches Microbiology, Enzyme Engineering, Synthetic Biology, Genetic Engineering for undergraduate students, and Advances in Biotechnology for graduate students in both Chinese and English. He has won 1 first prize, 2 second prizes, and 1 third prize of Jiangsu provincial-level achievements, and multiple honorary titles such as Jiangsu Excellent Graduate Thesis Advisor, Jiangsu University Excellent Teacher, and Outstanding Postgraduate Tutor. He is mainly working on synthetic biology including both basic and applied research, and has presided over 10 national-level research projects, including the National Key Research and Development Program of China, National Natural Science Foundation of China, and those of international cooperation. He has published over 200 papers in the major academic journals both at home and abroad. He has acted as Editor-in-Chief of 1 English book and co-author of 9 books. He holds over 10 authorized Chinese invention patents and 2 PCTs. He has served as a deputy Editor-in-Chief of many international journals and an evaluator of many key research projects of various countries.

齐向辉，教授、博士生导师

齐向辉，1998年毕业于中南林业科技大学，获农学学士学位；2006年毕业于广西大学，获理学博士学位。2007年至今在江苏大学食品与生物工程学院工作，担任微生物生物合成与智能制造团队负责人，江苏省高层次人才，福建省"闽江学者"特聘讲座教授。讲授本科生"微生物学""酶工程""合成生物学""基因工程"以及硕士研究生"生物技术进展"等中文与英文课程。获江苏省级成果一等奖1项、二等奖2项、三等奖1项，并获江苏省优秀研究生毕业论文指导教师、江苏大学优秀教师、优秀研究生导师等荣誉称号。主要从事合成生物学的基础与应用研究，主持国家重点研发计划、国家自然科学基金、国际合作等国家级课题10余项；在国内外重要学术期刊发表论文200多篇；主编英文专著1部，参编中英文专著9部；拥有10余件授权中国发明专利及2件PCT；担任多个国际期刊的副主编以及多个国家的重大项目评审专家。

Foreword

With the incoming of the third edition of *An Introduction to Molecular Biology with Chinese Translation*, we have to tell everybody with a heavy heart that: we have lost contact with Dr. Lapan since June 2019. After failure of many contacting attempts ever since, we were connected to the corresponding author of his previously published paper. Unfortunately, the answer is also being out of contact. As such, we could only wish everything of Dr. Lapan be alright!

In this new edition, "Molecular Identification" and "Gene Editing" sections are added into chapters 2 and 10 respectively. The former introduces application of molecular biology knowledge in social practice using nucleic acid testing and DNA fingerprinting as examples. The latter briefly outlines development stages of gene editing technology and functioning mechanism of the latest gene editing system. Due to absence of Dr. Lapan in this revision, no audio files are available for these two sections. Please make allowance for this.

Additionally, the following revisions have been made, including: addition of ten figures into chapters 5 to 7 and chapter 9, corrections/revisions to over sixty words/sentences, supplement of glossary into the book, and updating of the accessory teaching courseware. We sincerely hope that this new edition could help everybody better understand and learn knowledge in basic theory and practical application of molecular biology.

Best wishes,

Yong Wang
Xianghui Qi
March 2024

前言

在《英汉对照分子生物学导论》第三版即将出版之际，我们不得不怀着沉重的心情告诉大家：自2019年6月以来我们与勒潘博士失去了联系。在后续多次联系尝试失败后，我们联系到了他先前发表论文的通讯作者，很遗憾得知也是失联。如此，惟愿勒潘博士一切安好！

本次再版分别在第2章、第10章增加了"分子鉴定"与"基因编辑"两节内容。前者以核酸检测、DNA指纹分析为实例介绍分子生物学知识在社会实践中的应用；后者简要概述基因编辑技术的发展阶段及最新基因编辑系统的作用机理。由于勒潘博士缺席本次修订工作，因此这两节内容没有音频文件，希望大家谅解。

此外，还完成了下面几项修订工作，包括：增加了十幅插图到第5~7章及第9章中，修正了六十多处字句，增补了名词解释到书中，更新了配套教学课件。我们诚挚希望本书的再版能帮助大家更好地理解与学习分子生物学基础理论及其实际应用知识。

致以良好祝愿！

王 勇
齐向辉
2024年3月

Foreword to the First Edition

One of the wonderful things about molecular biology is that the field depends on constant communication. Important questions can only be answered through the collaboration of researchers from all over the world, who each contribute their piece of the puzzle and build on each other's work. But this lively interaction also poses challenges, particularly the problem of language. How can scientists collaborate if they speak and publish in different languages?

At current stage, most molecular biological researches are published in English, as are most international conferences. The language has become so central because a large number of scientists and research centers that have made important contributions to molecular biology were, and still are, based in England and the United States. The predominance of one language is not necessarily fair, but it is a reality that most professionals working in the field must face.

For Chinese students, learning biology in English can be very exciting, but also quite challenging. Chinese and English are very different languages, with completely different phonetics and writing systems. On top of this, biology textbooks and articles in English are especially difficult to understand, even for native English speakers. There is a lot of unique vocabularies, and there may be sentence structures particular to scientific writing.

In many cases, Chinese students are presented with a giant textbook only in English and instructors hope for the best. Our experience as teachers of molecular biology in China is that this approach

第一版前言

关于分子生物学的美好事物之一就是这一领域的发展依赖于不断的交流。重大问题只能通过全世界研究人员的合作才能找到答案,每个研究小组为解答难题贡献出他们的研究成果,并在他人研究成果的基础上开展更深入的研究。但是要开展这种密切合作也面临着许多挑战,特别是语言问题。如果科学家们说着不同的语言、用不同的语言发表文章,那他们之间怎么去开展合作呢?

现阶段,大多数分子生物学研究报告用英文发表,大多数国际会议也用英语发言。由于对分子生物学做出过重大贡献的许多科学家和研究中心都在英国和美国,现在也仍然是这些国家在发挥着主导作用,因此英语已经成为首要的语言。一种语言起主导作用不一定是公平的,但这是大多数在这一领域工作的专业人员必须面对的现实。

对中国学生来说,用英语学习生物学会是很令人兴奋的,但也会是颇具挑战性的。汉语和英语是很不一样的语言,具有完全不同的语音和语法体系。此外,生物学的英语教材和文章特别难以理解,即使对母语为英语的读者来说也是如此。书中有许多特殊的词汇,还会有一些特殊的科技写作方面的惯用句子结构。

许多时候,中国学生拿到手的是一本厚厚的英文教材,教师希望学生从中获得最大的收益。我们在中国讲授分子生物学的体会是:这样的做法是不

is unrealistic. Most students in China at this time, especially those who are on a course of scientific study, do not have enough training in English to understand textbooks written for native English speakers. Struggling with the language often comes at the expense of actually learning the concepts and principles of molecular biology.

We have put together this book with the hope of creating a more realistic, more sympathetic, and more specialized approach to learn molecular biology in English. Advanced students can learn by reading the English portion of the book, and refer to the Chinese translation to check their understanding, or for reference on individual words. Beginners in English can learn from the Chinese portion, and refer to the English for an introduction to vocabularies. With both languages present, students and instructors have more options for how to learn!

An added feature of this book is that all the English texts have been accompanied by audio files. They are especially helpful for those who intend to improve their listening comprehension of academic English. Meanwhile, the elaborately designed figures will enable readers to understand concepts and principles in molecular terms much more easily.

The book covers the material that we feel an advanced, undergraduate student of molecular biology should know. Unlike many textbooks used for molecular biology, this is not meant as a reference book, a large volume where one can look up any facts about the cell. Those kinds of volumes have their use, but we feel that it is unproductive to present students with such piles of information. If students can master all of the information present even in this relatively brief book, they will already be more informed than most students of the field, and even some professionals!

切合实际的。这一阶段的大多数中国学生，特别是那些还处于科学研究课程学习阶段的学生，并不具备足够的英语能力去理解为母语是英语的读者编写的教材。在语言上的些许进步实际上是以牺牲理解分子生物学概念与原理为代价的。

我们编写此书的目的是想为用英语学习分子生物学提供一种更实际、更让人喜欢和更专业的方法。程度好的学生可以阅读本书的英文部分，并参考中文译文以检查他们的理解情况，或参考个别单词的中文含义。英语程度差一点的学生可以从本书的中文部分学习，并参考英文而开始掌握专业词汇。使用这样的英汉对照教材，在如何学习上学生和教师都可以有更多的选择！

本书的另一个特点是：所有的英语课文都有配套的音频文件。它们对那些希望提高自身专业英语听力水平的人特别有帮助。同时，精心设计的插图能使读者更容易地从分子水平上理解相关的概念与原理。

本书包括的内容是我们认为一个分子生物学高年级本科生应该掌握的。与许多分子生物学教材不同，它不是一本参考书，不是一本让人查阅关于细胞各方面内容的大部头著作。那些大部头著作自然有它们的用武之地，而我们觉得向学生提供这么一大堆信息并不是很有成效的做法。如果学生能掌握这本相对来说比较简单的书中的所有内容，他们就已经是这一领域学生中出类拔萃的了，他们甚至可能比一些专业人员还要强！

It is with great excitement that we invite you to learn molecular biology with us in English. This is an important age for China and the rest of the world! As China becomes every day more prosperous, its research centers and companies are improving and have much to offer scientific communities around the world. At the same time, the continued development of China depends on the knowledge and skills that have been developed in the West and continue to flourish there. In molecular biology, there is much to gain if we can share and communicate with each other!

能邀请你与我们一道用英语来学习分子生物学让我们觉得很兴奋。中国和世界的其他国家正处于一个重要的时代！随着中国的日益进步与繁荣，中国的研究中心和公司正在取得更大的进展，它们正在为世界科学共同体提供更多的研究成果。同时，中国的持续发展需要了解西方已经拥有的和即将涌现的知识和技能。在分子生物学领域，如果我们能够共享知识与技能并互相交流，我们就都可以取得更大的成就！

Sincerely,
Sylvain W. Lapan
Yong Wang
October 2007

西尔维恩 W. 勒潘
王勇
谨识
2007 年 10 月

Foreword to the Second Edition

Dear Readers,

With this new edition of *An Introduction to Molecular Biology with Chinese Translation* we continue our mission that has captivated and motivated us from the beginning. Firstly, to convey fundamental concepts of how genetic information is replicated, expressed and modified. Secondly, to assist Chinese readers in learning words and phrases in the English language used to describe these processes. We have been honored by the positive feedback of students over the past decade, and used these comments to maintain and improve the qualities of this book that students value the most: the native English main text, accurate translation, clear figures and audio files narrated in American English.

The past decade has witnessed remarkable advancements in biotechnology — the reduction in cost of DNA sequencing and synthesis, the application of genome engineering techniques such as CRISPR, the use of transposases for generating DNA libraries and profiling chromatin, to name just a few. In each case, these advances would not have been made if not for basic research into genome structure, stability, and expression. We have an enormous amount still to gain from the basic study of molecular biology in a variety of organisms, as well as the innovative application of discoveries to solve challenges in biotechnology. It is our sincere hope that this new edition will advance the goal of introducing Chinese and English-speaking students to this exciting field and fostering global collaboration.

Best wishes,

Sylvain W. Lapan
Yong Wang
Jan 2018

第二版前言

亲爱的读者：

在新版《英汉对照分子生物学导论》中，我们继续努力达成促使与激励我们编写本书的初衷——首先，讲述遗传信息是如何进行复制、表达和改变的基本概念；其次，帮助中国读者学习描述这些生命过程的英语词汇与短语。我们很荣幸在过去十年里收到不少学生对本书的积极反馈，这些反馈表明了学生最关心本书的以下品质：地道的英语课文，准确的中文翻译，清晰的图解以及美式英语音频文件。在新版《英汉对照分子生物学导论》中，这些品质都得到了维护与提升。

过去十年里生物技术领域取得了令人瞩目的成就——DNA测序与合成成本显著降低、CRISPR等基因编辑技术的应用、使用转座酶生成DNA文库及分析染色质等等，不胜枚举。具体到某一方面，所有成就的取得都是建立在对基因组结构与稳定性以及基因表达特性的基础研究之上的。现今，分子生物学基础研究仍有待在许多不同生物中展开，生物技术领域所面临的挑战也需要创造性地应用分子生物学基础研究成果来应对。我们诚挚希望新版《英汉对照分子生物学导论》能有助于引导中国学生与英语国家学生进入这一令人兴奋的领域并培养全球合作精神。

致以良好祝愿！

西尔维恩·W. 勒潘
王勇
2018年1月

Contents / 目 录

Introduction 1	**绪论** 1
Chapter 1 Amino Acids to Proteins 5	**第 1 章 从氨基酸到蛋白质** 5
1.1 Protein Composition 5	1.1 蛋白质的组成 5
1.2 Protein Conformations 8	1.2 蛋白质的构象 8
1.2.1 Describing Protein Structure 9	1.2.1 描述蛋白质的结构 9
1.2.2 Chemical and Physical Basis for Protein Folding 13	1.2.2 蛋白质折叠的化学和物理基础 13
1.3 Protein Structure and Function: A Few Examples 16	1.3 蛋白质的结构与功能：几个例子 16
1.4 The Dynamics of Proteins 18	1.4 蛋白质动力学 18
1.5 Experiments 21	1.5 实验研究 21
Summary 21	小结 21
Vocabulary 22	词汇 22
Review Questions 23	习题 23
Exploration Questions 23	思考题 23
Chapter 2 Nucleic Acids 24	**第 2 章 核酸** 24
2.1 Properties of a Genetic Material 24	2.1 遗传物质的性质 24
2.2 Nucleic Acids and DNA structure 25	2.2 核酸与 DNA 结构 25
2.2.1 Nucleotides 25	2.2.1 核苷酸 25
2.2.2 General Structure of Nucleic Acids 27	2.2.2 核酸的一般结构 27
2.2.3 Structure of DNA 30	2.2.3 DNA 的结构 30
2.3 Molecular Identification 33	2.3 分子鉴定 33
2.3.1 Identification of DNA Molecules 33	2.3.1 DNA 分子鉴定 33
2.3.2 Identification of RNA Molecules 35	2.3.2 RNA 分子鉴定 35
2.4 DNA as the Genetic Material 37	2.4 DNA 作为遗传物质 37
2.5 DNA in the Cell 38	2.5 细胞中的 DNA 38
2.6 RNA (Ribonucleic Acid) 40	2.6 RNA（核糖核酸） 40
2.7 Experiments 42	2.7 实验研究 42
Summary 44	小结 44
Vocabulary 44	词汇 44
Review Questions 45	习题 45
Exploration Questions 46	思考题 46
Chapter 3 Transcription in Prokaryotes: Mechanism and Regulation 47	**第 3 章 原核生物转录：机理与调控** 47
3.1 Why Use an RNA Intermediate 47	3.1 为什么使用 RNA 作为中间物 47
3.2 Mechanism of Transcription 49	3.2 转录机理 49
3.2.1 Promoters 49	3.2.1 启动子 49

3.2.2 RNA Polymerase	3.2.2 RNA 聚合酶 ⋯ 51
3.2.3 Transcription Mechanism in Three Phases	3.2.3 转录机理的三个阶段 ⋯ 53
3.3 Regulation of Gene Expression in Prokaryotes	3.3 原核生物基因表达调控 ⋯ 58
3.3.1 Coordinate Regulation	3.3.1 协同调控 ⋯ 59
3.3.2 The *Lac* Operon	3.3.2 乳糖操纵子 ⋯ 61
3.3.3 The *Trp* Operon	3.3.3 色氨酸操纵子 ⋯ 64
3.3.4 *Ara* and *Gal* Operons	3.3.4 阿拉伯糖与半乳糖操纵子 ⋯ 70
3.4 Experiments	3.4 实验研究 ⋯ 72
Summary	小结 ⋯ 76
Vocabulary	词汇 ⋯ 77
Review Questions	习题 ⋯ 78
Exploration Questions	思考题 ⋯ 79

Chapter 4 Transcription in Eukaryotes: Mechanism and Regulation

第 4 章 真核生物转录：机理与调控 ⋯ 80

4.1 Eukaryotic RNA Polymerases	4.1 真核生物 RNA 聚合酶 ⋯ 80
4.2 Eukaryotic Promoters	4.2 真核生物启动子 ⋯ 82
4.3 General Transcription Factors and Initiation	4.3 通用转录因子与转录起始 ⋯ 83
4.3.1 TFⅡD	4.3.1 TFⅡD ⋯ 84
4.3.2 Other TFⅡs	4.3.2 其他 TFⅡs ⋯ 86
4.3.3 General Transcription Factors for RNA Polymerase Ⅰ and Ⅲ	4.3.3 RNA 聚合酶Ⅰ和Ⅲ的通用转录因子 ⋯ 86
4.4 Specific Transcription Factors and Transcriptional Regulation	4.4 特异转录因子与转录调控 ⋯ 88
4.4.1 Activators	4.4.1 激活蛋白 ⋯ 89
4.4.2 Repressors	4.4.2 阻遏蛋白 ⋯ 91
4.4.3 Enhancers and Silencers	4.4.3 增强子和沉默子 ⋯ 91
4.5 Structures of Specific Transcription Factors	4.5 特异转录因子的结构 ⋯ 93
4.5.1 DNA-Binding Motifs in Prokaryotes	4.5.1 原核生物 DNA 结合基序 ⋯ 94
4.5.2 DNA-Binding Motifs in Eukaryotes	4.5.2 真核生物 DNA 结合基序 ⋯ 95
4.6 Experiments	4.6 实验研究 ⋯ 97
4.6.1 RNA Polymerase Targets	4.6.1 RNA 聚合酶的目标 ⋯ 97
4.6.2 Modularity of Specific Transcription Factors	4.6.2 特异转录因子的模块化 ⋯ 98
Summary	小结 ⋯ 99
Vocabulary	词汇 ⋯ 100
Review Questions	习题 ⋯ 100
Exploration Questions	思考题 ⋯ 101

Chapter 5 mRNA Modifications in Eukaryotes

第 5 章 真核生物 mRNA 的修饰 ⋯ 102

5.1 Capping	5.1 加帽 ⋯ 102
5.2 Polyadenylation	5.2 聚腺苷酸化 ⋯ 104
5.3 Splicing	5.3 剪接 ⋯ 106
5.3.1 The Basic Splicing Reaction	5.3.1 基本的剪接反应 ⋯ 106
5.3.2 Proteins Involved in Splicing	5.3.2 在剪接中发挥作用的蛋白质 ⋯ 108

5.3.3 Self-Splicing	5.3.3 自我剪接	110
5.3.4 *Trans*-Splicing	5.3.4 反式剪接	111
5.3.5 Reasons for Introns	5.3.5 内含子存在的原因	113
5.4 mRNA Editing	5.4 mRNA 编辑	115
5.5 Experiments	5.5 实验研究	118
Summary	小结	120
Vocabulary	词汇	120
Review Questions	习题	120
Exploration Questions	思考题	121

Chapter 6 Translation / 第 6 章 翻译 — 122

6.1 The Genetic Code	6.1 遗传密码	122
6.2 Mechanism of Translation in Prokaryotes	6.2 原核生物翻译机理	125
6.2.1 Initiation	6.2.1 起始	125
6.2.2 Elongation	6.2.2 延伸	128
6.2.3 Termination	6.2.3 终止	132
6.3 Translation in Eukaryotes	6.3 真核生物翻译	132
6.4 tRNA Structure and Wobble	6.4 tRNA 的结构与摇摆	136
6.4.1 Anti-codons	6.4.1 反密码子	137
6.4.2 Wobble	6.4.2 摇摆	139
6.5 Experiments	6.5 实验研究	140
6.5.1 Deciphering the Genetic Code	6.5.1 破译遗传密码	140
6.5.2 Direction of Translation	6.5.2 翻译的方向	141
Summary	小结	142
Vocabulary	词汇	143
Review Questions	习题	143
Exploration Questions	思考题	144

Chapter 7 Regulation of Gene Expression in Eukaryotes / 第 7 章 真核生物基因表达调控 — 145

7.1 Histones and Transcriptional Regulation	7.1 组蛋白与转录调控	146
7.1.1 Histones and DNA Organization	7.1.1 组蛋白与 DNA 组织	146
7.1.2 Histones and Transcription	7.1.2 组蛋白与转录	149
7.1.3 Covalent Modification of Histones	7.1.3 组蛋白的共价修饰	151
7.1.4 Proteins that Recognize and Modify Histones	7.1.4 识别和修饰组蛋白的蛋白质	153
7.2 Post-Transcriptional Regulation	7.2 转录后调控	156
7.3 Nuclear Export	7.3 细胞核输出	158
7.4 RNA Stability	7.4 RNA 稳定性	160
7.4.1 mRNA Stability Regulation by Proteins	7.4.1 蛋白质调控 mRNA 稳定性	160
7.4.2 mRNA Stability Regulation by Small RNAs	7.4.2 小 RNA 调控 mRNA 稳定性	163
7.5 Translational Control	7.5 翻译调控	165
7.5.1 Global Control	7.5.1 全局控制	166
7.5.2 mRNA-Specific Control	7.5.2 mRNA 特异性控制	167
7.6 mRNA Localization	7.6 mRNA 定位	170

7.7　Protein Regulation
7.8　Experiments
　　7.8.1　Beads-on-a-string Structure
　　7.8.2　Repression of Gene Expression in Heterochromatin
Summary
Vocabulary
Review Questions
Exploration Questions

Chapter 8　DNA Replication
8.1　Semi-Conservative Replication
8.2　Initiation of Replication
8.3　Semi-Discontinuous Replication
8.4　Elongation of Replication and its Proteins
　　8.4.1　Helicase and SSBs
　　8.4.2　DNA Polymerases
　　8.4.3　Explanation for $3'\to 5'$ Synthesis
　　8.4.4　Primers
8.5　DNA Topology
8.6　DNA Replication in Eukaryotes
　　8.6.1　Initiation of DNA Replication in Eukaryotes
　　8.6.2　Telomeres
8.7　Experiments
Summary
Vocabulary
Review Questions
Exploration Questions

Chapter 9　Mutations and Mutation Repair
9.1　DNA Damage and Mutations
9.2　Point Mutations
　　9.2.1　Mismatched Base
　　9.2.2　Spontaneous Mutations
　　9.2.3　Induced Mutations
9.3　Insertions and Deletions
　　9.3.1　Strand Slippage
　　9.3.2　Transposons
　　9.3.3　Intercalating Agents
9.4　Large-Scale DNA Changes
9.5　Consequences of DNA Mutations
　　9.5.1　Consequences of Point Mutations
　　9.5.2　Consequences of Insertions and Deletions
　　9.5.3　Consequences of Translocations

7.7　蛋白质调控 …… 172
7.8　实验研究 …… 172
　　7.8.1　线珠结构 …… 172
　　7.8.2　异染色质中基因表达的抑制 …… 174
小结 …… 174
词汇 …… 175
习题 …… 176
思考题 …… 177

第 8 章　DNA 复制 …… 178
8.1　半保留复制 …… 178
8.2　复制的起始 …… 179
8.3　半不连续复制 …… 181
8.4　复制延伸及其相关蛋白 …… 184
　　8.4.1　解旋酶与 SSB …… 184
　　8.4.2　DNA 聚合酶 …… 185
　　8.4.3　关于 $3'\to 5'$ 合成 …… 188
　　8.4.4　引物 …… 191
8.5　DNA 拓扑学 …… 193
8.6　真核生物 DNA 复制 …… 195
　　8.6.1　真核生物 DNA 复制起始 …… 197
　　8.6.2　端粒 …… 199
8.7　实验研究 …… 200
小结 …… 203
词汇 …… 203
习题 …… 203
思考题 …… 204

第 9 章　突变与突变修复 …… 205
9.1　DNA 损伤与突变 …… 205
9.2　点突变 …… 206
　　9.2.1　错配的碱基 …… 206
　　9.2.2　自发突变 …… 207
　　9.2.3　诱发突变 …… 209
9.3　插入和缺失 …… 213
　　9.3.1　链滑动 …… 213
　　9.3.2　转座子 …… 214
　　9.3.3　嵌入剂 …… 215
9.4　大规模 DNA 变化 …… 216
9.5　DNA 突变的后果 …… 217
　　9.5.1　点突变的后果 …… 217
　　9.5.2　插入和缺失的后果 …… 219
　　9.5.3　易位的后果 …… 220

9.5.4 Mutation Hot Spots .... 9.5.4 突变热点 221
9.6 Mutation Repair .... 9.6 突变修复 222
    9.6.1 Direct Reversal ....   9.6.1 直接回复 223
    9.6.2 Mismatch Repair ....   9.6.2 错配修复 223
    9.6.3 Nucleotide Excision Repair ....   9.6.3 核苷酸切除修复 225
    9.6.4 Base Excision Repair ....   9.6.4 碱基切除修复 227
    9.6.5 Double-Stranded Break Repair ....   9.6.5 双链断裂修复 228
9.7 Experiments .... 9.7 实验研究 230
    9.7.1 Nucleotide Excision Repair and Human Disease ...   9.7.1 核苷酸切除修复与人类疾病 ... 230
    9.7.2 The Ames Test ....   9.7.2 埃姆斯测验法 231
Summary .... 小结 232
Vocabulary .... 词汇 232
Review Questions .... 习题 233
Exploration Questions .... 思考题 234

Chapter 10   Recombination   第 10 章   重组 235

10.1 Homologous Recombination .... 10.1 同源重组 235
    10.1.1 Mechanism for Crossing-Over ....   10.1.1 交换机理 236
    10.1.2 Mechanism for Double-Stranded Break Repair ...   10.1.2 双链断裂修复机理 ... 240
    10.1.3 The RecBCD Pathway ....   10.1.3 RecBCD 途径 241
    10.1.4 Gene Conversion ....   10.1.4 基因转换 241
10.2 Non-homologous Recombination .... 10.2 非同源重组 243
    10.2.1 Transposons ....   10.2.1 转座子 243
    10.2.2 Retrotransposons ....   10.2.2 反转录转座子 249
    10.2.3 Bacteriophage λ Integration ....   10.2.3 λ 噬菌体的整合 250
10.3 Gene Editing .... 10.3 基因编辑 251
    10.3.1 ZFNs System ....   10.3.1 ZFN 系统 251
    10.3.2 TALENs System ....   10.3.2 TALEN 系统 253
    10.3.3 CRISPR/Cas System ....   10.3.3 CRISPR/Cas 系统 254
Summary .... 小结 257
Vocabulary .... 词汇 257
Review Questions .... 习题 258
Exploration Questions .... 思考题 258

Answers to Review Questions   习题答案 259
Index（Chinese）   中文索引 262
Index（English）   英文索引 268
Glossary   名词解释 274
References   参考文献 302

Introduction

绪论

We say a cell is alive. But look inside a cell and all you will see are molecules, collections of atoms that are as inanimate as the paper of this book. How does a set of objects make a dynamic living organism? This question is the basis of the field of molecular biology.

The most fundamental features shared by all known life forms are the ability to reproduce and the ability to grow. If organisms were not able to reproduce, life would be impossible. Perhaps once in a billion years a precursor to life could arise spontaneously from a pool of chemicals; but if this form could not make more of itself, it would soon vanish. The necessity to grow is a direct result of the necessity to reproduce. As one cell makes two and two cells make four, for example, the mass of the original cell becomes more and more divided. If the offspring do not grow, after several generations they will be impossibly small.

So how can a set of molecules reproduce and grow? These functions are largely the work of large molecules called proteins, nature's tiny machines. For reproduction, proteins copy some parts of the cell, and pull one cell apart to make two cells. For growth, they can take in nutrients from the environment and put these nutrients together to make the structures of the cell. They also break down nutrients to provide the energy and molecular building blocks. Proteins are able to undertake so many tasks because of their large size, intricate three-dimensional structures, and complex chemical properties.

But proteins are not the end of the story. Proteins cannot directly reproduce themselves; therefore, they cannot be solely responsible for life. The formation of proteins depends on another kind of molecule called nucleic acids. Nucleic acids, mainly DNA, are especially well-suited for carrying the information to make proteins.

我们说一个细胞是活的。但是往细胞里面看，所有你看到的都是分子和各种各样的原子，它们就像本书的纸张一样没有生命。这样一些物体怎么会产生动态的生命呢？这是分子生物学领域所要研究的基本问题。

所有已知生命形态都具有的最基本特征是它们能够繁殖和生长。如果生物不能繁殖，生命就不可能存在。也许在十亿年里从一些化学物质中会自发出现生命的原始形式；但如果这一形式不能繁殖的话，它很快就会消失。生长的必要性是繁殖必要性的直接结果。例如，当一个细胞产生两个、两个细胞产生四个的时候，原始细胞的物质一次又一次地被分开。如果子代细胞不能生长的话，那么要不了几代细胞就会小到不可思议。

那么，这样一些分子是怎样繁殖和生长的呢？答案是：这些功能主要是由蛋白质这样的大分子来完成的，它们是自然界的微型生产机器。对繁殖来说，是蛋白质对细胞的某些部分进行了复制，并使一个细胞分裂产生两个细胞。对生长来说，是蛋白质把营养物质从环境中带进来，并用它们建造出细胞的各个组成部分。它们也分解营养物质产生能量和分子结构组件。蛋白质能够承担各种各样的任务，其原因就是它们具有较大的体积、精巧的三维结构和复杂的化学性质。

但是蛋白质并不是故事的结尾。蛋白质不能直接繁殖它们自身；因此，光有蛋白质还不足以实现所有的生命机能。蛋白质的形成依赖于另一种称为核酸的分子。核酸，主要是DNA，特别适合携带生产蛋白质的信息。DNA也从每个世

DNA is also passed down with each new generation, allowing offspring to completely reproduce the set of proteins present in the parent. If proteins are the cell's machines, DNA is the cell's head engineer—knowledgeable about how to make the machines, and able to pass this knowledge from generation to generation.

In order to make proteins, DNA is copied using another nucleic acid, called RNA. This process is called **transcription**. RNA is then used as a guide to put together the small molecules that make a protein, a process called **translation**. This general progression, DNA → RNA → Protein, is used in all life forms. Because it is so central to molecular biology, it has been termed the **central dogma**.

The general scheme of the central dogma is very basic, and many details need to be filled-in. The work of adding these details is spread between different fields of biology, of which molecular biology is just one. Molecular biology is most focused on the question of how nucleic acids and proteins are used to make more proteins. Other fields have different emphases. For example, genetics is focused particularly on the structure, function and inheritance of genes. Cell biologists may study how proteins and other molecules form the structures of the cell, how the cell divides, and how it interacts with the environment. Biochemistry has a classic focus on metabolic pathways in the cell and the enzymatic roles of proteins. Generally, these and other fields overlap to a great extent, and there is no saying where exactly one ends and one begins.

Although the topic of molecular biology as stated here may seem somewhat simple, many details must be mastered to truly understand the depth of the field. It is easy to get lost in these details, and so before we discover them, we lay out some guiding principles that can give a sense of order to many of these intricacies. Most important are the forces that shape life, including evolution, physics, and the availability of resources.

All life forms are the product of evolution. Life today is far too complicated to have arisen spontaneously from

代传递到下一世代中，使子代能够生产所有在亲本中出现的蛋白质。如果蛋白质是细胞的机器，那么细胞中的DNA就是总工程师——它们知道应该怎样生产机器，还能将这种知识一代一代地传递下去。

为了生产蛋白质，需要用到另外一种核酸即RNA来拷贝DNA。这一过程称为**转录**。之后，RNA用来作为引导者将小分子放在一起来产生蛋白质，这是一个称为**翻译**的过程。这种一般的进程，即DNA→RNA→蛋白质，被所有生命形式所采用。由于它对分子生物学是如此重要，以至于称它为**中心法则**。

中心法则的一般过程看上去很简单，但我们应该知道其中的许多细节。探询这些细节的工作分散在生物学的不同领域中，分子生物学是其中之一。分子生物学主要关注核酸和蛋白质如何用来生产更多的蛋白质。其他领域则各有其侧重点。例如，遗传学特别注重基因的结构、功能和遗传。细胞生物学家们可能会研究蛋白质和其他分子怎样形成细胞的结构、细胞怎样分裂、细胞如何与环境进行相互作用。生物化学则对细胞中的代谢途径和蛋白质的酶学作用更为关注。一般情况下，这些领域之间在很大程度上会有重叠，说不清某个领域的研究范围从哪儿开始、在哪儿结束。

虽然这里所说的分子生物学主题看起来有些简单，但若要真正理解这一领域深层次的东西，还是必须掌握许多细节内容。而一旦深入到细节中又很容易迷失方向，因此在开始学习之前，我们列出几条指导原则来帮助梳理许多这样的细节。这当中最重要的是那些塑造了生命的力量，包括进化、物理学和可利用资源。

所有生命形式都是进化的产物。今日的生命太复杂以至于不可能从简单的分子自发

simple molecules. Its present complexity can be understood by the accumulation of small advantageous changes over long periods of time. Because evolution is based on chance, new improvements in an organism generally arise as relatively modest changes to a previous state. Always remember, the mechanisms in a cell are not necessarily the best possible.

All life on Earth appears to have evolved from a common ancestor. One of the descendants of that ancestor had a life strategy so successful that it remained mostly unchanged in the evolution of millions of species over billions of years. That strategy has, at its core, the production of proteins from DNA using an RNA intermediate. The changes evolved by the millions of species that descended from that ancestor were, from a morphological perspective, enormous—leading to dinosaurs on the one hand, and bacteria on the other. From the perspective of molecular biology, however, many evolutionary changes have been, in a sense, icing on the cake; all cellular life forms retained the central dogma. As a consequence, in molecular biology we are able to make very relevant comparisons between distant life forms, like humans and bacteria.

In addition to sharing a common ancestor, all life on earth is subject to the same set of physical laws; like all matter, the molecules that make life are subject to the laws of physics. Motion and attraction of these molecules, for example, are guided and limited by strict rules. In the universe, all things tend to become more disordered, a property called entropy. Cells are highly ordered structures that must overcome the disorder of entropy by using energy. However, energy is always conserved in the universe. In order for a cell to obtain more energy, it must take energy from someplace else. The acquisition of energy and its efficient usage are necessities that fundamentally shape living things.

Available energy is only one example of a resource that is limited in the environment. Molecular building blocks of cellular components are another. The limited availability of

地出现在我们面前。它现在所具有的复杂性可以被理解成是在漫长的时光里一点一滴地积累了小的有利变化的结果。由于进化是有机会性的，一种生物中的新改进一般只是相对于先前的状态来说稍有变化而已。请总是记住一点：细胞中的某种机理不一定是所有可能性中最好的。

地球上所有的生命看起来都是从一个共同的祖先进化来的。那一祖先的后代之一具有一种生命策略，它是如此成功以至于在几十亿年的进化史中进化出来的几百万种生物中基本保持不变。作为这一策略的核心就是使用 RNA 作为从 DNA 生产蛋白质的中介物。从外观上看，来自共同祖先的几百万种生物在进化上是如此不同——它们大的像恐龙那样大，小的又像细菌那么小。而从分子生物学的角度看，进化所产生的许多改变从某种意义上来说只是蛋糕上的装饰而已；所有细胞生命形式都遵循中心法则。因此，在分子生物学中我们能够对亲缘关系很远的生命形式做出相关比较，例如我们可以比较人类和细菌之间的异同点。

除了拥有共同的祖先之外，地球上的所有生命都服从相同的物理学定律；像所有物质一样，组成生命的分子也服从物理学定律。例如，它们的分子运动与相互吸引受到物理学定律的严格指导和限制。在宇宙中，所有事物都倾向于变得更无序，这是一种称为熵的性质。细胞是高度有序的结构，必须利用能量去克服熵的无序。然而，能量在宇宙中总是守恒的。细胞为了能够有更多的能量可供使用，必须从别的地方获得能量。生物必须具备获得能量并加以有效使用的能力，这是生命现象的最基本体现。

可利用的能源只是环境中限制性资源的一个例子。细胞成分的分子部件是另一种限制性资源。资源的限制性意味着：

Introduction | 3

certain resources means that many cellular mechanisms are shaped by a need to be efficient. Cells cannot perform whatever task with whatever material. In some cases, better molecules are imaginable for the cellular tasks at hand, but these molecules are unattainable.

Within these guidelines, and many others, life on Earth has been able to find many solutions. Evolution, though slow and blind, has been a very powerful source of change over the course of billions of years of Earth's history. Within the laws of physics, life has been able to create order from a disordered mixture of chemicals—largely thanks to energy provided by the sun. Chemical building blocks, such as carbon, that are the basis for complex, functional structures are widely available on Earth. Also, the presence of liquid water on the planet has provided an ideal medium for chemical processes on which life is based.

In the following chapters of this book, we invite you to learn how inanimate molecules have come together to create life, steering between the limitations and the advantages provided by our planet and our universe.

许多细胞机理是为了有效地使用资源而塑造出来的。细胞并不能随心所欲地使用任何材料来执行任何任务。有的时候，细胞倒是想用更好的分子去完成手头的任务，但这些分子往往无法得到。

在诸如此类指导原则的框架内，地球上的生命找到了许多解决方案。进化虽然发生得极为缓慢而且盲目，但在地球存在的几十亿年历史过程中，它却是非常有效地获得解决方案的源泉。在物理学定律的限制下，生命从无序的化学物质混合物中创造出了有序的结构——这还应感谢太阳提供的能量。一些像碳原子这样的化学元素在地球上很容易获得，它们是组成具有复杂功能结构的基础。此外，地球上液态水的存在为生命赖以生存的化学过程提供了理想的介质。

在本书的后续章节中，我们邀请你来学习这些无生命的分子是如何组合在一起创造出生命的，它们又是如何在地球以及宇宙所提供的不利条件与有利条件之间曲折前行的。

Chapter 1 Amino Acids to Proteins

Life is most directly the work of **proteins**. Proteins allow organisms to grow and reproduce, the most fundamental properties of life. They provide shape and strength, and in many cases movement. They underlie cellular communication, but are also a key part of the boundaries that separate cells and organelles from their environment. In the cell, proteins are everywhere and do almost everything. In this chapter we examine the molecular composition of proteins, showing how the joining of small, simple molecules can produce large molecules with complicated shapes and extraordinary functions.

1.1 Protein Composition

Proteins are **polymers** of small molecules called **amino acids** (Figure 1.1). It is convenient to think of an amino acid as a carbon atom attached to four different chemical groups. Three of these are always the same: an **amino group**, a **carboxyl group**, and a hydrogen atom. The fourth group is generally termed the side chain, or **R group**, and varies between different amino acids. There are 20 different amino acids commonly used to make proteins, and all 20 have different R groups. R groups have various sizes and chemical properties (Figures 1.2 and Figures 1.3).

第 1 章 从氨基酸到蛋白质

生命几乎就是**蛋白质**的杰作。蛋白质让生物得以生长和繁殖,生长和繁殖是生命最基本的特征。蛋白质赋予了生物外形和力量,以及在许多情形中的运动功能。蛋白质还是细胞通讯的基础,也是细胞膜的组成部分(正是细胞膜将细胞和细胞器与它们所处的环境分隔开)。在细胞中,蛋白质无处不在,行使着几乎所有功能。本章我们将学习蛋白质的分子组成,了解简单小分子是如何连接在一起产生具有复杂形状和超乎寻常功能的大分子的。

1.1 蛋白质的组成

蛋白质是由称为**氨基酸**的小分子组成的**聚合物**(图1.1)。可以很方便地把氨基酸看成是由四个不同的化学基团连接到一个碳原子上而形成的结构。其中三个基团总是相同,即**氨基**、**羧基**和氢原子。第四个基团通常被称为侧链,或 **R 基团**,它随氨基酸的不同而有变化。通常有20种不同的氨基酸被用来组成蛋白质,它们均具有不同的 R 基团。R 基团大小不一,其化学性质也不同(图1.2 和图 1.3)。

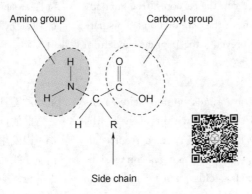

Figure 1.1 General amino acid structure

图 1.1 氨基酸的一般结构

Hydrophobic R groups	疏水的 R 基团		
Glycine	甘氨酸	Gly	G
Alanine	丙氨酸	Ala	A
Isoleucine	异亮氨酸	Ile	I
Leucine	亮氨酸	Leu	L
Methionine	甲硫氨酸	Met	M
Phenylalanine	苯丙氨酸	Phe	F
Tryptophan	色氨酸	Trp	W
Proline	脯氨酸	Pro	P
Valine	缬氨酸	Val	V

Hydrophilic R groups	亲水的 R 基团		
Serine	丝氨酸	Ser	S
Threonine	苏氨酸	Thr	T
Asparagine	天冬酰胺	Asn	N
Glutamine	谷氨酰胺	Gln	Q
Cysteine	半胱氨酸	Cys	C
Tyrosine	酪氨酸	Tyr	Y

Acidic R groups	酸性的 R 基团		
Aspartic acid	天冬氨酸	Asp	D
Glutamic acid	谷氨酸	Glu	E

Basic R groups	碱性的 R 基团		
Arginine	精氨酸	Arg	R
Histidine	组氨酸	His	H
Lysine	赖氨酸	Lys	K

Figure 1.2 The twenty amino acids and their abbreviations

图 1.2 20 种氨基酸以及它们的缩写表示法

Amino acids are joined to each other by combining the amino end of one with the carboxyl end of another (Figure 1.4). Because all amino acids have these two ends, any amino acid can join to any other amino acid. The polymer that results from these combinations is linear, meaning that there are no branchpoints. Proteins can be composed of any combination of the twenty amino acids, in any number, attached in any order. In fact, this flexibility in composition is necessary to produce the wide variety of proteins that are used in nature.

氨基酸可以通过一个氨基酸上的氨基与另一个氨基酸上的羧基结合而互相连接起来（图 1.4）。由于所有氨基酸都具有这两个基团，因此任何氨基酸都可以与任何其他氨基酸相连接。这种结合形成的聚合物是线性的，意味着它们没有分支。蛋白质可以由 20 种氨基酸以任何组合、任何数目和任何顺序组成。事实上，这种在组成上的灵活性对于产生在自然界中用到的种类繁多的蛋白质是必需的。

The bond that joins two amino acids in a protein is called a **peptide bond**. It is a kind of amide bond. The peptide bond is quite strong and rigid, and does not allow rotation. This is because the double-bond joining the carbon and oxygen is also distributed between the same carbon and the adjacent nitrogen. The redistribution of electron density gives a partial double-bond character to the carbon-nitrogen bond, which is the core component of the peptide bond. This **partial double-bond** character prevents the peptide bond from rotating easily.

在蛋白质中连接两个氨基酸之间的键叫作**肽键**。它是一种酰胺键。肽键很强并具有刚性，不允许旋转。这是因为连接碳和氧的双键也在同一个碳和邻近的氮之间进行分配。电子密度的重新分布使得碳-氮键具有部分双键的特性，而碳-氮键是肽键的核心成分。这种**部分双键**特性防止了肽键发生自由旋转。

Although peptide bonds are rigid, amino acid chains are flexible because other bonds within each of the amino acids can rotate (Figure 1.5). As a result, although proteins are linear, they are not one-dimensional. The linear molecule bends, folds, and twists to form complicated three-dimensional structures. We explore protein structures in the next section.

虽然肽键具有刚性，但是氨基酸链还是容易弯曲的，因为位于氨基酸里面的其他键都可以发生旋转（图 1.5）。结果，虽然蛋白质是线性的，但它们并不是一维的。线性的分子会发生弯曲、折叠和扭曲从而形成复杂的三维结构。我们在下一节中去探寻蛋白质结构方面的知识。

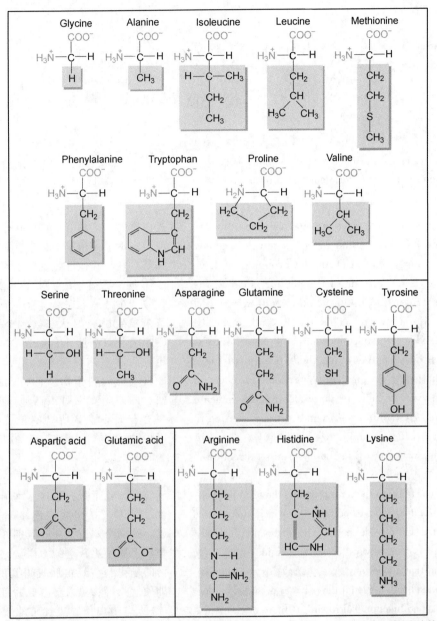

Figure 1.3　Structures of the 20 common amino acids　　　图 1.3　20 种常见氨基酸的结构

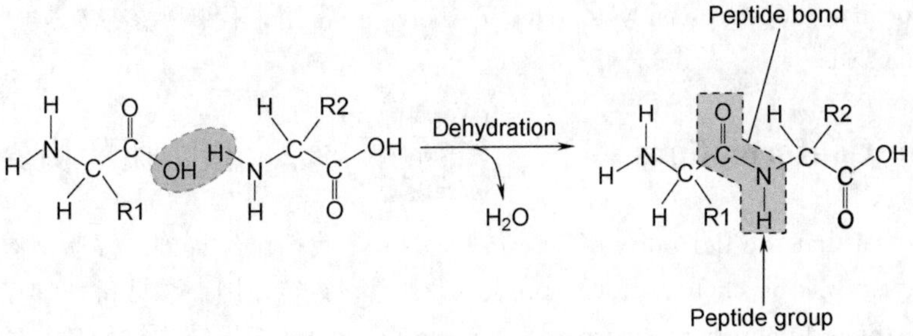

Figure 1.4　The formation of a peptide bond　　　图 1.4　肽键的形成

1.1　Protein Composition

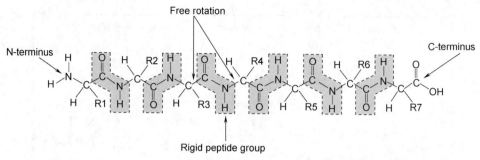

Figure 1.5 Rigid and flexible bonds in an amino acid polymer, and location of N-terminus and C-terminus

图 1.5 氨基酸聚合物中的刚性键和柔性键以及 N-末端和 C-末端的位置

The ends of a protein have unique chemical properties. On one end, an amino group is left unreacted, unlike the other amino acids in the protein that have been polymerized. This end of the peptide is called the **N-terminus** (Figure1.5). On the other end, a carboxyl group remains unreacted. This end is called the **C-terminus**. Under cellular conditions, the N-terminus is usually positively charged, and the C-terminus is negatively charged. A protein's termini provide a convenient way to indicate directionality. To indicate the position of an amino acid A in a protein relative to amino acid B, for example, we might say, "A is 5 amino acids away from B in the N-terminal direction."

蛋白质的末端具有独特的化学性质。在一个末端上保留了一个没有发生反应的氨基，它与其他已经发生了聚合的氨基酸不同。这个末端叫作 **N-末端**（图1.5）。在另一个末端上有一个羧基保持着未发生反应的状态。这个末端叫作 **C-末端**。在细胞中，通常 N-末端带有正电荷而 C-末端带有负电荷。蛋白质的末端为说明蛋白质的方向提供了方便。例如，要说明蛋白质中氨基酸 A 相对于氨基酸 B 所处的位置，我们可以说："A 位于 B 的 N-末端方向距离 5 个氨基酸之处。"

Note that in addition to the term protein, the term **polypeptide** can be used to refer to polymers in which amino acids are connected by peptide bonds. The distinction between these terms is not always clear. Polypeptide is a more general term that applies any time amino acids are polymerized by peptide bonds, even if the polymer is made synthetically and has no function. 'Protein' is more often used to describe one or more polypeptides that have a function in nature. The term **peptide** may also be used instead of 'polypeptide' or 'protein', most frequently to refer to short polymers of amino acids.

请注意，除了蛋白质这一术语外，**多肽**也可以用来指由肽键连接起来的聚合物。这两个术语之间的区别并不是很清楚。多肽是更普遍的术语，它适用于任何时候氨基酸经由肽键得到的聚合物，即使此聚合物是人工合成并且是没有功能的。"蛋白质"更经常地用来描述自然界中具有功能的一种或多种多肽。在描述氨基酸的短聚合物时，术语**肽**也常常用来代替"多肽"或"蛋白质"。

1.2 Protein Conformations

1.2 蛋白质的构象

We have mentioned that the flexibility of certain bonds in a polypeptide allows the chain to fold, producing a wide variety of three-dimensional structures, or **conformations**. These shapes, along with the chemical

我们已经提到，多肽中有些化学键可以旋转，因而多肽链能够发生折叠，产生多种三维结构或**构象**。这些形状与组成它们的氨基酸的化学性质一起，使得蛋

properties of the amino acids that compose them, are what give proteins their specialized functions.

1.2.1 Describing Protein Structure

To simplify discussion of protein conformations, four levels of protein structure have been defined (Figure 1.6). **Primary structure** simply refers to the sequence of amino acids in a polypeptide. **Secondary structures** are the most basic forms that regions in a polypeptide make by folding. **Tertiary structure** is the larger, more complicated conformation of the whole polypeptide, and is mostly composed of various secondary structures. Finally, **quaternary structure** refers to the shape of a protein that has multiple polypeptides, each with its own tertiary structure.

1.2.1 描述蛋白质的结构

为了简化对于蛋白质构象的讨论，通常把蛋白质结构划分成四个水平（图1.6）。**一级结构**就是指多肽链中氨基酸的顺序。**二级结构**是多肽中的区域通过折叠产生的最基本的结构形式。**三级结构**是整条多肽链的更大更复杂的构象，通常由不同的二级结构组成。最后，**四级结构**指由几条多肽链组成的蛋白质形状，每条多肽链具有自身的三级结构。

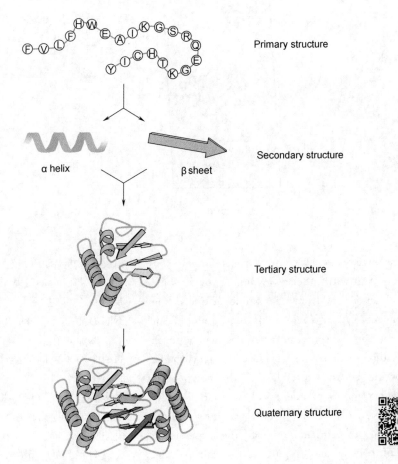

Figure 1.6　Four levels of protein structure

图 1.6　蛋白质结构的四个水平

The secondary structures of peptides usually have one of two forms, **α helix** or **β sheet** (The Greek letters are pronounced alpha and beta, respectively). The α

肽的二级结构通常有 **α 螺旋**和 **β 折叠**两种（希腊字母分别读作阿尔法和贝塔）。α 螺旋是多肽链的螺旋形弯曲，其中一

helix is a helical twisting of the polypeptide in which the N—H of an amino acid interacts with a carbonyl oxygen three amino acids away (Figure 1.7). This kind of bond is called a hydrogen bond, and is described in the next section. The R groups of the amino acids stick out from the helix. There are 3.6 amino acids for every complete turn of the helix.

个氨基酸上的 N—H 与距离它三个氨基酸的羰基上的氧发生相互作用（图 1.7）。这种键被称为氢键，将在下一节里进行描述。氨基酸的 R 基团指向螺旋的外面，螺旋的每一圈包含了 3.6 个氨基酸。

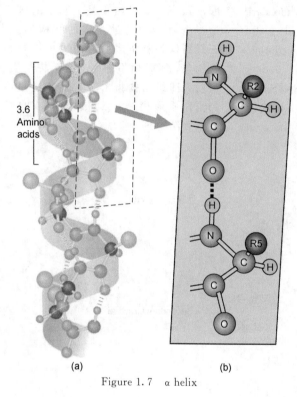

(a) Helix structure(Source:National Institute of Health)
(b) Detail of interaction in α helix structure

(a) 螺旋结构（来源：美国国家卫生研究院）
(b) α 螺旋内相互作用的细节

Figure 1.7　α helix

图 1.7　α 螺旋

The β sheet structure, by contrast, is relatively flat (Figure 1.8). As with α helices, binding involves interaction of an N—H from one amino acid and the double-bonded oxygen from a different amino acid. However, in the β sheet, the two atoms may be much more than three amino acids apart in the polypeptide. Binding occurs between stretches of the polypeptide that lie approximately parallel to each other. β sheets are often classified as either parallel or anti-parallel, depending on the orientation of the interacting stretches of polypeptide (Figure 1.9). Anti-parallel sheets usually form when a stretch of polypeptide folds backs on itself several times, causing neighboring regions to face each other in opposite orientations.

β sheets and α helices each can be formed from many

相反，β 折叠相对较为扁平（图 1.8）。与 α 螺旋相同，氨基酸之间的结合也牵涉到一个氨基酸上的 N—H 与另一个不同的氨基酸上的双键氧发生相互作用。然而，在 β 折叠中，发生相互作用的两个原子之间的距离可以大大超过多肽链中三个氨基酸。这种相互作用发生在多肽链中大致保持相互平行的不同区段之间。依据多肽链中发生相互作用区段的方向不同，β 折叠常被分为平行的或反向平行的（图 1.9）。反向平行折叠常常由多肽链的一个区段自身往返折叠数次形成，使得邻近区域的肽链以相反的方向相互面对。

β 折叠和 α 螺旋均可在许多不同的氨基

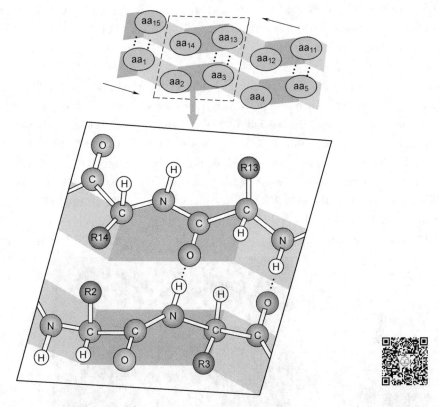

Figure 1.8 Structure of β sheet　　　图 1.8 β 折叠的结构

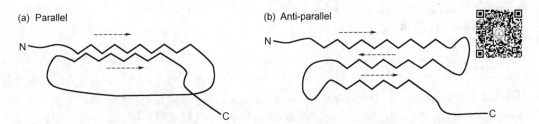

Figure 1.9 Parallel (a) and anti-parallel β sheets (b)　　图 1.9 平行 (a) 与反向平行 (b) 的 β 折叠

different amino acid sequences. Although the structures do not vary greatly, the chemical and physical properties of each structure are very different depending on the specific combination of amino acids that compose it. An α helix rich in leucine amino acids, for instance, will have a different reactivity and a different position in the protein than α helix rich in glutamine.

Some proteins are entirely made of β sheet or α helix, but most proteins contain a mix of both structures. The structures fold together in complicated ways to make the larger, tertiary structure of the peptide. Whereas

酸序列中形成。虽然结构上并没有很大的不同，但由于组成某种结构的氨基酸组合的特殊性，每种结构的化学性质和物理性质会有很大不同。例如，一个富含亮氨酸的 α 螺旋与一个富含谷氨酰胺的 α 螺旋会具有不同的反应活性，在蛋白质中也会出现在不同的位置。

有些蛋白质全部由 β 折叠或 α 螺旋组成，但绝大多数蛋白质由两种结构混合组成。它们在一起以复杂的方式折叠产生更大的三级结构。二级结构只限于少

secondary structure is limited to only several forms, like helices and sheets, tertiary structure may describe a vast number of different forms.

Many tertiary structures superficially look like globs, but they actually have shapes that are very specific for their functions, down to the last atom (Figure 1.10). The glob may, for example, have a surface with just the right charge distribution and folded shape to bind tightly to one kind of molecule out of the thousands of different proteins in the cell. We will see many more examples of the relationship between a protein's shape and function in a following section.

数几种，如螺旋和折叠，而三级结构的数目则可以是一个巨大的数字。

许多三级结构表面上看起来就像是一些小团，但实际上它们具有行使特定功能所需要的非常特殊的形状，这种结构甚至要精细到最终的原子水平（图1.10）。例如，就是这样的一小团东西需要具备一个拥有正确的电荷分布、以正确的形状折叠的表面，以便与细胞中几千种蛋白质中的一种发生紧密结合。在下一节中我们将看到更多的蛋白质形状与功能之间相互关系的例子。

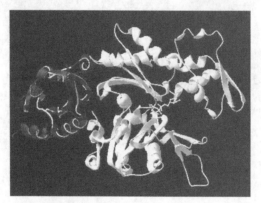

Figure 1.10　Example of a globular protein binding a small molecule (Source：http://www.wikipedia.org)

图1.10　一个与小分子结合的球状蛋白的例子（来源：http://www.wikipedia.org）

It is common for several polypeptides, each with its own tertiary structure, to come together to form a complex. The complex is still called a protein, even though it is made of more than one polypeptide. For example, the protein hemoglobin, which carries oxygen in the blood, is made of four polypeptides, two of one kind, and two of another (Figure 1.11). In this case, each polypeptide is called a **subunit** of the protein. The term quaternary structure is used to describe how subunits come together, in specific stoichiometries and orientations, to form a larger protein.

常见几种具有各自三级结构的多肽聚集在一起，形成复合体。这样的复合体仍旧被叫作蛋白质，虽然它由不止一条多肽链组成。例如，在血液中输送氧气的血红蛋白就是由四条多肽链组成，其中的两条属于一种多肽链，另外两条属于另一种多肽链（图1.11）。在这种情形下，每条多肽链叫作蛋白质的**亚基**。术语四级结构用来描述亚基怎样在一起、以什么特殊的化学组成和方向来形成更大的蛋白质。

It is also appropriate here to introduce the term **protein domain**. A protein domain is a functional region of a protein. It is not tertiary structure, because it usually does not comprise the whole polypeptide, but it is not as simple or as general as a secondary structure. An example of a protein domain might be the region that

在此有必要介绍**蛋白质域**这一术语。蛋白质域就是蛋白质的功能区域。它不是三级结构，因为它通常并不包含整个多肽，但又不仅仅像二级结构那样简单或普遍。举例来说，能使蛋白质结合到DNA上去的区域就是一个蛋

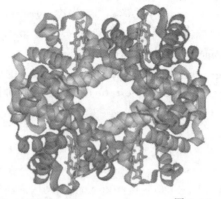

Figure 1.11 Hemoglobin quaternary structure (Source: http://www.wikipedia.org)

allows a protein to bind to DNA. Some protein domains appear almost unchanged in many different proteins in the cell, serving a similar function in each. These domains are called **modules**. When protein domains share such a closely related structure and function they usually have a common evolutionary history.

Protein structures are often represented by simplified, schematic diagrams. These images help the viewer to appreciate certain aspects of the protein's structure, such as its general shape, without getting overwhelmed by too many details. In the most popular representation of protein structures, α helices are represented as ribbons and β sheets are shown as flat arrows, but individual amino acids cannot be discerned. While viewing proteins in this way, it is easy to forget the importance of individual amino acids to protein function. Remember that on each ribbon or arrow there are amino acids with R groups sticking out. These groups may be charged, or hydrophobic, big or small. Their properties determine how the protein is folded and how it will function.

1.2.2 Chemical and Physical Basis for Protein Folding

Loosely speaking, a polypeptide with a given sequence always folds into the same shape. Often the peptide folds automatically, without any help from other molecules. This is a truly wonderful property. It means that instead of having to hold information about the complicated three-dimensional shapes of each protein, the cell just has to put together a primary sequence, and the polypeptide takes care of the rest! In some cases, proteins called **chaperones** help the folding process. But chaperones never do all the work of folding a protein; usually they help guide the protein toward one of several possible conformations.

Peptide folding is a complicated process and scientists are still trying to understand how it occurs. It is not possible at this time to look at the sequence of amino acids in a protein and know the protein's structure. However, the types of physical interactions that are involved in protein folding are understood. The most important are covalent bonding, hydrogen bonding, ionic bonding, Van der Waals forces, and hydrophobic interaction (Figure 1.12).

白质复杂的三维形状信息，多肽链自身会做好后面的所有事情！在有些情况下，被称为**侣伴蛋白**的蛋白质也参与折叠过程。但是侣伴蛋白并不能做完折叠蛋白质的所有工作；通常它们只是引导蛋白质向几种可能构象中的某一种进行折叠。

肽链的折叠是一个复杂的过程，科学家们仍然在努力探寻它是如何发生的。想要从一种蛋白质的氨基酸顺序上知道蛋白质的结构现在还不可能。然而，在蛋白质折叠中牵涉到的物理相互作用类型已经弄清楚。最重要的是共价键、氢键、离子键、范德瓦耳斯力和疏水相互作用（图1.12）。

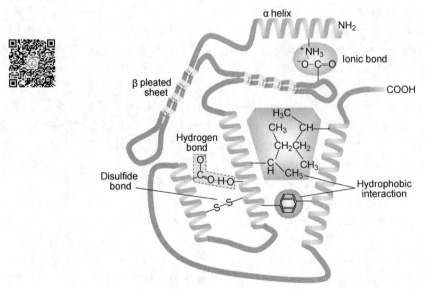

Figure 1.12　The molecular interactions of protein folding

图1.12　蛋白质折叠的分子相互作用

A **covalent bond** is the strongest bond that can join atoms. Covalent bonding occurs when a pair of electrons is shared between two atoms. Amino acids are held together in their primary structure by covalent bonds. In one case, covalent bonds can form between amino acids that are not adjacent in the polypeptide. This is only possible between cysteine amino acids, which have an —SH group. Two —SH groups react to form a covalent bond, called a **disulfide bond**, that can hold distant parts of the protein together tightly.

共价键是能连接原子的最强的键。共价键是在一对电子被两个原子共享时形成的。氨基酸在它们的初级结构中就是由共价键连接在一起的。在一种情形下，共价键可以在多肽中并不相邻的两个氨基酸之间形成。这只有在半胱氨酸之间才有可能，半胱氨酸含有—SH基团。两个—SH基团反应形成共价键，称为**二硫键**，它能把蛋白质中相距很远的部分紧密地连接在一起。

Hydrogen bonds are much weaker than covalent bonds, but they are abundant, and strong enough to greatly affect protein folding. Hydrogen bonding occurs with hydrogen atoms that are bound to electronegative atoms like oxygen or nitrogen. These change electron density in the hydrogen, giving it a partial positive charge. An atom with a partial negative charge, like a nitrogen in an amine, or an oxygen in a carbonyl group, is attracted to the partially charged hydrogen. This attraction causes what is called a hydrogen bond.

Ionic bonds are interactions that occur because of attraction between two ions. Ions are atoms that have gained or lost one or more electrons and therefore have full positive or negative charges. Opposite charges attract each other; accordingly, ions with opposite charges are drawn to each other, forming a kind of bond. This is also called an electrostatic interaction. Unlike covalent bonds and hydrogen bonds, electrostatic interactions can act at a distance, attracting or repelling ions that are not immediately adjacent in space. In proteins most ionic interactions involve amino acids with ionic R groups, such as lysine and glutamic acid.

Relative to the bonds described above, **Van der Waals forces** are extremely weak. However, unlike the bonds already mentioned, these interactions can occur between any two atoms. Van der Waals forces occur because the cloud of electrons around an atom is sometimes shifted, causing the atom to have a little bit of positive charge in one area, and negative charge in another area. This induces similar shifts in neighboring atoms. Areas with slightly different charge concentrations can then become attracted to each other, causing transient interactions.

Hydrophobic interaction is the tendency for non-polar molecules to gather near each other, away from water molecules (Figure 1.13). Non-polar molecules and water do not interact strongly with each other. Therefore, when non-polar molecules mix with water they force the water molecules into relatively ordered structures that surround the non-polar molecules. However, according to the second law of thermodynamics, systems always tend toward increasing disorder. Thus, it

氢键比共价键弱得多，但它们数量很大，因而足够强，从而对蛋白质的折叠产生很大影响。氢键发生在与具有电负性的氧原子或氮原子连接的氢原子上。这种氧原子或氮原子会改变氢原子的电子密度，使氢原子带部分正电荷。一个具有部分负电荷的原子，比如氨基上的氮或羰基上的氧，会被吸引到带部分正电荷的氢上。这种吸引造就了氢键。

离子键是因为两个离子之间的吸引而发生的相互作用。离子是获得或失去了一个或多个电子的原子，它们具有完全的正电荷或负电荷。异性电荷互相吸引；相应地，具有相反电荷的离子也会被吸引到一起，形成一种键。这也称为静电相互作用。与共价键和氢键不同，静电相互作用能够远距离发挥作用，可以使空间上不是紧邻的离子发生吸引或排斥。在蛋白质中，绝大多数离子相互作用发生在具有离子性质的R基团之间，比如赖氨酸和谷氨酸。

相对于上述几种键而言，**范德瓦耳斯力**是非常弱的。然而，与已经提及的几种键不同的是，这些相互作用可以发生在任意两个原子之间。范德瓦耳斯力发生的原因是围绕一个原子的电子云有时候会发生变化，使原子在某一区域带有一小部分正电荷，而在另一区域带有一小部分负电荷。这会诱使邻近的原子发生类似的变化。具有轻微不同电荷密度的区域可以因此互相吸引，引起短暂相互作用。

疏水相互作用是非极性分子远离水分子而互相聚集在一起的一种倾向（图1.13）。非极性分子和水不会发生强烈的相互作用。因此，当非极性分子与水混合时，疏水相互作用迫使水分子在非极性分子周围形成相对较为有序的结构。然而，根据热力学第二定律，系统总是倾向于增加无序性。因此，对于这两种分子来说，采取互相回避的策略更

is more realistic for the two kinds of molecules to simply avoid each other. The apparent result is that the non-polar molecules are attracted to each other. In reality, they are forced together because they do not mix well with water. For proteins that exist in water, the effect of this hydrophobic interaction is that non-polar amino acids are forced into the center of the protein, away from the watery environment of the cell.

为现实。表观的结果是非极性分子被相互吸引到一起。真实情形是它们被挤到了一起，因为它们不能与水很好地混合。对于水中的蛋白质而言，疏水相互作用的效果是：非极性氨基酸被挤到了蛋白质的中心——一个细胞中远离水环境的区域。

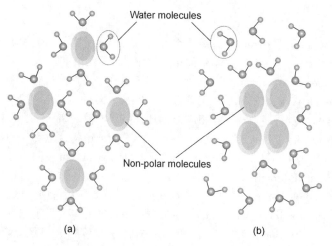

(a) Water molecules form relatively ordered structures around non-polar molecule
(b) Non-polar molecules are forced together by water molecules

(a) 水分子在非极性分子周围形成较为有序的结构
(b) 非极性分子被水分子挤到了一起

Figure 1.13 Hydrophobic interaction

图 1.13 疏水相互作用

1.3 Protein Structure and Function: A Few Examples

1.3 蛋白质的结构与功能：几个例子

Depending on which amino acids are polymerized and in what order, proteins can have many different sizes and shapes. Some proteins are big, some are small, dense or hollow, long or straight. Structure plays a major part in determining how proteins function. In this section we introduce a few proteins to illustrate this principle.

依赖于由哪些氨基酸以什么顺序聚合，蛋白质会具有许多不同的尺寸和形状。有些蛋白质很大，有些很小，有些紧密，有些中空，有些细长，有些挺直。结构在决定蛋白质如何发挥作用方面起主要作用。本节我们介绍几种蛋白质，来说明这一原理。

Under the physical conditions of the cell, many important chemical reactions occur very slowly. Proteins called **enzymes** act as catalysts, accelerating reactions so that they occur quickly enough to be useful. An example of an enzyme is carbonic anhydrase (Figure 1.14). Carbonic anhydrase catalyses the reaction $CO_2 + H_2O \longrightarrow H_2CO_3$, an important reaction that allows CO_2 to dissolve in blood. It is remarkably efficient, causing this reaction to occur 10 000 000 times faster than normal!

在细胞的物理条件下，许多重要的化学反应发生得相当缓慢。被称为**酶**的蛋白质能起催化剂的作用，它们可以加速反应以便反应发生得足够快从而变得更为有用。碳酸酐酶是一个例子（图 1.14）。它催化 $CO_2 + H_2O \longrightarrow H_2CO_3$ 这一反应，这是一个使 CO_2 得以溶解在血液里的重要反应。碳酸酐酶效率非常高，能使反应的速度比普通情况快 10 000 000 倍！

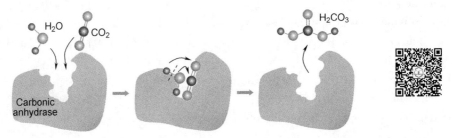

Figure 1.14 Schematic of carbonic anhydrase with active site

图 1.14 碳酸酐酶活性位点示意图

In order for enzymes like carbonic anhydrase to function, substrate molecules bind to the protein in an area called the **active site**. This can only happen if the active site has a structure that tightly accommodates the molecules. The site must also be shaped so that it rejects molecules that are not involved in the reaction. Within the active site, amino acids with specific chemical properties are positioned just right so that their atoms make contact with the atoms of the substrates. These interactions are crucial for stabilizing intermediate structures in the reaction.

为了使碳酸酐酶这样的酶能发挥作用，需要底物分子结合到它的一个称为**活性位点**的区域。这只有当活性位点具有能很好适应底物分子的结构时才能实现。这个位点还必须具有适当的形状，这样才能将不参与反应的分子排斥在外。在活性位点内部，具有特殊化学性质的氨基酸被置于正确的位置，使它们的原子与底物的原子接触。在反应中，这些相互作用对保持中间结构的稳定是非常重要的。

While certain protein structures are particularly well suited for catalyzing reactions, many structures are better suited for other roles. **Actin** is a relatively small and simple protein, but it can join with other actin proteins to form long, thin structures called **actin filaments** (Figure 1.15). These filaments line the inside of cells and are often used to provide support, some-

一些蛋白质结构特别适合起催化反应的作用，而更多结构更适合于发挥其他作用。**肌动蛋白**是较小、较简单的蛋白质，但它能与其他肌动蛋白形成长的、细的结构，称为**肌动蛋白丝**（图1.15）。肌动蛋白丝排列在细胞的内部，通常用来提供支撑，有些像撑起帐篷的柱子。

(a)
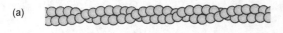
○ An actin monomer

(b)

(a) The actin filament
(b) Actin filaments in a cell as seen by fluorescence microscopy
  (Source: http://www.wikipedia.org)

(a) 肌动蛋白丝
(b) 荧光显微镜下所看到的肌动蛋白丝
  （来源： http://www.wikipedia.org）

Figure 1.15 Actin filament

图 1.15 肌动蛋白丝

what like poles holding up a tent. The structure of the actin protein underlies its function on two levels. First, each individual actin protein has a surface structure that allows it to bind to other actin proteins, so that filaments can be formed. Second, the long and rigid structure of the filaments that form from joining actin proteins is well-suited for providing support to the cell.

We have already seen several protein functions that are determined by structure, from catalysis, to protein binding, to cellular support. In a class of proteins called motor proteins, three-dimensional structures dynamically change, allowing the polypeptide to move and exert mechanical forces. The structure of a motor protein called **kinesin** even includes a pair of legs! Kinesin is used to carry cellular components, like organelles, from one end of a cell to another (Figure 1.16). The cargo is bound to central head region, while the two legs attach to microtubules. A reaction in a catalytic domain of kinesin provides energy which is used to move the legs. As a result, the protein walks along the microtubule toward its destination.

肌动蛋白的结构在两个层面上说明了它的功能。第一，每个肌动蛋白分子具有一个可以与其他肌动蛋白结合的表面结构，这样就能形成丝状结构。第二，由肌动蛋白连接而成的长而坚硬的丝状结构很适合为细胞提供支撑。

我们已经看到了几种蛋白质结构决定其功能的例子，包括催化作用、蛋白质结合作用以及细胞支撑作用。在一类叫作动力蛋白的蛋白质中，其三维结构可发生动态的改变，使多肽发生位移并施加机械力。一种叫作**驱动蛋白**的动力蛋白结构中甚至包含了两条腿！驱动蛋白能够将像细胞器那样的细胞组分从细胞的一端运送到另一端（图1.16）。被运送的货物绑在中央的头部区域，两条腿则附着在微管上。在驱动蛋白的催化功能域发生的反应提供了移动双腿所需要的能量。结果，驱动蛋白沿着微管向它的目的地行进。

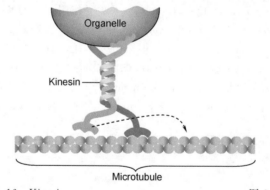

Figure 1.16　Kinesin

图 1.16　驱动蛋白

There are many proteins in the cell, and a huge array of functions and structures. We have given only a few examples from among these to illustrate the importance of structure for the function of a protein. We will encounter more with every chapter of the book.

在细胞中的许多蛋白质具有极其多种多样的功能和结构。我们已经给出了少数几个例子来说明蛋白质结构对功能的重要性。在本书的每一章中，我们会遇到更多这样的例子。

1.4　The Dynamics of Proteins

1.4　蛋白质动力学

The example of kinesin movement illustrates an important point that can be easily forgotten when learning

关于驱动蛋白运动的例子让我们看到了一点：蛋白质是动态的，这一重点在学

about proteins; they are dynamic. The structure of a protein is not frozen in place, contrary to what pictures in a book would lead you to think. Firstly, the atoms in a protein are always moving a little bit and colliding with their environment. The amount of movement in a region of a protein depends on how tightly the amino acids in that region are bound to each other. α helices and β sheets do not move very much, but the unstructured loops in between may move quite a lot.

Aside from the small random movements within a protein, large movements can occur that significantly alter protein structure, especially during catalysis. For instance, the enzyme hexokinase adds a phosphate group to glucose. Binding of glucose to this enzyme causes a shift in the structure of the protein which brings the glucose molecule into close contact with a terminal phosphate of ATP that is initially bound at a distant site. A catalysis mechanism in which binding of substrate causes such a large structural change that promotes catalysis is called an **induced-fit mechanism**. In some cases catalysis occurs without a change in structure at the active site; this is called a **lock-and-key mechanism** (Figure 1.17).

习蛋白质的时候很容易被忘记。与书中图片会让你认为的那样相反，蛋白质的结构并不是一成不变的。首先，蛋白质中的原子总是在做运动而与环境中的分子发生碰撞。在蛋白质的某一区域中这种运动强烈与否取决于这一区域中氨基酸之间互相结合的紧密程度。α螺旋和β折叠运动幅度不大，但将它们连接起来的环因为没有固定结构运动幅度就很大。

除了蛋白质中小的随机运动以外，也会发生能显著改变蛋白质结构的大幅运动，尤其在催化作用中。例如，已糖激酶催化将磷酸基团加到葡萄糖上的反应。葡萄糖与该酶的结合会引起蛋白结构的变化，这一变化使葡萄糖分子接近ATP的末端磷酸基团，该基团原先位于远离葡萄糖分子的位置。这种由于底物的结合引起较大结构变化从而促进催化作用的机理称为**诱导-契合机理**。在有些情形下，活性位点的结构不发生变化，这种催化机理被称为**锁-钥机理**（图1.17）。

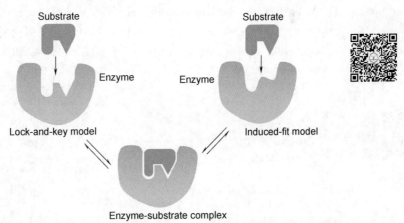

Figure 1.17 Induced-fit vs. lock-and-key models

图1.17 诱导-契合与锁-钥模型

The change in shape of the active site is still relatively minor relative to the structure of the entire protein. Very large changes in protein structure are possible too, as we saw with the example of kinesin. Many other examples exist of proteins that undergo major structural movements. The protein ATP-synthetase, for example, has a collection of subunits that spin when the

活性位点形状的改变相对于整个蛋白质结构来说还是较小的变化。但如我们从驱动蛋白例子中看到的那样，蛋白质结构发生较大变化也是可能的。还有许多蛋白质结构可以发生更大的变化。例如，ATP合成酶具有一组亚基，当它暴露在质子梯度中时，这些亚基会发生

protein is exposed to a gradient of protons. RNA polymerase, a protein we will be familiar with in future chapters, is a collection of several subunits that resembles a crab claw. When it first binds to DNA, the claw is open, but before beginning to make RNA it snaps shut.

Just as it is inaccurate to think of proteins as having a very fixed structure, it is also unrealistic to think of them as having a fixed place. Most proteins do not move about the cell like kinesin, using ATP to take steps in a chosen direction. Many proteins move around by random motion, colliding with other randomly moving molecules in the cell.

If a protein's structure and location can change, it should not be surprising that a protein's function can change as well. The activity of some proteins, for example, can be regulated by the binding of a small molecule to a regulatory site on the protein. This is called **allosteric regulation** (Figure 1.18). An example of allosteric regulation is seen in a protein called the *lac* repressor, that we will encounter in Chapter 3. This protein normally binds to DNA. However, when a small molecule called lactose binds to the regulatory site on this protein, it can no longer bind to DNA. This occurs because binding of lactose in the regulatory domain causes a structural change in the DNA-binding domain of the protein, which affects the protein's function.

旋转。在后续章节中我们将会非常熟悉的 RNA 聚合酶具有由多个亚基组合成的与蟹钳相似的形状。当它初次结合 DNA 时，钳是张开的，但在开始产生 RNA 之前，它会及时关闭。

正如把蛋白质想象成具有非常固定的结构是不准确的那样，认为蛋白质会待在固定的地方也是不现实的。大多数蛋白质并不像驱动蛋白那样能做相对于细胞的运动（驱动蛋白利用 ATP 来朝着选定的方向迈步）。许多蛋白质以随机方式移动，与细胞中其他随机运动的分子发生碰撞。

如果蛋白质的结构和位置可以发生改变，那么它的功能也可以发生变化就不足为奇了。例如，某些蛋白质的活性会因为结合到它的调控位点上的小分子而发生改变。这叫作**别构调节**（图 1.18）。别构调节的例子见于 *lac* 阻遏蛋白，我们将在第 3 章遇到它。这种蛋白质一般情况下会与 DNA 结合。然而，当像乳糖这样的小分子结合到它的调控位点上时，它就不再与 DNA 结合了。之所以发生这种变化是因为乳糖在调控区域的结合引起了蛋白质的 DNA 结合域结构发生变化，从而影响了蛋白质的功能。

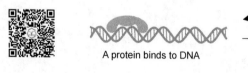

Figure 1.18　Allosteric regulation

图 1.18　别构调节

In many cases, protein function is altered by association with other proteins. One protein can have several functions depending on which combinations of proteins it is associated with. There are also examples of protein function being changed by altering the composition of the protein. A chemical group can be covalently attached to a protein, for example, to regulate its activity. We will see many more examples of how proteins are regulated throughout the course of the book.

在很多情形下，蛋白质通过与其他蛋白质相联合而改变功能。一种蛋白质可以具有几种功能，这取决于它与其他蛋白质形成什么样的组合。也有蛋白质通过改变它们自身的组成而改变功能的例子。例如，可以将化学基团共价连接到蛋白质上以调节它的活性。在本书的后续章节中我们会看到更多有关蛋白质调控的例子。

1.5 Experiments

One of the important principles of this chapter is that the amino acid sequence of a protein, its primary structure, is enough to determine its full three-dimensional conformation. This was demonstrated in the 1960s by a simple experiment now known as the '**Anfinsen experiment**', after the scientist Christian Anfinsen (Figure 1.19).

1.5 实验研究

本章的重点原理之一是：蛋白质的一级结构，即它的氨基酸顺序，足以决定它的全部三维构象。这一原理在20世纪60年代被一个简单的实验所证实，即根据科学家克里斯蒂安·安芬森的名字命名的"**安芬森实验**"（图1.19）。

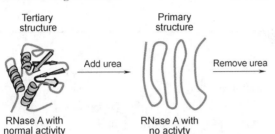

Figure 1.19　The Anfinsen experiment

图1.19　安芬森实验

A protein was put in solution and its activity was demonstrated relative to a substrate. The protein was an enzyme called RNase A that cleaves substrate RNA. Next, a high concentration of the molecule urea was added to the solution. Urea forms bonds with the amino acids in the protein, disrupting their normal interaction with each other. Although this does not affect the primary structure of the protein, it undoes its normal three-dimensional conformation. This is called **denaturing** the protein. The effect of the urea on the structure of the protein was evident because the RNase A protein was no longer able to cleave RNA.

一种蛋白质被放在了一种溶液中，其活性也用一种底物得到了证实。该蛋白质是一种称为RNase A的酶，它能切割底物RNA。然后，向溶液中加入高浓度的尿素。尿素与蛋白质中的氨基酸形成化学键，破坏了它们之间的正常相互作用。虽然这没有影响到蛋白质的一级结构，但它破坏了蛋白质正常的三维构象。这称为蛋白质的**变性**。尿素对蛋白质结构的影响得到了证实，因为该RNase A蛋白不能再切割RNA。

Finally, urea was removed from the solution, and the original solution around the protein was restored. Remarkably, the protein's activity, which had previously been eliminated, was also restored and it could cleave RNA. If the protein is functional after having been denatured, that means its normal three-dimensional structure has been restored. The only way this is possible is if the unfolded, primary structure of the protein that was produced during the urea treatment was able to refold automatically into its normal tertiary structure.

最后，从溶液中去掉尿素，这样，原先在蛋白质周围的溶液得到了恢复。值得注意的是，刚才被消除了的蛋白质活性也被恢复了，它又能切割RNA了。如果蛋白质在被变性之后又具备了功能，这说明它的正常三维结构已经恢复了。使这种现象成为可能的唯一方式是：经过尿素处理产生的蛋白质解折叠的一级结构能够自动地重新折叠成它正常的三级结构。

Summary

Most cellular activity is performed by proteins. The variety of amino acids and their chemical properties

小结

绝大多数细胞活动是通过蛋白质来实现的。不同氨基酸种类和它们的化学性质

allow the formation of countless different polypeptides, each with unique, complicated three-dimensional structures. These structures, and the chemical properties of the amino acids arranged in the structures, allow proteins to take on the multitude of functions required for life to exist. Although protein structures are highly specified for function, they should not be thought of as frozen. Proteins structures can move, and proteins themselves can move. As a result, proteins can even have dynamic functions.

导致了无数不同的多肽的形成，每种多肽都拥有独特的、复杂的三维结构。这些结构和形成这些结构的氨基酸的化学性质一起使蛋白质能够具备生命生存所需要的功能多样性。虽然蛋白质结构在功能上具有高度特异性，但不能认为它们具有固定的模样。蛋白质结构可以发生变化，它们自身也能移动，这使得蛋白质甚至具有动态的功能。

Vocabulary 词汇

acidic [əˈsidik]	酸性的	ionic bond [aiˈɔnik]	离子键
actin [ˈæktin]	肌动蛋白	isoleucine [ˌaisəuˈluːsiːn]	异亮氨酸
actin filament [ˈfiləmənt]	肌动蛋白丝	kinesin [kaiˈnesin]	驱动蛋白
alanine [ˈæləniːn]	丙氨酸	leucine [ˈluːsiːn]	亮氨酸
allosteric regulation [ˌæləˈsterik]	别构调节	lock-and-key model	锁-钥模型
amino acid [ˈæminəu] [ˈæsid]	氨基酸	lysine [ˈlaisiːn]	赖氨酸
amino group [ˈæminəu]	氨基	methionine [meˈθaiəniːn]	甲硫氨酸
arginine [ˈɑːdʒinin]	精氨酸	negative charge	负电荷
asparagine [əsˈpærədʒiːn]	天冬酰胺	non-polar molecule [ˈnɔnˈpəulə] [ˈmɔlikjuːl]	非极性分子
aspartic acid [əˈspɑːtik]	天冬氨酸		
ATP synthetase [ˈsinθiteis]	ATP 合成酶	N-terminus [ˈtəːminəs]	氨基末端
basic [ˈbeisik]	碱性的	organism [ˈɔːgənizəm]	生物体
carbonic anhydrase [kɑːˈbɔnik] [ænˈhaidreis]	碳酸酐酶	partial double bond	部分双键
		partial positive charge	部分正电荷
carbonyl group [ˈkɑːbənil]	羰基	peptide bond [ˈpeptaid] [bɔnd]	肽键
carboxyl group [kɑːˈbɔksil]	羧基	phenylalanine [ˌfenəlˈæləniːn]	苯丙氨酸
conformation [ˌkɔnfɔːˈmeiʃən]	构象	polymer [ˈpɔlimə]	聚合物
covalent bond [kəuˈveilənt]	共价键	positive charge	正电荷
C-terminus	羧基末端	primary structure	一级结构
cysteine [ˈsistin]	半胱氨酸	proline [ˈprəuliːn]	脯氨酸
disulfide bond [daiˈsʌlfaid]	二硫键	protein [ˈprəutiːn]	蛋白质
dynamics [daiˈnæmiks]	动力学	protein folding	蛋白质折叠
free rotation	自由旋转	quaternary structure [kwəˈtəːnəri]	四级结构
glutamic acid [ˈgluːtæmik]	谷氨酸	secondary structure	二级结构
glutamine [ˈgluːtəmiːn]	谷氨酰胺	serine [ˈseriːn]	丝氨酸
glycine [ˈglaisiːn]	甘氨酸	β sheet	β折叠
α helix [ˈhiːliks]	α 螺旋	side chain	侧链
histidine [ˈhistidiːn]	组氨酸	tertiary structure [ˈtəːʃəri]	三级结构
hydrogen bond [ˈhaidrədʒən]	氢键	threonine [ˈθriːəniːn]	苏氨酸
hydrophilic [ˌhaidrəuˈfilik]	亲水的	tryptophan [ˈtriptəfæn]	色氨酸
hydrophobic [ˌhaidrəuˈfəubik]	疏水的	tyrosine [ˈtirəsiːn]	酪氨酸
hydrophobic interaction	疏水相互作用	valine [ˈvæliːn]	缬氨酸
induced-fit model	诱导-契合模型	Van der Waals force	范德瓦耳斯力

Review Questions 习题

Ⅰ. True/False Questions（判断题）

1. Proteins may have different amino acids, but are all of the same length.
2. There are twenty major proteins in the cell.
3. Arginine is usually found on the inside of folded cytoplasmic proteins.
4. All amino acid R groups are acidic.
5. Hydrophobic effect occurs because hydrophobic molecules are attracted to each other.
6. The N-terminus of a protein is usually positively charged.
7. An α helix can be formed from any polypeptide.
8. Proteins can only bind to other proteins.
9. Kinesin is another word for motor protein.
10. A protein always begins to function as soon as it is made by the ribosome.

Ⅱ. Multiple Choice Questions（选择题）

1. Which of the following amino acids is not hydrophobic?
    a. valine
    b. leucine
    c. lysine
    d. isoleucine
    e. glycine

2. Which atom is not found in the 20 common amino acids?
    a. carbon
    b. nitrogen
    c. oxygen
    d. sulfur
    e. fluorine

3. How many different polypeptides of 100 amino acids is it possible to make?
    a. 100
    b. 1000
    c. 33333
    d. 100^{20}
    e. 20^{100}

4. Which of the following is not an amino acid?
    a. glucine
    b. glutamine
    c. leucine
    d. tryptophan
    e. tyrosine

5. Which of the following conditions does not usually denature proteins?
    a. neutral pH
    b. high temperature
    c. high salt concentration
    d. presence of urea
    e. high concentration of acid

Exploration Questions 思考题

1. Do you think the Anfinsen experiment would have worked with most proteins in the cell? Why or why not?
2. What are some of the changes that might happen to a protein after it is translated?
3. Do you think that two proteins could act exactly the same without having exactly the same amino acid sequence and composition?
4. Do you know of any proteins that were not mentioned in this chapter? How does their structure and chemistry help them to achieve their function?

Chapter 2　Nucleic Acids

Proteins are extremely complex molecules, well-suited for their functions. Something must hold the information for making proteins, because there is no chance that they could spontaneously arise in each new generation. Indeed, molecules called **nucleic acids** hold the information for how to make all of the proteins in the cell. In this chapter we examine nucleic acids and how they are suited for their functions.

2.1　Properties of a Genetic Material

What holds the information for making proteins? Since proteins produce the traits of an organism, this is similar to asking: What determines the traits of an organism? Vaguely speaking, this material is called the **genetic material.**

Using macroscopic observations of life, we know that the genetic material must have certain properties.

(1) **It must be able to hold information for how to make proteins**

(2) **It must be stable**

Fish give birth to fish, and humans give birth to humans. The necessary traits of organism are passed down from generation to generation with very little change, often for millions of years. There are fish alive today that have essentially the same traits as fish **100** million years ago. Thus, the genetic material must be stably inherited by offspring from parents. It must also be stable for long periods of time within one generation. The traits of an organism usually stay unchanged during its entire lifetime, even if it is very long lived. Again, this indicates that the genetic material that directs these traits must be stable.

(3) **It must have some capacity for change**

Although we have said that traits can be very stably inherited for a long period of time, some change in traits must occur, at least on a large time scale. This is clear because otherwise evolution would not be possible. Human ancestors, millions of years ago, were ape-like animals. Many of our traits have remained the same since then, but clearly many have also changed, such as hairiness and intelligence. Therefore, the genetic material, which directs such traits, must have also changed since that time.

(4) It must be able to be copied

Two parents can give birth to many children. Each child is a complete organism, with almost all the same components as the parents. This means that each child must receive his own complete set of the genetic material. If this is to happen generation after generation, the genetic material must be able to be copied.

2.2 Nucleic Acids and DNA structure

Fortunately for life on earth, a class of molecules does exist with all four necessary properties of a genetic material. These molecules are called nucleic acids. We first describe the structure and composition of nucleic acids, and then return to explain how they are suited for their roles as the carriers of traits, holding information for how to make proteins.

2.2.1 Nucleotides

Like proteins, nucleic acids are long polymers of smaller molecules. However, instead of using amino acids as monomers, nucleic acids contain small molecules called **nucleotides** (Figure 2.1). Nucleotides are often understood as a collection of three parts: a **nitrogenous base**, a sugar, and a triphosphate. The sugar is at the center of the nucleotide and has five carbons. It forms a pentagonal ring, with four carbons and one oxygen. The fifth carbon of the sugar sticks out from the ring. Carbons in the sugar are numbered by convention, as in the figure. The triphosphate group, consisting of three covalently bonded phosphates, is located on the fifth carbon.

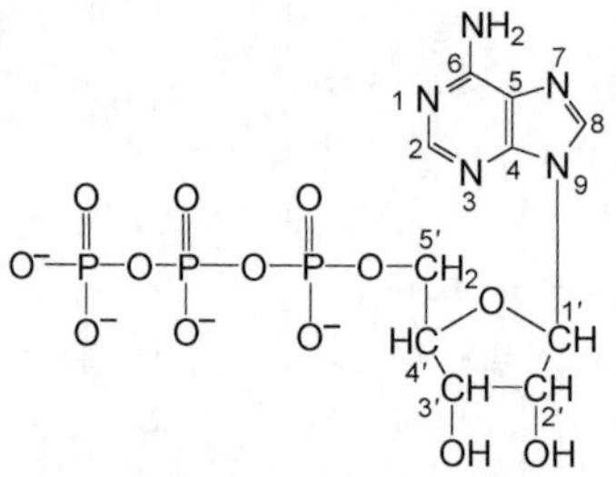

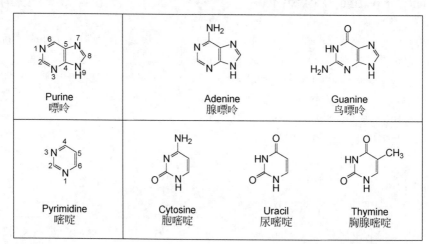

Figure 2.1　General structure of a nucleotide

图 2.1　核苷酸的一般结构

The **nitrogenous base** of a nucleotide is connected to the 1′ carbon of the sugar by an *N*-glycosidic bond. Each nucleotide has one base. Bases are ringed structures (Figure 2.2). They can contain one or two rings. Bases with one ring are called **pyrimidines**; bases with two rings are called **purines**. The name nitrogenous base comes from the fact that each ring has two nitrogen atoms. A variety of different chemical groups, such as amino and carbonyl groups, may be attached to the rings to give each base its unique chemical properties.

核苷酸的**含氮碱基**通过 *N*-糖苷键连接到糖的 1′碳上。每一个核苷酸含有一个碱基，碱基具有环状结构（图 2.2）。它们含有一个或两个环。含有一个环的碱基叫作**嘧啶**；含有两个环的碱基叫作**嘌呤**。含氮碱基的名称来自每个环都含有两个氮原子这样的事实。一些不同的化学基团，如氨基和羰基，可以连接到这些环上，从而赋予每种碱基独特的化学性质。

Figure 2.2　Bases of DNA and RNA

图 2.2　DNA 和 RNA 中的碱基

Most nucleotides can be broadly divided into two groups according to which kind of five-carbon sugar they contain (Figure 2.3). Nucleotides containing the sugar ribose are called **ribonucleotides**. Those containing a deoxyribose are called **deoxyribonucleotides**. Deoxyribose and ribose are exactly the same except that the former does not have a hydroxyl group attached to 2′ carbon.

根据所含五碳糖的种类不同，大多数核苷酸可以被大体上分为两组（图 2.3）。含有核糖的核苷酸叫作**核糖核苷酸**，含有脱氧核糖的核苷酸叫作**脱氧核糖核苷酸**。脱氧核糖除了在 2′碳上没有羟基以外与核糖完全相同。

Four deoxyribonucleotides are commonly found in cells.

在细胞中有四种常见的脱氧核糖核苷

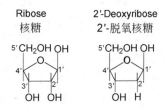

Figure 2.3　Ribose vs. Deoxyribose

图 2.3　核糖与脱氧核糖

Each one is made using a different base：**Adenine**（A），**Guanine**（G），**Cytosine**（C），or **Thymine**（T）. These bases are very frequently referred to using the first letter of the name. A and G are purines，while C and T are pyrimidines. Nucleotides that are made using these bases are respectively called adenylate, guanylate, cytidylate, and thymidylate.

酸，每种分别包含不同的碱基：**腺嘌呤**（A）、**鸟嘌呤**（G）、**胞嘧啶**（C）或**胸腺嘧啶**（T）。这些碱基常常用它们名称的第一个字母来表示。A 与 G 是嘌呤，C 与 T 是嘧啶。由这些碱基构成的核苷酸分别叫作腺苷酸、鸟苷酸、胞苷酸和胸苷酸。

Similarly，four different ribonucleotides are commonly found. A，G，and C are all used in ribonucleotides，but T is replaced by a slightly different base called **uracil**（U）. U is a pyrimidine（Figure 2.4）. The ribonucleotide containing U is called uridylate.

类似地，有四种常见的核糖核苷酸。A、G 和 C 均在核糖核苷酸中使用，但 T 被一种稍微不同的碱基即**尿嘧啶**（U）取代。U 是一种嘧啶（图 2.4）。含有 U 的核糖核苷酸叫作尿苷酸。

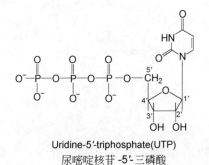

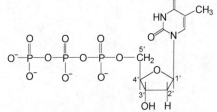

Figure 2.4　Uridylate and thymidylate，a ribonucleotide and a deoxyribonucleotide

图 2.4　尿苷酸和胸苷酸，一种核糖核苷酸和一种脱氧核糖核苷酸

2.2.2　General Structure of Nucleic Acids

Nucleotides are polymerized to form nucleic acids. If deoxy-ribonucleotides are used，the polymer is called **deoxyribonucleic acid**，or just DNA for short. If ribonucleotides are used，the polymer is called **ribonucleic acid**，or RNA for short. We will begin by discussing the structure of DNA，after which the structure of RNA will be easy to understand.

When nucleotides are polymerized，the triphosphate

2.2.2　核酸的一般结构

核苷酸聚合起来产生的就是核酸。如果使用的是**脱氧核糖核苷酸**，产生的聚合物是脱氧核糖核酸，或简称为 DNA。如果使用的是核糖核苷酸，产生的聚合物是**核糖核酸**，或简称为 RNA。我们先来讨论 DNA 的结构，之后就能较容易地理解 RNA 的结构了。

当核苷酸发生聚合的时候，一个核苷酸

group of one nucleotide reacts with the 3′-OH (hydroxyl) group of another (Figure 2.5). In this reaction, two phosphates are removed from one nucleotide, and the two nucleotides are connected by the single remaining phosphate. This is called a **phosphodiester bond**. The energy for forming the phosphodiester bond comes from breaking the high energy links between the first and second, and second and third phosphates in the triphosphate group. Actually, nucleotides have almost the exact same structure and reactivity as the cell's most common energy carrier, ATP.

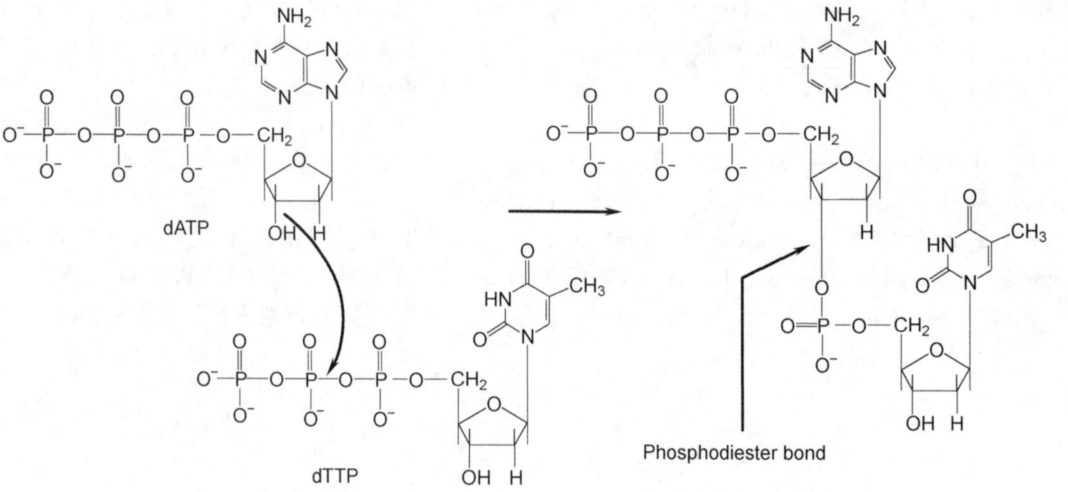

Figure 2.5 Polymerization reaction of nucleotides, creation of a phosphodiester bond

Both sides of each nucleotide can react with other nucleotides; the 3′-OH of nucleotide A can react with the 5′-triphosphate of nucleotide B, and the 3′-OH of nucleotide B can react with the 5′-triphosphate of nucleotide C, and so forth. This allows for the formation of polymers (Figure 2.6). DNA polymers can be quite long, sometimes millions of nucleotides in length. DNA can include any of the four common deoxyribonucleotides attached in any order.

Like peptides, the nucleic acid polymer is linear, meaning it has no branches. Stretched out, it has a simple structure. On one end of the polymer is the 3′-OH group of a nucleotide. This end is called the **three prime end** and is written 3′ end. On the other end is a

的三磷酸基团与另一个核苷酸的 3′-OH（羟基）发生反应（图 2.5）。在这一反应中，两个磷酸基团从一个核苷酸上移去，之后两个核苷酸由留下的单磷酸连接在一起，形成的键叫作**磷酸二酯键**。形成磷酸二酯键所需的能量来自打断三磷酸基团中第一个和第二个，以及第二个和第三个磷酸之间的高能键。实际上，核苷酸具有与细胞中最普遍的能量载体 ATP 几乎完全相同的结构和反应活性。

图 2.5 核苷酸的聚合反应——磷酸二酯键的产生

每个核苷酸的两端都能与其他核苷酸发生反应，核苷酸 A 的 3′-OH 可以与核苷酸 B 的 5′-三磷酸反应，之后核苷酸 B 的 3′-OH 又可以与核苷酸 C 的 5′-三磷酸反应，依次进行，导致聚合物的形成（图 2.6）。DNA 聚合物可以很长，有时长达几百万个核苷酸。DNA 可以由四种常见的脱氧核糖核苷酸以任何顺序连接在一起形成。

与肽一样，核酸聚合物是线性的，意味着它没有分支。伸展开来的话，它具有一种简单的结构。在聚合物的一个末端是 3′-OH。这一末端叫作**3′末端**。另一个末端是另一个核苷酸的 5′-三磷酸基

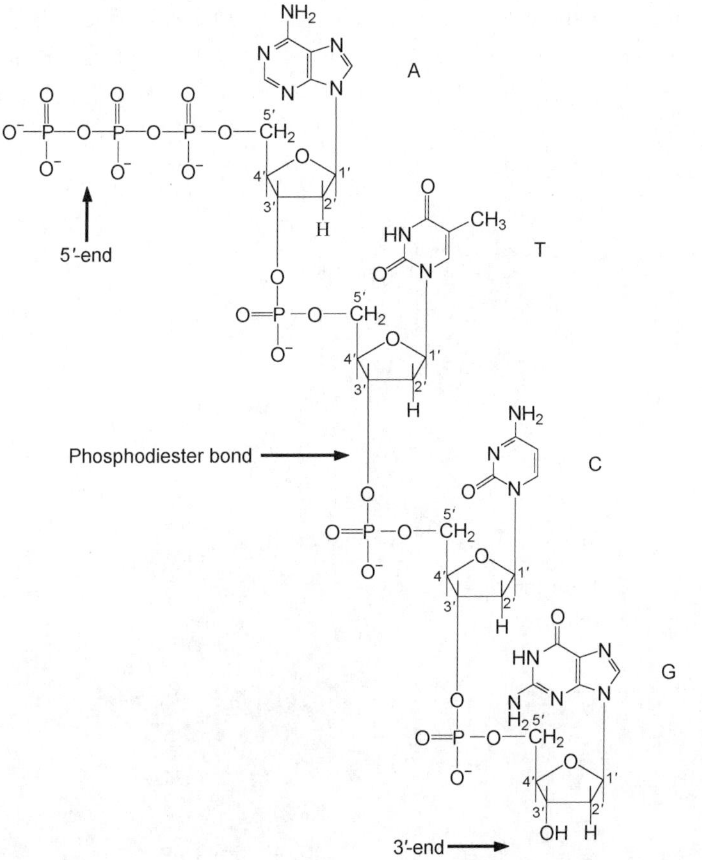

Figure 2.6 DNA, a polymer of deoxyribonucleotides. Labeling of 5′ and 3′ ends

图 2.6 DNA——脱氧核糖核苷酸的聚合物。标明了 5′ 和 3′ 末端

5′-triphosphate group of a different nucleotide. That is called the **five prime end**, and is written 5′ end.

If we ignore the bases on each nucleotide, between the two ends of the DNA is a simple, repeating structure of sugars connected by phosphodiester bonds. This structure is often called the **sugar-phosphate backbone** (Figure 2.7). The backbone structure does not depend on the identity of the bases in the DNA. The backbone has many negative charges due to the phosphodiester bonds that hold together the nucleotides. Phospodiester bonds are acidic; although oxygen in these bonds has the potential to interact with protons (H^+), at normal cellular pH values (around pH7), these oxygen atoms are typically not associated with protons, and display a negative charge as a result. It is the acidic nature of this bond that accounts for the word 'acid' in nucleic acid. Acids can be defined in several ways; the two most common definitions are the 'Bronsted-Lowry' definition and the 'Lewis' defini-

团。这一末端叫作 **5′ 末端**。

如果我们忽略每个核苷酸上的碱基，那么在两个末端之间 DNA 是简单的、重复的由磷酸二酯键连接起来的糖。这种结构常常被叫作**糖-磷酸骨架**（图2.7）。糖-磷酸骨架的结构与 DNA 由哪些碱基组成无关。骨架上带有许多负电荷，这些负电荷来自把核苷酸连接在一起的磷酸二酯键。磷酸二酯键是酸性的；虽然这些键中的氧原子能够与质子（H^+）发生相互作用，但在正常细胞 pH 值（约 pH7）条件下，这些氧原子不会与质子发生结合，结果使该键带上负电荷。正是磷酸二酯键的酸性特征说明了核酸中"酸"字的来源。对酸的定义有多种，两种最常见的定义是"布劳恩斯特德-劳沃里"定义和"刘易斯"定义。

2.2 Nucleic Acids and DNA structure

tion. The former states that acids are molecules or chemical groups that tend to lose protons (usually by donating them to water) at pH7. The Lewis definition states that acids are molecules or chemical groups that tend to accept pairs of electrons.

前者称，酸是一些在 pH7 条件下倾向于失去质子的（通常将质子贡献给水分子）的分子或化学基团。刘易斯定义中酸是一些倾向于获取电子对的分子或化学基团。

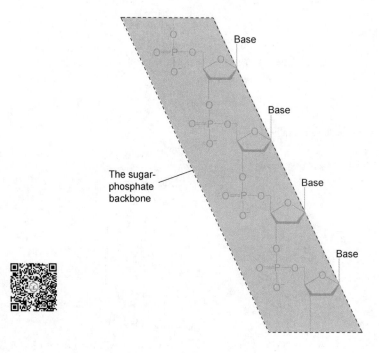

Figure 2.7　The sugar-phosphate backbone

图 2.7　糖-磷酸骨架

2.2.3　Structure of DNA

We have described DNA so far as a linear arrangement of covalently bonded nucleotides. We call this arrangement a **DNA strand**, or **single-stranded DNA**. In the cell, DNA strands rarely exist alone. Each strand is stuck to another strand, and the arrangement is called **double-stranded DNA**.

The two strands in a double-stranded DNA molecule are held together by hydrogen bonds between their bases (Figure 2.8). These bonds can only form correctly between adenine on one strand and thymine on the other, or between guanine on one strand and cytosine on the other (Figure 2.9). Bases that can correctly bond to each other are said to be **complementary**. Two hydrogen bonded nucleotides, each one located on a different strand, is called a **base pair**. Note that A-T base pairs are connected by two hydrogen bonds and G-C base pairs are connected

2.2.3　DNA 的结构

到现在为止，我们把 DNA 描述成是由共价键连接起来的核苷酸的线性排列。我们称这种排列为一条 **DNA 链**，或**单链 DNA**。在细胞中，DNA 很少以单链形式存在。每条链与另一条链黏附在一起，这种排列方式叫作**双链 DNA**。

在双链 DNA 分子中，两条链的碱基之间形成氢键从而将它们黏附在一起（图 2.8）。这些键只能在一条链的腺嘌呤和另一条链的胸腺嘧啶之间，或一条链的鸟嘌呤和另一条链的胞嘧啶之间正确形成（图 2.9）。能正确形成这种键的碱基是**互补**的。分别位于两条链上的能形成氢键的碱基叫作**碱基对**。注意，A-T 碱基对由两个氢键连接，G-C 碱基对由三个氢键连接。虽然氢键是较弱的化学

by three hydrogen bonds. Although hydrogen bonds are relatively weak, there are so many of them in a double-stranded DNA molecule that the two strands stay together quite strongly under physiological conditions.

键，但它们在双链 DNA 分子中数量是如此之多以至于在生理条件下能使两条链紧紧地结合在一起。

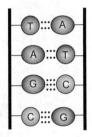

Figure 2.8　Double-stranded DNA　　　　　图 2.8　双链 DNA

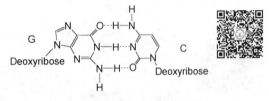

Figure 2.9　Pairing between complementary bases　　图 2.9　互补碱基之间的配对

The orientation of the two strands relative to one another is **anti-parallel** (Figure 2.10). Imagine a double-stranded DNA with one strand called X and the other strand called Y. On one end of the double-stranded DNA, strand X will have a 5′ end and strand Y will have a 3′ end. On the other end of the molecule, strand X will have a 3′ end and strand Y will have a 5′ end. This is what is meant by the term anti-parallel. The reason for this arrangement is that it puts bases in the right orientation to bind to each other. Parallel DNA strands cannot bind to each other.

两条链的方向相对于各自来说是**反向平行**的（图 2.10）。设想一条双链 DNA 由一条 X 链和另一条 Y 链组成。在双链 DNA 的一个末端，X 链具有 5′末端而 Y 链具有 3′末端。在它的另一个末端，X 链具有 3′末端而 Y 链具有 5′末端。这就是反向平行的含义。之所以采用这种排列方式是因为这样能使碱基处于正确的方向以互相发生结合。同向平行的 DNA 单链之间不能互相结合。

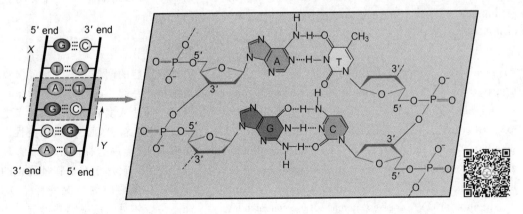

Figure 2.10　Antiparallel DNA strands　　　　图 2.10　反向平行的 DNA 链

The model described thus far may appear somewhat like a ladder, with paired bases bridging two straight strands. In reality, this ladder is twisted into a shape called a **double helix** (Figure 2.11). The double helix is a very compact structure. Each base pair lies parallel and very near to the base pair above and below. This is called **base-stacking**. The stacked bases face the inside of the molecule, while the sugar-phosphate backbone lines the outside.

到现在为止，我们描述出来的模型可能有些像梯子——由配对的碱基把两条伸直的链连接在一起。实际上，梯子是扭转成**双螺旋**形状的（图 2.11）。双螺旋是一种非常紧密的结构。每个碱基对平行伸展并且与上面的和下面的碱基对非常靠近。这一现象叫作**碱基堆积**。堆积起来的碱基朝向分子的内部，而糖-磷酸骨架则排列在外面。

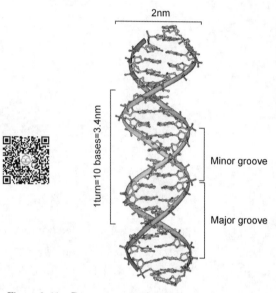

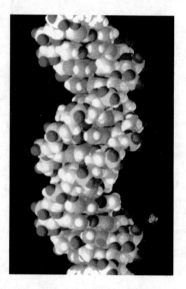

Figure 2.11 Representation of the DNA double helix (Source: http://www.wikipedia.org)

图 2.11 DNA 双螺旋图像 （来源：http://www.wikipedia.org）

The outer edges of the DNA molecule have grooves. The larger grooves are called **major grooves**, and the smaller ones are called **minor grooves**. Most proteins that bind DNA bind at the major groove. Every complete turn of the double helix is 3.4 nanometers long, and comprises approximately 10 base pairs. The double helix is roughly 2 nanometers wide.

DNA 分子的外边缘有沟，较大的沟叫作**大沟**，较小的沟叫作**小沟**。大多数与 DNA 结合的蛋白质结合在 DNA 的大沟上。双螺旋的每一圈长 3.4nm，由大约 10 个碱基对组成，双螺旋的宽度大约是 2nm。

Although the strands of the double helix are held tightly together by numerous hydrogen bonds, under special conditions such as high alkaline concentration, low salt concentration, or high heat, the two strands may come apart fully or partially. This is called **denaturation**. Denaturation also occurs as part of some normal cellular processes like DNA replication. As with proteins, denaturation of DNA does not damage the covalent bonds within each strand. Experimentally, denaturation is de-

虽然双螺旋的两条链被无数的氢键紧紧地维持在一起，在特殊的条件下，比如强碱、低盐浓度或高温下，两条链会完全或部分地被分开。这叫作**变性**。变性在像 DNA 复制这样的正常细胞过程中也会发生，它是正常细胞过程中的一个环节。与蛋白质的情况一样，DNA 变性不会破坏每条链中的共价键。实验时变性是通过测定样品在 260nm 处的光

termined by measuring the absorbance of light at 260nm. Single-stranded DNA absorbs light well at this wavelength, but double-stranded DNA does not. Therefore, measuring absorbance can indicate to what extent a DNA molecule is denatured (Figure 2.12). A DNA molecule that has been denatured can return to its previous, double-stranded state if the denaturing conditions are reversed. This is called **renaturation**.

吸收来检查的。单链 DNA 在此波长下吸收很多光，而双链 DNA 吸收较少的光。因此，测定样品的光吸收值便可以显示 DNA 分子变性到了什么程度（图 2.12）。如果变性条件被逆转的话，已经变性的 DNA 分子可以回复到先前的双链状态。这一过程称为**复性**。

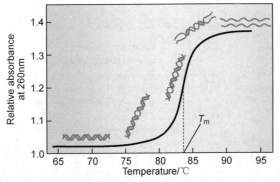

Figure 2.12 Denaturation of DNA. Absorbance and denaturation as a function of temperature

图 2.12 DNA 变性。光吸收和变性作为温度的函数

2.3 Molecular Identification

While renaturation means that the two reunited single strands are from the same DNA, **hybridization** refers to formation of a double-stranded molecule from two single strands of different origin. This property allows the development of various techniques for identifying specific DNA and RNA molecules.

2.3.1 Identification of DNA Molecules

A DNA molecule can be identified by using a labeled short DNA piece (called "**probe**") which is complementary to certain region of it. Such identification is widely used not only for biological studies but also in social activities. Human DNAs are fairly different with each other in location and number of "**minisatellites**", which are short repeated nucleotide segments. These differences constitute **DNA fingerprints** of different individuals.

To obtain DNA fingerprints, DNA samples are firstly treated with restriction endonuclease to yield nucleotide segments containing those with minisatellites (Figure 2.13).

2.3 分子鉴定

复性意味着两条重新结合的单链来自相同的 DNA，而**杂交**指的是不同来源的两条单链形成双链分子。这一特性导致针对特异性 DNA 和 RNA 分子研发出了不同的鉴定技术。

2.3.1 DNA 分子鉴定

DNA 分子可以使用标记的短 DNA 片段（称为"探针"）进行鉴定，探针与所需鉴定 DNA 分子的一部分是互补的。这样的鉴定不仅在生物学研究中应用广泛，在社会活动中也频繁使用。人类个体的 DNA 由于"小卫星"（即：短的重复核苷酸片段）位置与数量的不同而呈现差别。这些差别组成了不同个体的 **DNA 指纹**。

为获得 DNA 指纹，首先使用限制性核酸内切酶处理 DNA 样本以产生含有小卫星的核苷酸片段（图 2.13）。经凝胶电

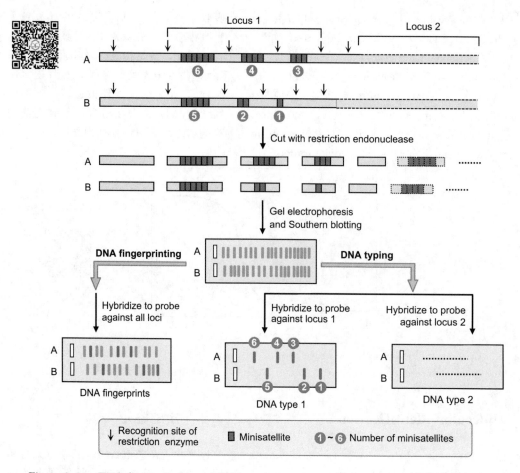

Figure 2.13　DNA fingerprinting and DNA typing

图 2.13　DNA 指纹分析与 DNA 分型术

After gel electrophoresis and **Southern blotting**, a whole set of nucleotide segments are obtained. Then, through using a probe that hybridizes to all minisatellite loci, DNA fingerprints containing complex bands are obtained. This process is termed **DNA fingerprinting**. Although the difference of DNA fingerprint between each sample can be seen, it is not sufficient for practical application especially when it is to be used as forensic evidence.

In actual practice, **DNA typing** is used because it presents a much less number of DNA bands. It uses three to five probes, each of which only hybridizes to a single minisatellite locus. Based on such **DNA type** that displays only a few bands, it is much easier to discriminate different individuals. As shown in figure 2.13, DNA type 1 of individual A is clearly different with that of individual B, because they (assuming that they have no genetic relationship) have three bands containing 6, 4,

泳和 **Southern 转印**后得到全套核苷酸片段。之后，通过使用与所有小卫星位点杂交的探针，得到含有复杂条带的 DNA 指纹，这一过程称为 **DNA 指纹分析**。虽然可以看出不同样本 DNA 指纹之间的差别，但这样的结果还不能满足实际应用，尤其是需要把它作为法庭证据的时候。

在实际应用中，使用的是显示很少 DNA 条带的 **DNA 分型术**。它使用三到五个探针，每个探针只会与一个小卫星位点杂交。根据这种只显示了少数条带的 **DNA 类型**，区分不同个体就容易多了。如图 2.13 所示，个体 A 的 DNA 类型 1 与个体 B 明显不同，因为他们（假设他们之间没有亲缘关系）在位点 1 分别含有 6、4、3 和 5、2、1 个小卫

3 and 5, 2, 1 minisatellites at locus 1, respectively. However, judgement based on only one DNA type could have an error rate of around 1%, meaning that one in a hundred people of random population will have the same DNA type with individual A or B. To solve this problem, three or more probes are used to obtain more DNA types for accurate judgement. For instance, if three probes are used in a parentage test and all three DNA types from individuals A and B are identical, it will be judged that the probability of parentage between them is considered to be 99.9999%.

2.3.2 Identification of RNA Molecules

An RNA molecule can be identified through firstly reverse-transcribing it into cDNA (**complementary DNA**) and then using specific primers to amplify the target fragment. If a test sample has the target fragment with which the specific primers can hybridize, this fragment will be amplified and detected with the help of a labeling agent. **Nucleic acid testing** against SARS-CoV-2 (severe acute respiratory syndrome coronavirus 2) presents a good example of this identification strategy.

SARS-CoV-2 has a genome of single-strand RNA with nearly 30,000 nucleotides. A test sample will be first extracted for viral RNA, then reverse-transcribed to obtain first-strand cDNA, and finally subjected to qPCR (quantitative polymerase chain reaction) for determination of **virus load** in the sample (Figure 2.14). The virus load is proportional to fluorescence produced from the labeling agent, which contains a fluorescent reporter dye (R) and a fluorescent quencher dye (Q). When coupled by a short nucleotide segment, the quencher blocks emission of fluorescence due to proximity. After the coupling nucleotides are degraded by polymerase (in synthesizing the second-strand cDNA), reporter dye emits fluorescence to display existence of the target fragment.

Cycle threshold value (**Ct value**) has been used to determine how many amplification cycles are needed for a test sample to cross the threshold of fluorescent emission. This value is inversely related to virus load of the sample. For example, sample A in figure 2.14 has a

星。然而，只根据一种DNA类型做出的判断有约1%的错误率，意味着随机人群里每一百人中有一人会具有与个体A或个体B相同的DNA类型。为解决这一问题，需要使用三个或更多探针，以获得更多DNA类型来做出准确判断。举例来说，如果在亲子鉴定中使用了三个探针，并且个体A与个体B的三种DNA类型都相同，那就会判断他们之间存在亲子关系的可能性是99.9999%。

2.3.2 RNA分子鉴定

RNA分子可以通过先反转录成cDNA（**互补DNA**）然后用特异引物扩增目标片段的方法进行鉴定。如果样本中含有能与特异引物杂交的目标片段，该片段就会被扩增，借助标记物便可以显现其存在。新型冠状病毒（严重急性呼吸综合征冠状病毒2）**核酸检测**即为该检测策略的良好实例。

新型冠状病毒的基因组是一条由近30000个核苷酸组成的单链RNA。测试样本先要用于提取病毒RNA，然后反转录获取cDNA第一链，最后使用qPCR（定量聚合酶链式反应）测定样本中的**病毒载量**（图2.14）。病毒载量与标记物所产生的荧光量成正比。标记物含有荧光报告染料（R）和荧光猝灭染料（Q）。当两者被短核苷酸片段联结在一起时，由于距离靠近，猝灭染料会阻碍荧光释放。当起联结作用的核苷酸片段被聚合酶（在合成cDNA第二链时）降解后，报告染料就能释放荧光，显示存在目标片段。

循环临界值（**Ct值**）被用来确定样本需要多少个扩增循环来跨越荧光释放的临界值。该值与样本中的病毒载量成反比。例如，图2.14中样本A的Ct值是20，意味着它的病毒载量是比较高的，

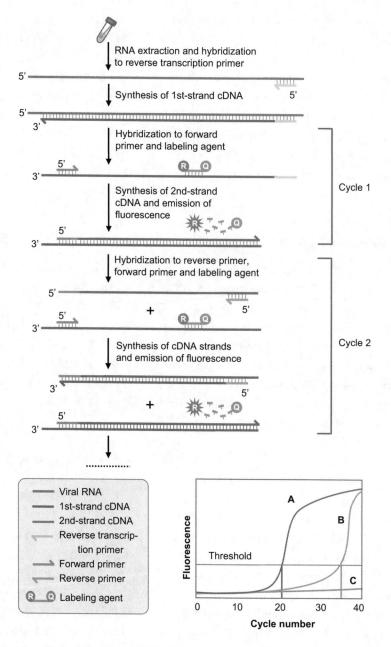

Figure 2.14　Nucleic acid testing against SARS-CoV-2

图 2.14　新型冠状病毒核酸检测

Ct value of 20, meaning that its virus load is relatively high because only 20 amplification cycles are needed for its fluorescent emission to cross the threshold. Sample B has a low virus load, because 35 amplification cycles are needed for its fluorescent emission to cross the threshold. There is no virus in sample C, because its fluorescent emission does not reach the threshold from the beginning to the end.

因为只需经过 20 个扩增循环它的荧光释放就超过了临界值。样本 B 中病毒载量较低，因为它的荧光释放需要经过 35 个扩增循环才能超过临界值。样本 C 中不含病毒，因为它的荧光释放自始至终都没有达到临界值。

2.4　DNA as the Genetic Material

2.4　DNA 作为遗传物质

By now we can begin to understand how DNA can be responsible for the functions necessary of a genetic material.

现在我们可以开始理解为什么 DNA 会具有作为遗传物质所必须具备的功能了。

① The sequence of bases in a DNA strand can be used to hold information about the sequence of amino acids that make proteins. Each amino acid in a protein is represented by a group of three bases in the DNA. We will learn much more about this in the following chapters.

① DNA 链中的碱基序列可以用来保存生产蛋白质的氨基酸序列信息。蛋白质中的每一个氨基酸都由 DNA 中的一组三个碱基表示。我们将在后续的章节中学到更多关于这方面的知识。

② The structure of DNA provides the stability required of a genetic material. The bases, which hold the information, are packed tightly together inside the helix, away from molecules in the environment that might harm them. The sugar-phosphate backbone, which does not hold information, lines the outside of the molecule. The covalent bonds within the double helix are all quite strong. The hydrogen bonds between the two strands are numerous, cementing the two strands together.

② DNA 的结构提供了作为遗传物质所需要的稳定性。保存信息的碱基被紧密地包装在螺旋的里面，远离环境中可能会破坏它们的分子。而没有保存信息的糖-磷酸骨架排列在分子的外面。在双螺旋中的共价键也都很强。两条链之间的氢键数量非常非常多，使两条链紧紧地黏附在一起。

The structure of DNA also allows some kinds of damage to be repaired quite easily. In a double-stranded molecule, every strand is an exact complement of the other strand. This means that if a base is lost on one strand, the cell can easily determine which base should replace it by looking at the base at that position on the complementary strand.

DNA 的结构也使得它很容易地对某些类型的损伤进行修复。在双链分子中，每条单链都是另一条单链的完全互补链。这就是说，如果一条单链上丢失了一个碱基，细胞可以通过检查互补链对应位置的碱基而很容易地确定应该用哪个碱基来替补。

③ Although the double helix is quite strong and is often easily repaired, it is not invincible. DNA is fairly intricate, and possesses many possible targets for damage. This is especially true for the chemical groups of nitrogenous bases. Also, the two DNA strands can be separated by other molecules or under special conditions of denaturation. This can expose the bases to damage. The occasional susceptibility of DNA to damage is important because it causes changes in DNA base sequence that allows evolution.

③ 虽然双螺旋很稳定，也很容易对损伤进行修复，但它并不是牢不可破的。DNA 的结构相当复杂，它上面含有许多可能的破坏目标。这对于含氮碱基上的化学基团来说更是如此。此外，两条 DNA 链可以被其他分子或其他特殊的变性条件分开，这样就会把碱基暴露出来而使它们易于遭受破坏。这种 DNA 对损伤偶尔表现出的脆弱性很重要，因为它能引起 DNA 碱基序列的变化，这样才使进化得以实现。

④ The double-stranded nature of DNA allows it to be easily copied. Each strand can only bind to a perfectly

④ DNA 的双链特性使它很容易被复制。每条链只能与另一条完全互补的单

complementary strand. We have already seen that this allows certain kinds of DNA damage to be repaired. More importantly, this means that each strand can be used as a template to build a whole new double-stranded molecule. Thus, copying of DNA, at least in theory, is relatively simple. This satisfies our final requirement for the genetic material.

链结合。我们已经看到这可以使某些类型的 DNA 损伤得到修复。更重要的是，它意味着每条单链都可以被用作模板去建造另一条全新的双链分子。这样，至少从理论上来说复制 DNA 是较为简单的工作。这满足了作为遗传物质所需具备的最后一个要求。

2.5 DNA in the Cell

2.5 细胞中的 DNA

Cells contain thousands of different proteins. Accordingly, cells must contain huge amounts of DNA to hold the information for how to make these proteins. The information for making each protein is held in a stretch of DNA that is hundreds to thousands of bases long. This stretch of DNA is called a **gene** (Figure 2.15). The sequence of DNA bases within the gene holds information about what sequence of amino acids should be used to make the protein. We say that these bases 'code' for protein. The DNA also holds other information, such as how much of each protein should be made, where, and when.

细胞含有几千种不同的蛋白质，相应地，它应该含有很大数量的 DNA 来保存如何生产这些蛋白质的信息。生产每种蛋白质的信息被保存在一段长几百个到几千个碱基的 DNA 片段中，这样的 DNA 片段叫作**基因**（图 2.15）。基因中 DNA 碱基的顺序保存着用哪些氨基酸序列去生产蛋白质的信息。我们说，这些碱基为蛋白质“编码”。DNA 也保存了一些其他信息，比如每种蛋白质应该生产多少、在哪里生产以及什么时间生产。

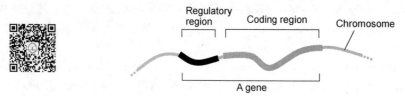

Figure 2.15 Simple representation of a gene

图 2.15 一个基因的简单图像

In prokaryotes, which are relatively simple organisms with relatively few proteins, the amount of DNA is small enough that it can all fit in one, continuous molecule of several million base pairs. The molecule is called a **chromosome**. Prokaryotic chromosomes are circular, meaning they have no beginning or end (Figure 2.16). However, the circle is usually twisted and its circular shape is not always apparent. The chromosome is attached to the interior wall of the bacterial cell.

原核生物是相对较为简单的生物，相对来说含有较少的蛋白质，它们的 DNA 量也较小，可以全部放入一条连续的几百万碱基长的分子中，这一分子叫作**染色体**。原核生物的染色体是环状的，意味着它们没有开头和结尾（图 2.16）。然而，环状分子常常发生扭转，使得环的形状并不总是那么明显。染色体附着在细菌细胞的内壁上。

Eukaryotes have much more DNA than prokaryotes. For example, the human genome has approximately 3 billion base pairs of DNA. The human genome is so large that if the DNA from one human cell is stretched out, it is ap-

真核生物的 DNA 比原核生物的多很多。例如，人类基因组含有约 30 亿个碱基对的 DNA。人类基因组是如此庞大，以至于如果把一个人类细胞中的

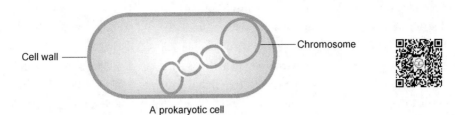

Figure 2.16 Prokaryotic chromosome

图 2.16 原核生物的染色体

proximately 2 meters long! All of this DNA requires organization in order to fit into a compartment of approximately 10 micrometers in length.

DNA in a eukaryotic cell is also organized in structures called chromosomes. However, eukaryotic chromosomes are linear, not circular, and eukaryotes almost always have more than one chromosome (Figure 2.17). All of the chromosomes are kept inside a special compartment of the cell, the **nucleus.** Within each chromosome, proteins called histones wind the eukaryotic DNA, like thread around a spool, as a means of organizing and protecting it.

DNA 全部伸展出来的话，它们约有 2m 长！所有这些 DNA 需要得到良好的组织以便放入约 10μm 长的小空间内。

真核细胞中的 DNA 也被组织成染色体的结构。然而，真核生物的染色体是线性的，不是环状的，并且几乎总是具有一条以上染色体（图 2.17）。所有的染色体被放在细胞中的特殊区域即**细胞核**中。在每条染色体内，叫作组蛋白的蛋白质缠绕真核生物的 DNA，像是细线缠绕线轴一样，这既对 DNA 进行了组织又为它提供了保护。

Figure 2.17 Eukaryotic chromosomes in the nucleus during prophase of mitosis

图 2.17 细胞核中处于有丝分裂前期的真核生物染色体

Eukaryotic chromosomes are often depicted as having a compact H-shape, or perhaps a butterfly shape, as in figure 2.17. This representation is based on chromosomes during mitosis, during which DNA is packaged unusually tightly so that it can be easily maneuvered. Also, each chromosome during mitosis is paired to an exact copy of itself. At most times in the cell, however, chromosomes are very long, thin double-stranded DNA molecules, with a much more random structure, like a noodle. Also, outside of mitosis or meiosis, eukaryotic chromosomes are not paired to copies of themselves.

真核生物染色体常被描绘成如图 2.17 所示具有紧密的 H 形，或有点像蝴蝶那样的形状。这种表示方法是基于染色体在有丝分裂过程中，这时 DNA 会被包装得非常紧密以方便对它进行调遣。还有，处于有丝分裂中的每条染色体会与它本身的完全复制品进行配对。而在细胞中的大多数时间里，细细的双链 DNA 分子具有更随机的结构，看起来就像是一团面条。另外，未进行有丝分裂或减数分裂的真核生物染色体并不与它们自身的拷贝配对。

Some eukaryotic DNA is also found inside organelles called mitochondria and chloroplasts. These pieces of DNA are circular, like prokaryotic chromosomes. A popular theory called the **endosymbiotic hypothesis** suggests that these organelles were once free-living prokaryotes that came to live inside other cells. Although most of their genes have since been lost or incorporated into the DNA of the nucleus, some of the original DNA of these one-time prokaryotes remains in the organelle.

2.6 RNA (Ribonucleic Acid)

Ribonucleotides and deoxyribonucleotides polymerize to form RNA and DNA, respectively. The main difference between the two nucleic acids is that DNA is missing an oxygen on the $2'$ carbon of each sugar. In spite of this very small molecular difference, DNA and RNA have very different functions in the cell. DNA is the genetic material in essentially all organisms. RNA is used as an intermediate copy of the DNA to be used for making proteins, and has other functions discussed later in the text. The most important reason that DNA is used as the genetic material is that it is much more stable than RNA. Although the molecular difference between DNA and RNA is small, the effect of this difference for stability is large. The $2'$-OH of RNA reacts easily with phosphodiester bonds to create ruptures in the RNA strand (Figure 2.18). RNA is also less stable than DNA because the base uracil is easily converted into the cytosine. This creates mutations that are very difficult for the cell to repair. By using thymine instead of uracil, DNA can avoid this type of mutation.

The only example of RNA being used as the genetic material is in viruses. In this case, the relative instability of RNA may actually be favorable. Viruses are usually produced in very large numbers, and depend on rapid evolution for survival and dispersal. The ease with which alterations occur to RNA would allow viruses to evolve more rapidly than if they used DNA. At the same time, the large numbers of viruses allow populations to accommodate the high rates of virus 'death' that come from having such a high mutation rate. Using RNA also allows viruses to make

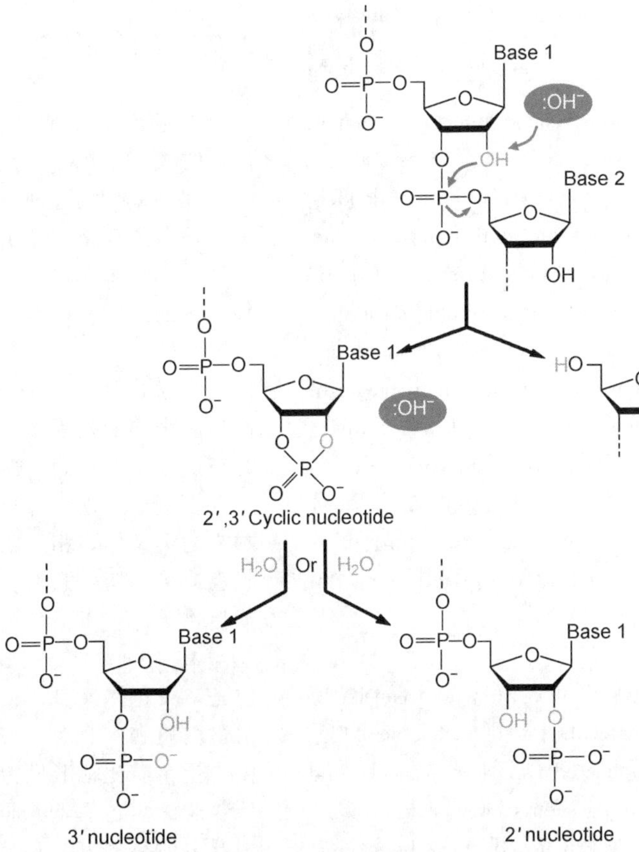

Figure 2.18　Reaction that causes instability of RNA

图 2.18　造成 RNA 不稳定性的反应

proteins directly from the genetic material, without having to coordinate an intermediate step like transcription.

RNA should not be thought of as a cheap imitation of DNA. The relative instability of RNA is very important in its role as a copy of the genetic material, because these copies need to be degraded for the cell to have a dynamic physiology. Furthermore, RNA can fold into a wide range of complex three-dimensional shapes, similar to a peptide. This means that RNA can act as an enzyme, and can have other functions far beyond its role as a genetic messenger. Ribosomes, which are highly structured complexes that manufacture proteins, are largely composed of RNA. According to a widely-accepted theory, in early life forms RNA even took on the roles of DNA and protein! We will see quite a few examples of RNA structures that serve a function throughout this book.

物质上直接产生蛋白质，而不必去协调诸如转录这样的中间步骤。

RNA 不应该被想象成是 DNA 的廉价仿制物。RNA 的相对不稳定性对用它来作为遗传物质的拷贝是很重要的，因为这些拷贝需要被降解掉以使细胞具有动态的生理学特性。而且，RNA 可以折叠成许许多多与肽相似的复杂三维形状。这意味着 RNA 可以起到酶的作用，也可以具有除了担负遗传信使角色以外更多的其他功能。在蛋白质生产中起作用的结构复杂的核糖体主要由 RNA 组成。根据一种被广泛接受的理论，在早期生命形式中，RNA 甚至扮演着 DNA 和蛋白质的角色！在本书中我们将看到不少 RNA 结构行使某种功能的例子。

2.7 Experiments

The identity of the genetic material became more clear in the 19th century thanks to two sources, genetics and microscopy. Geneticists at the time, notably Gregor Mendel, performed breeding experiments to determine laws for how traits are passed from parents to offspring. Around the same period, microscopists were examining structures called chromosomes inside cells. By comparing conclusions from these fields, it was later evident that the behavior of chromosomes is very consistent with the laws of inheritance determined by geneticists. For example, animal cells contain chromosomes from each parent. Likewise, the traits of an animal are inherited from both parents. This strongly suggested that the genetic material lies on chromosomes.

But what is a chromosome? It was known to consist of protein and of a kind of molecule called nucleic acid. Because protein is more complicated than nucleic acid, and the role of the genetic material seems likewise very complicated, most scientists first believed that the genetic material was protein. But much evidence, including two famous experiments, ultimately proved that in fact nucleic acid, specifically DNA, is the genetic material.

In a series of experiments begun in 1944 by Avery, MacLeod and McCarthy, it was shown that a trait could be passed between two bacteria by simply transferring DNA from one bacteria to another (Figure 2.19). Specifically, one strain of bacteria was lethal if

2.7 实验研究

得益于遗传学和显微术的贡献,在19世纪时关于遗传物质是什么的问题已经弄得比较清楚了。那一时期的遗传学家,著名的有格雷高尔·孟德尔,开展了育种实验来研究性状从亲本传递到子代中去的规则。基本上在同一时期,显微镜学家对细胞中的染色体结构进行了检查。通过比较从这些领域得到的结论,后来弄清楚了一点,即染色体的行为与遗传学家得到的性状遗传规则非常一致。例如,动物的细胞含有从每个亲本来的染色体,相应地,一种动物的性状也遗传自两个亲本。这强烈地意味着遗传物质存在于染色体上。

但是,染色体又是什么呢?已经知晓它由蛋白质和一种称为核酸的分子组成。因为蛋白质比核酸更为复杂,并且从遗传物质的角色看起来同样是很复杂的,因此多数科学家一开始都相信遗传物质是蛋白质。但是更多的证据,包括两个著名实验的结果,最终证明了实际上核酸尤其是 DNA 才是遗传物质。

从 1944 年开始进行的一系列实验中,艾弗里、麦克劳德和麦卡锡观察到只需通过转移 DNA 就可以将性状在两种细菌中进行传递(图 2.19)。详细地说,一种菌株如果被注射到小鼠体内的话会

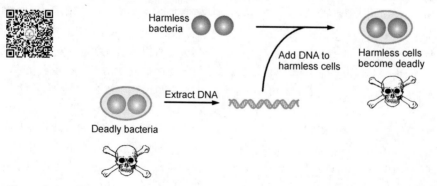

Figure 2.19　Experiment showing that DNA alone can carry a trait

图 2.19　说明仅 DNA 就能携带某一性状的实验

injected into a mouse, whereas the other strain was harmless. If pure DNA was extracted from the lethal bacteria and added to the harmless bacteria, the harmless bacteria became lethal to mice. This meant that the genetic information for lethality was carried by DNA.

Another experiment, performed in 1950, showed that DNA alone can hold the information to make a whole virus (Figure 2.20). It was known that viruses called bacteriophages replicate with the help of bacteria. The bacteriophages inject some material into the bacteria, and that material is used to make new viruses. This experiment showed that when bacteriophages replicate, they only need to inject their DNA into host cells; no protein is transmitted.

导致小鼠死亡，而另一种菌株则是无害的。如果从致死细菌中提取出纯的DNA并加到无害细菌中去的话，原先无害的细菌就变成了能导致小鼠死亡的致死性菌株。这意味着致死性的遗传信息是由DNA携带的。

在1950年有另一项实验显示，单单DNA就能保存用于产生整个病毒的信息（图2.20）。已经知晓，被称为噬菌体的病毒能在细菌的帮助下进行复制。噬菌体注射某种物质到细菌的体内，这种物质被用于产生新的病毒。这一实验显示，当噬菌体复制的时候，它们只需要将它们的DNA注射到宿主细胞中，并没有进行蛋白质的传递。

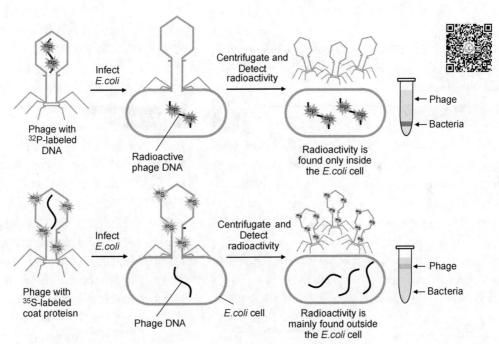

Figure 2.20 The Hershey-Chase experiment, showing that DNA carries the information to make a whole new virus

图2.20 Hershey-Chase实验，说明DNA携带了产生完整新病毒的信息

This was shown by making two preparations of virus. One was grown using radioactive phosphorous (^{32}P). The radioactivity could be used as a label that would allow certain molecules to be tracked in the experiment. Because only DNA contains phosphorous, not protein, the radioactive medium specifically labeled DNA in the bacteriophages. In another preparation, viruses were

这是通过制备两种病毒样本得到证明的。一种病毒样本的制备采用放射性磷（^{32}P）做标记。放射性可以用作标记对实验中的某些分子进行追踪。由于只有DNA含有磷，蛋白质不含磷，因此这种放射性培养基特异性地对噬菌体中的DNA进行标记。另一种病毒样本的制

2.7 Experiments | 43

grown using radioactive sulfur (^{35}S). Proteins contain sulfur, but DNA does not, so this specifically labeled protein in these viruses.

Next, each preparation was used to infect bacteria. After some elapsed time, the mixtures were spun at high speeds to separate the bacteria from the viruses. In the mixture infected by viruses with radioactive phosphorous, radiation was detected mostly inside the bacteria. This meant that DNA was injected into the bacteria before separation. In the experiment that used viruses with radioactive sulfur, radiation was found only outside the bacteria, in the original viruses. This means that viral proteins did not enter the bacteria. Because whatever material injected by viruses into bacteria is used to make new viruses, this shows that the DNA carries all the information for how to make a new virus, and that protein carries none of the information. Alfred Hershey and Martha Chase won the Nobel Prize for performing this work, and the experiment is now commonly known as the **Hershey-Chase experiment**.

备采用放射性硫（^{35}S）做标记。蛋白质含有硫，而DNA不含硫，所以这种培养基特异性地标记蛋白质。

下一步，每种病毒样本都用来侵染细菌。侵染了一段时间后，将得到的混合物高速离心而使细菌和病毒分开。在用放射性磷做标记的病毒进行侵染得到的混合物中，放射性主要存在于细菌体内，意味着DNA在分离前被注射到了细菌体内。在用含有放射性硫的病毒所做的侵染实验中，放射性只存在于细菌外面，位于原先那些病毒体内。这意味着病毒的蛋白质并没有进入细菌内部。由于不管是什么物质被病毒注射到了细菌体内，它都被用来产生新的病毒，这说明，DNA携带着如何产生新病毒的所有信息，而蛋白质没有携带任何信息。阿尔佛雷德·赫尔希和玛撒·蔡斯因为此项工作而获得了诺贝尔奖，他们所做的实验现在通常称为**赫尔希-蔡斯实验**。

Summary

Nucleic acids, notably DNA, are used to store information about how to make all of the proteins needed in an organism. DNA also possesses three other properties required of a genetic material: it is stable, it has some capacity for change so as to allow evolution, and it can be copied. The structure of DNA is largely to thank for these properties: the sequence of bases can hold information, while the double helix structure gives a high level of stability, ease of repair, and a convenient means of replication. RNA, though very slightly different in structure from DNA, generally performs quite a different role in cells. It can provide a copy of the DNA for use in making proteins, and can have catalytic as well as other functions.

小结

核酸，主要是DNA，被用于储存一种生物如何生产所有它所需蛋白质的信息。DNA也拥有作为遗传物质所需要的其他三项性质：稳定，具有一定的容纳变化的能力以使进化得以发生以及能够被复制。具备这些性质主要得益于DNA的结构：碱基序列可以保存信息、双螺旋结构带来了高水平的稳定性、易于修复以及复制的便利性。虽然与DNA在结构上只有细微的不同之处，一般来说RNA却在细胞中担负相当不同的角色。它可以在生产蛋白质中作为DNA的一个拷贝，也可以具有催化功能和一些其他功能。

Vocabulary 词汇

absorbance [ab'sɔːbəns]	吸收	base	碱基
adenine ['ædəniːn]	腺嘌呤	chromosome ['krəuməsəum]	染色体
anti-parallel [ˌænti'pærəlel]	反向平行的	complementary [ˌkɔmplə'mentəri]	互补的
bacteriophage [bæk'tiəriəfeidʒ]	噬菌体	cytosine ['saitəsiːn]	胞嘧啶
bacterium [bæk'tiəriəm]	细菌	denaturation [diːneitʃə'reiʃən]	变性
base-stacking [ˌbeis'stækiŋ]	碱基堆积	denature [diː'neitʃə]	变性

deoxyribonucleic acid [diːˈɔksiˌraibəuˈnjuːkliːik]	脱氧核糖核酸（DNA）	nucleus [ˈnjuːkliəs]	细胞核
		phage	噬菌体
deoxyribonucleotide [diːˈɔksiˌraibəuˈnuːkliətaid]	脱氧核糖核苷酸	phosphate [ˈfɔsfeit]	磷酸根
		phosphodiester bond [ˌfɔsfədaiˈiːstə]	磷酸二酯键
deoxyribose [diːˈɔksiˈraibəus]	脱氧核糖	phosphorus [ˈfɔsfərəs]	磷
double helix [ˈhiːliks]	双螺旋	physiology [ˌfiziˈɔlədʒi]	生理学
double-stranded DNA	双链 DNA	prokaryote [prəuˈkæriət]	原核生物
eukaryote [juˈkæriət]	真核生物	prokaryotic [ˌprəuˈkæriˈɔtik]	原核生物的
eukaryotic [juːˌkæriˈɔtik]	真核生物的	purine [ˈpjuəriːn]	嘌呤
five prime end	5′末端	pyrimidine [ˌpaiəˈrimidːn]	嘧啶
genetic material	遗传物质	ribonucleic acid [ˌraibəuˈnjuːkliik]	核糖核酸（RNA）
guanine [ˈgwɑːniːn]	鸟嘌呤		
hydroxyl group [haiˈdrɔksil]	羟基	ribonucleotide [ˌraibəuˈnuːkliətaid]	核糖核苷酸
inherit [inˈherit]	遗传，继承	ribose [ˈraibəus]	核糖
intricate [ˈintrikit]	错综复杂的	single-stranded DNA	单链 DNA
invincible [inˈvinsəbl]	不可战胜的	sugar-phosphate backbone	糖-磷酸骨架
major groove [gruːv]	（DNA）大沟	sulfur [ˈsʌfə]	硫
minor groove	（DNA）小沟	three prime end [praim]	3′末端
nanometer [ˈneinəˌmiːtə]	纳米	thymine [ˈθaimiːn]	胸腺嘧啶
nitrogenous base [naiˈtrɔdʒinəs]	含氮碱基	triphosphate [traiˈfɔsfeit]	三磷酸（基团）
nucleic acid [ˈnjuːkliik]	核酸	uracil [ˈjuərəsil]	尿嘧啶
nucleotide [ˈnjuːkliətaid]	核苷酸	virus [ˈvaiərəs]	病毒

Review Questions

Ⅰ. True/False Questions（判断题）

1. Deoxyribonucleotides have two more oxygen atoms than ribonucleotides.
2. Once a DNA molecule is denatured it cannot be renatured.
3. All RNA molecules are the same length, but DNA molecules can have different length.
4. The strongest bonds in a DNA molecule occur between the bases of the two strands.
5. Phosphorous, and not sulfur, is common in RNA.
6. Purines always form base pairs with purines.
7. Each chromosome is made of an unbroken double-stranded DNA.
8. RNA uses thymine instead of uracil.
9. In RNA, bases are connected to the 5′ carbon of ribose.
10. RNA is only used by the cell when DNA is unavailable.

Ⅱ. Multiple Choice Questions（选择题）

1. Proteins usually bind to DNA at the _____.
   a. minor grooves
   b. major grooves
   c. 5′ end
   d. 3′ end
   e. purines

2. RNA is less stable than DNA in part because it _____.
   a. is shorter
   b. has three-dimensional folded structures
   c. is used to make protein
   d. has no phosphodiester bonds
   e. none of the above

3. 260nm wavelength light _____.

a. is absorbed better by double-stranded DNA than single-stranded DNA
   b. is absorbed better by single-stranded DNA than double-stranded DNA
   c. denatures double-stranded DNA
   d. denatures single-stranded DNA
   e. breaks phosphodiester bonds
4. DNA is a good genetic material for the following reasons, except:
   a. it is easily copied
   b. it is easily repaired
   c. it can never be damaged
   d. it holds information
   e. all of the above
5. A DNA molecule is found to be 30％ C and 20％ A. What percent of the molecule is T?
   a. 30％
   b. 20％
   c. 50％
   d. 10％
   e. There is no way to predict.

Exploration Questions   思考题

1. Biologists knew that chromosomes carried the genetic material long before they knew that the genetic material was DNA. Why do you think it took them so long to discover the importance of DNA?
2. Many biologists theorize that in early life forms, there was very little DNA and protein. What molecule could have been present instead? Why is it a good candidate for these functions?
3. Discovery of the structure of DNA by James Watson and Francis Crick is considered one of the most important discoveries in the history of biology. Why do you think this might be the case?
4. The concept of a 'gene' is not simple, and we have only just begun to discover what a gene is or can be. What is your current understanding of what a gene is?

Chapter 3 Transcription in Prokaryotes: Mechanism and Regulation

第 3 章 原核生物转录：机理与调控

In order to make proteins, the base sequence of a gene must be read by ribosomes and translated into proteins. In theory, translation might occur using DNA, with ribosomes reading genetic information directly from a chromosome. In reality, a copy of the DNA is first made using RNA, in a process called transcription (Figure 3.1). The RNA copy holds the base sequence that is directly translated into protein. In this chapter we examine why transcription occurs, and its mechanism in prokaryotes. In chapter 4 we examine transcription in eukaryotes.

为了产生蛋白质，基因的碱基序列必须被核糖体阅读后翻译成蛋白质。理论上，翻译应该可以直接使用DNA——通过核糖体直接从染色体上读取遗传信息实现。而实际上却是通过称为转录的过程用RNA为DNA制作了一个拷贝（图3.1）。RNA拷贝保存着用于直接翻译成蛋白质的碱基序列。本章我们学习转录为什么会发生以及原核生物的转录机理。在第4章中再学习真核生物转录方面的知识。

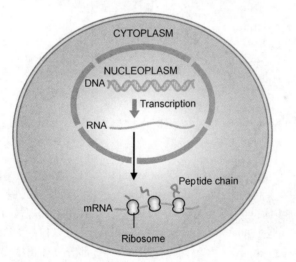

Figure 3.1 The Central Dogma. Protein is made from an RNA copy of the DNA

图 3.1 中心法则。蛋白质是从DNA的拷贝RNA上生产出来的

3.1 Why Use an RNA Intermediate

3.1 为什么使用RNA作为中间物

Chromosomes are extremely large structures, usually containing millions of bases. Reading sequence directly from DNA might simply be too cumbersome using ribosomes, which are relatively large complexes in their own

染色体是极其庞大的结构，通常含有几百万个碱基。用核糖体直接从DNA读取序列会很容易成为缓慢而缺乏效率的工作，因为相对来说核糖体本身已经是

3.1 Why Use an RNA Intermediate | 47

right. More importantly, the genetic material must be very stable over time. Even small changes in genes can cause major malfunctions in protein production in a cell and its descendants. Direct translation of the DNA, which would require frequent separation of DNA strands, could greatly increase the exposure of the genetic material to damage.

By using transcription, ribosomes can read an RNA copy, and therefore never have to come in contact with the chromosome. Transcription does indeed require other proteins to contact the chromosome and separate DNA strands; however, because one RNA molecule can be used to make many proteins (Figure 3.2), much less contact occurs than if translation were direct from the DNA, separating the double helix each time a protein had to be made.

很大的复合体。更重要的是，遗传物质必须在长时间内保持稳定。在细胞及其后代中，基因即使发生很小的变化也可能导致蛋白质生产中出现很大的问题。对DNA进行直接翻译需要将DNA链频繁地分开，这大大增加了遗传物质暴露出来并遭受损伤的机会。

通过采用转录过程，核糖体可以从RNA拷贝上读取信息，因此根本不需要与染色体接触。转录确实需要其他的蛋白质与染色体接触以便分开DNA链；然而，由于一个RNA分子可以用来生产许多蛋白质（图3.2），这就比从DNA直接翻译所发生的接触要少得多，如果从DNA直接翻译，每生产一个蛋白质分子时都需要将双螺旋分开。

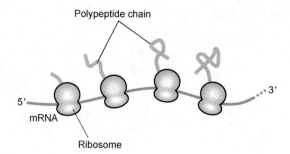

Figure 3.2　One mRNA can be used to make many proteins

图3.2　一条mRNA可以被用来生产许多蛋白质

As an analogy, imagine a rare and important book at the library. Because of the book's value, it cannot be taken out and only the librarian is allowed to touch it. The librarian can, however, make photocopies of whichever part of the book students might want to read. Therefore, the information in the book can be accessed easily while the book remains protected.

Generally speaking, a process with more steps allows more opportunities for regulation. Transcription introduces a major step in the process of protein production, and indeed it is frequently regulated. By deciding which genes to transcribe into RNA, in what quantity and when, the cell can exercise very tight control over the set of proteins it produces. The relative instability of RNA is probably beneficial in this

打个比方，在图书馆里有一本很珍贵、很重要的图书。由于它的价值所在，只有图书管理员可以去碰它。但图书管理员可以根据需要将学生想要阅读的任何部分复印出很多份。这样，读者从书中获得了所需的信息，同时珍贵的图书也得到了保护。

一般来说，具有多个步骤的过程提供了更多的对它进行调控的机会。转录是蛋白质生产过程中的一个主要步骤，它也确实频繁地处于被调控之中。通过决定哪些基因在什么时间要转录成多少RNA，细胞可以对它要生产的蛋白质种类与数量行使非常严密的控制。RNA的相对不稳定性在这种情形下或许是有

context. As RNA is constantly degraded, new RNA must constantly be made. This dynamic turnover gives the cell many opportunities to change its program of transcription.

3.2 Mechanism of Transcription

The mechanism of transcription is based on the fact that ribonucleotides can form base pairs with deoxyribonucleotides. Another way to say this is that RNA **hybridize**s to DNA. This should not be too surprising because the two kinds of molecules are chemically and structurally similar.

During transcription, the two strands of DNA at a gene are separated (Figure 3.3). One of these strands is used as template to make RNA. Either strand of DNA can be used as template for transcription, although only one strand is used for any given gene. Once the template strand has been selected and isolated, each DNA base is paired to a complementary ribonucleotide by an enzyme called **RNA polymerase**. This enzyme also connects the ribonucleotides to each other to make an RNA strand. The final RNA molecule is a perfect complement of the template strand. It is therefore a perfect copy of the other DNA strand, except that the RNA contains uracil instead of thymine.

3.2 转录机理

转录的机理是基于这样一个事实：核糖核苷酸能与脱氧核糖核苷酸形成碱基对。另一种表述方式是：RNA 可以与 DNA **杂交**。这不应该让我们觉得特别奇怪，因为这两种分子在化学性质和结构上都是相似的。

发生转录时，某一基因 DNA 的两条链被分开（图 3.3）。其中的一条链用来作为生产 RNA 的模板。DNA 两条链中的任何一条都可以用来作为转录的模板，而对任何给定的基因来说只使用其中的一条链。一旦模板链被挑选出来并被分隔开来，每个 DNA 碱基就会通过称为 **RNA 聚合酶**的酶来与互补的核糖核苷酸配对。这种酶也能将核糖核苷酸互相连接起来产生 RNA 链。最终的 RNA 分子是模板链的一个完美的互补物，它也是另一条 DNA 链的完美复制品，不同之处在于 RNA 中含有尿嘧啶而不是胸腺嘧啶。

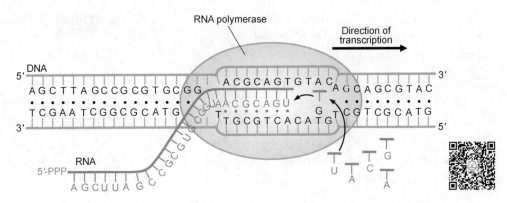

Figure 3.3 The transcription bubble allows ribonucleotides to pair to the template strand

图 3.3 转录泡使核糖核苷酸可以与模板链配对

3.2.1 Promoters

Genes contain more than just DNA sequences that are transcribed to make proteins; they also contain non-

3.2.1 启动子

基因不仅具有转录以后用于生产蛋白质的 DNA 序列，它们还含有不编码的序

coding sequences that carry crucial information. One such region is called the **promoter**, which partly is responsible for denoting the beginning of genes. Essentially every gene in *E. coli* is under the control of a promoter.

There is a lot of DNA in a genome, much of which does not contain genes and should not be transcribed (Figure 3.4). Furthermore, when transcription occurs, it must begin at a precise location at the beginning of the gene. An RNA that is too long or too short may be unable to make functional proteins. The promoter tells RNA polymerase which pieces of the genome to transcribe and precisely where to begin transcription. Promoters also recruit RNA polymerase to genes, and therefore exercise control over the rate of transcription.

列，这些序列携带了至关重要的信息。其中有一个这样的区域叫作**启动子**，它也部分地起到标记基因起始位置的作用。基本上大肠杆菌的每个基因都处于相应启动子的控制之下。

在一个基因组中有很多 DNA，其中有许多并不含有基因，因此不会被转录（图 3.4）。还有，当转录发生的时候，它必须在基因起始部分的准确位置开始。一条过长或过短的 RNA 很可能产生不了有功能的蛋白质。启动子能告诉 RNA 聚合酶在基因组的哪些地方以及在哪个确切的位置开始进行转录。启动子还能把 RNA 聚合酶召集到靠近基因的位置，因此又能对转录速率行使控制。

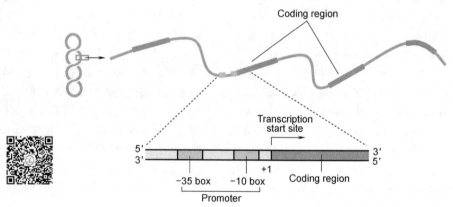

Figure 3.4　Much of the DNA in a genome does not code for protein. Promoters mark the beginning of coding regions

图 3.4　基因组中的许多 DNA 并不编码蛋白质。启动子标明了编码区域的起始位置

Transcription occurs in the $3'\rightarrow 5'$ direction relative to the DNA. Therefore, the beginning of a gene is generally considered to the $3'$ end of the template strand. The promoter is located on the $3'$ side of the first base to be transcribed. This region, coming before the first transcribed nucleotide, is often called the **upstream region**.

相对于 DNA 来说，转录以 $3'\rightarrow 5'$ 的方向进行。因此，一个基因的开头一般被认为是模板链的 $3'$ 端。启动子位于需要转录的第一个碱基的 $3'$ 那一侧。这一区域，即第一个被转录核苷酸的前面，常被称为**上游区域**。

E. coli promoters contain two main elements, the **−10 box** and the **−35 box**. These short DNA sequences are named according to their approximate location in the gene. By convention, the first transcribed base is labeled +1, and bases upstream of this are given negative labels according to their distance from this first base. Thus, the

大肠杆菌启动子含有两个重要元件，即**−10 框**和**−35 框**。这两段短的 DNA 序列是根据它们在基因中所处的大概位置命名的。按照惯例，第一个被转录的碱基标为 +1，这一位置的上游根据它们离开第一个碱基的距离用负号标记。这样一

−35 base is 35 bases upstream of the first transcribed base. The −35 box is so named because its sequence of bases is approximately centered at the −35 base. The −10 box is similarly named.

The −10 box and the −35 box have the **consensus sequences** TATAAT and TTGACA, respectively. The term consensus sequence refers to a generalized sequence from which most actual sequences differ very little or not at all. In other words, most of the −10 boxes in *E. coli* will have a sequence that differs by no more than one or two base pairs from the consensus sequence TATAAT.

A −10 box may have the sequence TATAAA and function perfectly well. However, if it deviates much more from the consensus sequence, its function will be weakened. The promoter will be less able to attract **RNA polymerase**, and transcription of the gene will occur less frequently. For genes that are not meant to produce very much protein, this may in fact be desirable. Thus, we can see that promoters do more than just mark the beginnings of genes; they are actively involved in the process of initiation of transcription and help to determine the rate of transcription.

3.2.2 RNA Polymerase

RNA polymerase is the main enzyme of transcription (Figure 3.5). Prokaryotes have only one RNA polymerase, and it has four subunits: α, β, β′ and σ. The α, β and β′ subunits together are called the **RNA polymerase core**. This collection of subunits is able to perform transcription; however, it is unable to recognize promoters. Using only the core polymerase, transcription begins at random locations, and the RNA produced is not suitable for making proteins.

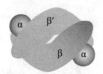
RNA polymerase core

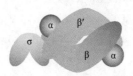
RNA polymerase holoenzyme

Figure 3.5 Subunits of RNA polymerase

In order to recognize promoters, and start transcription in the correct location, the σ subunit must join the core polymerase. This full collection is called the **RNA polymerase holoenzyme**. The σ subunit makes almost all of the contacts with DNA at the promoter. The protein's structure and amino acid composition allows it to recognize specific base sequences in the DNA and bind to them.

The full structure of RNA polymerase somewhat resembles a crab claw. At first the claw is open and can accommodate a double-stranded DNA. Afterwards, the two DNA strands are locally denatured and the claw closes, creating a channel through which only the template strand passes (Figure 3.6). Notice how the structure of the polymerase is well suited for its function, and also how its function depends on dynamic structural changes in the protein.

在一起。这种完整的亚基组合叫作 **RNA 聚合酶全酶**。σ亚基几乎完成了所有与启动子区域 DNA 结合的工作，它的结构和氨基酸组成使它能够识别 DNA 中的特异碱基序列并与之结合。

RNA 聚合酶的完整结构有些类似于螃蟹的钳。开始的时候，钳是张开的，它能容纳一条双链 DNA。之后，两条 DNA 链发生局部变性，钳随之关闭，创造出一个只允许模板链通过的通道（图 3.6）。请注意聚合酶的结构是怎样很好地适合行使它的功能的，并且它的功能是如何依赖于蛋白质结构发生动态变化的。

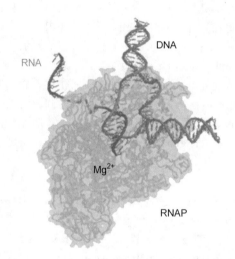

Figure 3.6 Catalytic area of RNA polymerase. Crab craw shape of enzyme not apparent
(Source: http://www.wikipedia.com)

图 3.6 RNA 聚合酶的催化区。酶的蟹钳形状在图中不明显
（来源：http://www.wikipedia.com）

The process of polymerizing ribonucleotides occurs within the channel that holds the template strand. The catalytic site contains a magnesium ion, Mg^{2+}. It is fairly common for catalytic domains of proteins to contain metal ions, as these atoms have special chemical properties, like high affinity for electrons, which are not found in amino acids. The metal ion is not covalently bound to the protein, but is held in place by strong bonds with amino acids at the catalytic center.

使核糖核苷酸聚合的过程发生在保留有模板链的通道内。催化位点含有一个镁离子（Mg^{2+}）。蛋白质的催化域含有金属离子是相当普遍的事，因为这些原子具有特殊的化学性质，比如对电子的高亲和性，这些特性是氨基酸所不具备的。金属离子并不与蛋白质共价连接，但能通过与催化中心的氨基酸形成很强的键而被保留在对应的位置上。

3.2.3 Transcription Mechanism in Three Phases

The mechanism of transcription is often considered to have three different phases: initiation, elongation, and termination.

(1) Initiation

Initiation begins with the binding of the holoenzyme to the promoter in the upstream region of a gene. From here, initiation is divided into four stages (Figure 3.7). At first, the holoenzyme has bound to double-stranded DNA at the promoter. The strands of DNA have not been separated yet, so this is called the '**closed promoter complex**'. Next, the two strands are separated by the holoenzyme in order to expose a template strand. The claw of the polymerase closes around the template strand. This is called the '**open promoter complex**'.

3.2.3 转录机理的三个阶段

转录的机理常被认为由三个阶段组成：起始、延伸和终止。

(1) 起始

起始开始于全酶结合到一个基因上游区域的启动子上。从这以后，起始被分为四个步骤（图3.7）。第一，全酶已经结合到双链DNA的启动子上，但DNA链还没有被分开，因此它被叫作"**闭合启动子复合体**"。第二，两条链被全酶分开以便露出模板链。聚合酶的钳围绕着模板链关闭，形成"**开放启动子复合体**"。

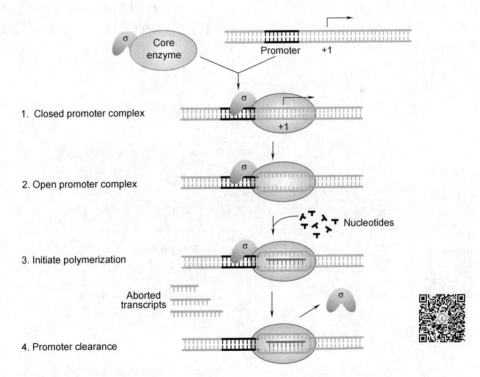

Figure 3.7 Four steps of transcription initiation

图 3.7 转录起始的四个步骤

The third step is required for polymerase to get its footing. Short pieces of RNA are made by the polymerase,

第三，聚合酶取得立足点。聚合酶先产生一些小的RNA片段，但它们通常不

but usually these are not stable enough to be the beginning of a long RNA, and are discarded. Finally, in the fourth step, the RNA polymerase creates an RNA molecule that hybridizes well to the DNA and the protein can begin transcribing the rest of the gene. Because the polymerase moves away from the promoter at this step, it is called '**promoter clearance**'. The σ subunit, which is required during initiation, probably separates from the polymerase after this fourth step and can be re-used elsewhere.

(2) Elongation

By locally separating the two DNA strands at the promoter, RNA polymerase creates an opening called the **transcription bubble**. The bubble allows RNA polymerase access to a single strand of DNA to use as a template. After initiation, RNA polymerase and its transcription bubble move down the length of the gene, exposing the template DNA strand 10~20 bases at a time (Figure 3.8). The two DNA strands are kept apart at the bubble by a piece of the polymerase called the **rudder**, which inserts itself between the strands.

够稳定，不能作为生产长链 RNA 的开始，因此会被弃掉。最后，在第四步，RNA 聚合酶生产出了一个 RNA 分子，这个 RNA 分子与 DNA 很好地杂交在一起，之后 RNA 聚合酶就可以开始转录基因的剩下部分了。因为聚合酶在这一步骤中离开了启动子，因此称为"**启动子清空**"。第四步之后，在起始阶段起作用的 σ 亚基从聚合酶上脱离下来，然后又可以在别的地方重新使用。

(2) 延伸

通过在启动子的位置局部分开两条 DNA 链，RNA 聚合酶创造了一个开口，称为**转录泡**。转录泡使 RNA 聚合酶接近 DNA 的单链并把它作为模板。在起始后，RNA 聚合酶和它的转录泡沿着基因序列向下移动，每次暴露模板 DNA 链的 10~20 个碱基（图 3.8）。转录泡中两条 DNA 链被聚合酶上叫作**方向舵**的结构分隔开，方向舵插入到了两条链之间。

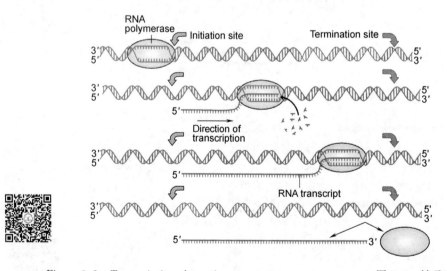

Figure 3.8 Transcription elongation 　　图 3.8 转录延伸

For each new base of template that is exposed, RNA polymerase finds a complementary base and adds it to the growing RNA strand. The first new ribonucleotide has a 5′-triphosphate on one side, and 3′-OH on the other side. During the polymerization reaction, the 3′-OH of

对被暴露出来的模板上的每个新碱基而言，RNA 聚合酶会找到一个互补的碱基并把它加到生长中的 RNA 链上。第一个新的核糖核苷酸一端带有 5′-三磷酸，另一端是 3′-OH。在聚合反应过程中，第一

this first ribonucleotide reacts with the 5′-triphosphate of an incoming ribonucleotide. In the reaction, two phosphates are lost from the triphosphate group, providing energy for the reaction. A new phosphodiester bond is formed with the one remaining phosphate.

The newly added ribonucleotide now provides the growing RNA strand with a new 3′-OH end. Because the first ribonucleotide retains its 5′ end (including the triphosphate group), and all additional ribonucleotides add a new 3′ end, transcription is said to create RNA in the 5′→3′ direction. Recall from chapter 2 that in double-stranded DNA, the single strands are always anti-parallel. This is also the case for the DNA-RNA hybrid formed during transcription. Thus, in order for 5′→3′ RNA to grow on a DNA template, polymerase must move down the DNA template in the 3′→5′ direction.

RNA polymerase sometimes incorrectly incorporates a base that is not complementary to the template. There are two **proofread**ing mechanisms, and both are stimulated by the fact that incorrect base pairing causes the RNA polymerase to favor reversing along the DNA template rather than proceeding. In **pyrophosphorolysis**, the polymerization reaction is simply reversed. A pyrophosphate tethered by Mg^{2+} ion in the active site attacks the phosphodiester bond leading to release of an NTP from the 3′ terminal. This occurs simultaneously with the backtracking of the polymerase by one position on the DNA template (Figure 3.9). The other mechanism is **hydrolysis**. In this mechanism, the triphosphate of the mis-incorporated nucleotide stimulates binding of a second Mg^{2+} ion within the active site. This positions a water molecule for nucleophilic attack that can cleave the phosphodiester bond. The hydrolytic reaction may remove a small stretch of nucleotides that includes the mis-incorporated nucleotide (Figure 3.10).

When RNA is first synthesized, it is hybridized to the single-stranded DNA template. Because the transcriptional bubble remains approximately 10 bases long, a limited length of RNA can be bound to the DNA at one

个核糖核苷酸的 3′-OH 与另一个新来的核糖核苷酸的 5′-三磷酸发生反应。在这一反应中，两个磷酸从三磷酸基团失去来为反应提供能量，而留下的那个磷酸形成了一个新的磷酸二酯键。

现在，新加上去的核糖核苷酸为生长中的 RNA 链提供了一个新的 3′-OH 末端。由于第一个核糖核苷酸保留了它的 5′末端（包括三磷酸基团），并且每个新加进来的核糖核苷酸都带来了一个新的 3′末端，因此转录被说成是以 5′→3′方向产生 RNA。回忆一下第二章中所讲的，在双链 DNA 中单链之间总是反向平行的，在转录中形成的 DNA-RNA 杂种分子中也是同样的情形。这样的话，为了在 DNA 模板上沿 5′→3′方向合成 RNA，聚合酶必须沿 DNA 模板以 3′→5′方向往前移动。

RNA 聚合酶有时会把与模板不互补的碱基加到新合成的链上。RNA 聚合酶具有两种**校正**机理，它们都基于这样一种事实：不互补碱基配对使 RNA 聚合酶更倾向于沿 DNA 模板后退而不是前进。在**焦磷酸解作用**中，聚合反应被逆转了：与 Mg^{2+} 结合在一起的焦磷酸根在活性位点攻击磷酸二酯键，导致从 3′末端释放出一个 NTP。这是与聚合酶在 DNA 模板链上后退一个碱基的距离同时发生的（图 3.9）。另一种校正机理叫作**水解**。在这一机理中，错误整合进来的核苷酸上的三磷酸基团刺激第二个 Mg^{2+} 结合到活性位点，从而将一个水分子置于可发生亲核攻击的位置，水分子的亲核攻击导致磷酸二酯键断裂。这种水解反应可能会去除一小段核苷酸，其中包含错误整合进来的核苷酸（图 3.10）。

当 RNA 刚被合成出来的时候，它是与单链 DNA 模板杂交在一起的。因为转录泡大约保持在 10 个碱基的范围，因此每次只有较短的 RNA 能与 DNA 结

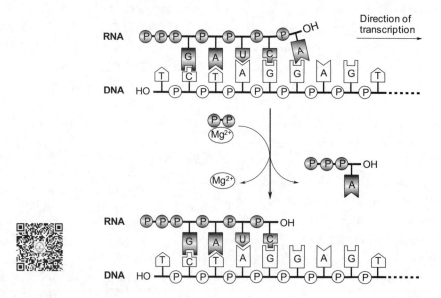

Figure 3.9　Proofreading through pyrophosphorolysis

图 3.9　通过焦磷酸解作用进行校正

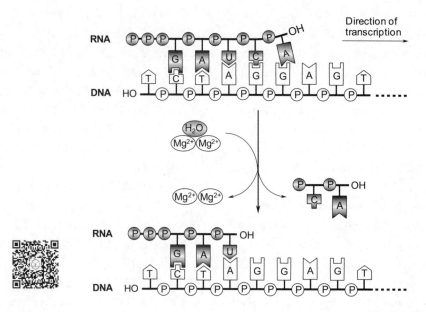

Figure 3.10　Proofreading through hydrolysis

图 3.10　通过水解进行校正

time. As the RNA polymerase moves down the length of the gene, and the RNA grows longer and longer, the RNA bases that were added earlier must detach from the DNA. Thus, during transcription most of the newly produced single-stranded RNA dangle from the transcription bubble.

合。在 RNA 聚合酶沿着基因序列移动且 RNA 链越来越长的时候，之前加上去的 RNA 碱基必须从 DNA 上脱离。这样，在转录过程中大多数新合成的单链 RNA 会挂在转录泡下。

56　│　Chapter 3　Transcription in Prokaryotes: Mechanism and Regulation

(3) Termination

When RNA polymerase reaches the end of the gene, most bases of the new RNA strand have already been detached from the DNA template. However, there are still the approximately 10 hybridized bases in the transcription bubble as well as RNA polymerase itself that must all come apart from the DNA. Prokaryotes employ two different strategies to terminate transcription: **intrinsic termination** and **ρ-dependent termination**.

Intrinsic termination only relies on nucleotide sequence to terminate transcription, not protein (Figure 3.11). The sequence is an inverted repeat, as shown in the figure. When this sequence is transcribed into RNA, the two complementary repeats bind to each other by base pairing. The resulting two-dimensional RNA structure is called a **hairpin loop** because it looks something like a hairpin.

(3) 终止

当 RNA 聚合酶到达基因的末尾时，新 RNA 链的大多数碱基已经从 DNA 模板上脱离。然而，仍然有大约 10 个杂交的碱基以及 RNA 聚合酶自身还处在转录泡位置，所有这些都必须与 DNA 分开。原核生物采用两种不同的策略来终止转录：**内在型终止**和 **ρ 依赖型终止**。

内在型终止仅仅依靠核苷酸序列来终止转录，它不需要蛋白质参与（图 3.11）。这个序列是如图所示的反向重复序列。当这个序列被转录成 RNA 时，两段互补的重复序列通过碱基配对互相结合在一起，得到一种叫作**发夹环**的二维 RNA 结构，因为它看起来有点像发夹。

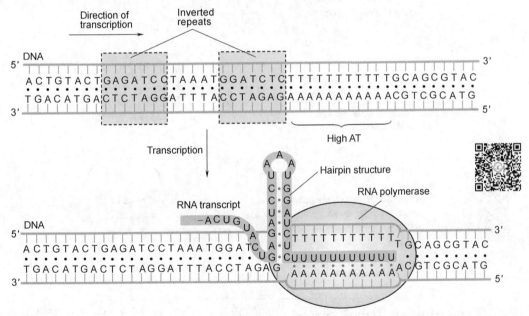

Figure 3.11 Intrinsic termination sequences and formation of the hairpin loop

图 3.11 内在型终止序列以及发夹环的形成

The clumsy hairpin shape destabilizes the RNA-DNA hybrid, allowing it to come apart more easily. It is aided by a long string of A's just downstream in the DNA, which are transcribed into U's in the hybridized RNA. These A-U base-pairs holding the RNA and DNA together slow the RNA polymerase. They also allow the RNA to separate easily from the DNA

这种笨拙的发夹形状使 RNA-DNA 杂交分子变得很不稳定，使它们更容易分开。这一机理还得益于 DNA 下游正好有一长串的 A，它们在杂交的 RNA 中被转录成了 U。这些使 RNA 和 DNA 保持在一起的 A-U 碱基对与发夹环一起使 RNA 聚合酶前进的速度减慢。它

because of their weak pairing; like A-T base pairs, A-U pairs are connected by only two hydrogen bonds, rather than three for G-C pairs.

ρ-dependent termination also employs a hairpin loop to terminate transcription (Figure 3.12). However, it requires a protein called ρ to aid in termination. ρ binds to the growing RNA before RNA polymerase has reached the inverted repeat and follows the polymerase. Once the hairpin has formed, RNA polymerase stops and ρ catches up with it. Finally, ρ separates the RNA from the DNA, and transcription ends. No string of A's is required in the DNA for ρ-dependent termination.

们之间弱的配对能力也使 RNA 更容易从 DNA 上脱离下来；因为与 A-T 碱基对相似，A-U 碱基对只由两个氢键连接，而不是 G-C 碱基对中的三个氢键。

ρ 依赖型终止也采用发夹环来终止转录（图 3.12）。然而，它需要一种叫作 ρ 的蛋白质来帮助。ρ 在 RNA 聚合酶到达反向重复序列之前就结合到了生长中的 RNA 上，并一直跟在 RNA 聚合酶后面。一旦 RNA 上形成了发夹，RNA 聚合酶便停下来，ρ 就会赶上它。最后，ρ 使 RNA 从 DNA 上分开，转录完全停止。ρ 依赖型终止不需要 DNA 上有一连串 A。

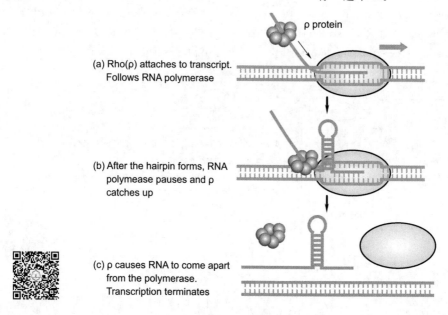

Figure 3.12   ρ-dependent termination of transcription

图 3.12   ρ 依赖型转录终止

3.3 Regulation of Gene Expression in Prokaryotes

3.3 原核生物基因表达调控

We have seen that genes are first transcribed into mRNA before translation in proteins. Not all genes in a cell are transcribed and translated all of the time. Among the thousands of genes in a prokaryote, some are always needed and are constantly expressed. But other proteins are only needed at certain time or in certain environment. To make these proteins under the

我们已经看到，基因在翻译成蛋白质之前需要先转录成 mRNA。细胞中并不是所有基因在所有的时间里都需要转录和翻译的。在原核生物的几千个基因中，有些基因总是需要的，它们一直在表达。而其他蛋白质只在一定的时间里或一定的环境条件下才需要。如果在错误的情况下制造这

wrong circumstance would be at best inefficient, and at worst, lethal.

The production of protein from a gene is called **gene expression**. Control of gene expression is important in prokaryotes in part because these are tiny, single-celled organisms with little capacity for independent movement. The environments in which they land can be quite varied and can change quickly—from wet to dry, hot to cold, nutritious to deserted. Instead of choosing their environments, like a more motile organism, they often just make the best use of wherever they land. Doing so requires a flexible physiology, which in turns depends on a flexible program of gene expression.

Prokaryotes mainly regulate gene expression by controlling transcription of genes (Figure 3.13). If a gene is not transcribed, no protein can be made from it; by contrast, high production of a particular protein can be achieved by transcribing a gene at a high rate. In prokaryotes, protein production is rarely controlled above the level of transcription.

3.3.1 Coordinate Regulation

In prokaryotes, genes that share a related function are often regulated together. For instance, the five genes that make the amino acid tryptophan in *E. coli* are expressed as a group. These five peptides work together in a pathway, and all are needed to make tryptophan. There is never any reason to express one without the others, so they are always expressed together.

Genes that are subject to such **coordinate regulation** are organized in structures called **operons** (Figure 3.14). In an operon the related genes are located adjacent to each other on the chromosome. Instead of each having their own promoters, all of the genes are under the control of one regulatory region. This region includes one promoter and another element called the **operator**.

RNA polymerase transcribes all of the genes of the operon in one piece, producing a long mRNA. This

些蛋白质，则最好的结果是造成低效率生产，最糟的结果是导致生物死亡。

从基因开始生产出蛋白质的过程叫作**基因表达**。在原核生物中对基因表达进行调控非常重要，部分原因是：原核生物是微小的单细胞生物，具有很小的独立运动能力。它们所处的环境可能是很不相同的并且变化很快——从潮湿到干燥、炎热到寒冷、营养丰富到贫乏。由于不能像具有更强运动能力的生物一样去选择环境，它们通常只能对任何它们所处的环境做最大限度的利用。要做到这一点需要有灵活的生理活动，而这依赖于基因表达的灵活性。

原核生物主要通过控制基因的转录来调控基因的表达（图3.13）。假如一个基因不被转录，则不会有蛋白质从这一基因产生出来；相反，通过高速率转录某一基因，该基因所编码的蛋白就可以被大量生产出来。在原核生物中，蛋白质生产很少是在转录之上的水平进行调控的。

3.3.1 协同调控

在原核生物中，功能相关的基因通常一起被调控。例如，在大肠杆菌中，合成色氨酸的五个基因作为一组基因进行表达。这五种肽在代谢途径中一起发挥作用，并且在生产色氨酸中都需要。从来没有任何理由去表达其中的一个基因而不表达其他基因，因此它们总是在一起表达。

这样一些被**协同调控**的基因组织起来的结构叫作**操纵子**（图3.14）。在操纵子中，一些相关的基因在染色体上处于相互靠近的位置。它们并不是各自具有自己的启动子，而是所有基因都处于同一个调控区的控制之下。这一调控区包含一个启动子和另一个叫作**操纵基因**的元件。

RNA聚合酶将操纵子上的所有基因转录到一条链上，产生一条长链mRNA。

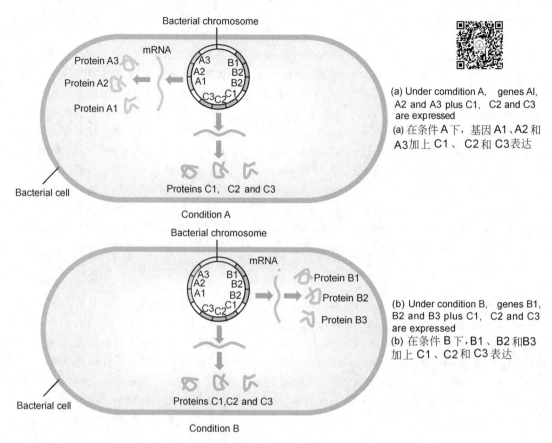

Figure 3.13 Not all genes are transcribed. By controlling transcription, different sets of proteins are made under different conditions

图 3.13 不是所有基因都会被转录，通过控制转录，在不同的条件下产生出不同的蛋白质

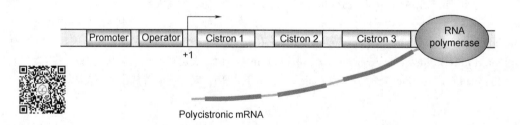

Figure 3.14 An operon being transcribed to produce a polycistronic mRNA

图 3.14 一个正在转录产生多顺反子 mRNA 的操纵子

mRNA is called **polycistronic mRNA** because it codes for multiple cistrons, which is another word for genes. Between each cistron in a polycistronic mRNA there are short untranslated regions that contain signals to stop and start translation. This allows the genes to be translated as separate proteins, instead of one long protein.

这一长链 mRNA 被称为**多顺反子 mRNA**，因为它编码了多个顺反子。顺反子是基因这一名称的另一种叫法。在多顺反子 mRNA 中，顺反子之间含有短的非翻译区域，这些区域含有翻译的终止和起始信号，它们使基因被翻译成各自的蛋白质，而不是一条长链蛋白质。

3.3.2 The *Lac* Operon

The best studied operon is the **lac operon** of *E. coli*, which contains genes for metabolizing the sugar lactose. It is a good example of how a prokaryote can alter its set of proteins to adapt to nutrient availability in the environment.

(1) The Conditions of Lactose Metabolism

E. coli generally uses the simple sugar glucose as a source of energy. Glucose is often abundant, and is relatively simple to metabolize. However, if *E. coli* finds itself in an environment without glucose, it can adapt to use other sources of energy. One possible source is **lactose**.

Lactose is a disaccharide composed of one monomer of glucose and one monomer of galactose (Figure 3.15). In order to metabolize lactose, *E. coli* expresses three extra proteins: **lactose permease**, to bring lactose into the cell; **β-galactosidase**, to separate lactose into glucose and galactose; and **galactoside transacetylase**, which has an unknown function. Lactose permease is the first example we have encountered of a membrane protein. Its position within the plasma membrane allows lactose, which otherwise could not cross the membrane, to enter the cell.

3.3.2 乳糖操纵子

研究得最详细的操纵子是大肠杆菌的**乳糖操纵子**，它含有代谢乳糖的基因。它是原核生物如何调节其多种蛋白质产物以适应环境营养成分变化的很好的一个例子。

(1) 乳糖代谢的条件

大肠杆菌通常使用葡萄糖作为能量来源。葡萄糖通常较为丰富，也相对容易被代谢。然而，如果大肠杆菌发现它处于一个没有葡萄糖的环境中时，它也能适应使用其他能量来源。一种可能的来源是**乳糖**。

乳糖是一种双糖，由一个葡萄糖单体和一个半乳糖单体组成（图3.15）。为了代谢乳糖，大肠杆菌需要表达三种额外的蛋白质：**乳糖渗透酶**，用于将乳糖送进细胞里；**β-半乳糖苷酶**，用于把乳糖分离成葡萄糖和半乳糖；**半乳糖苷转乙酰基酶**，其功能尚不清楚。乳糖渗透酶是我们遇到的第一个膜蛋白的例子。它位于质膜中，帮助乳糖进入细胞，如果没有它的话乳糖就不能越过细胞膜。

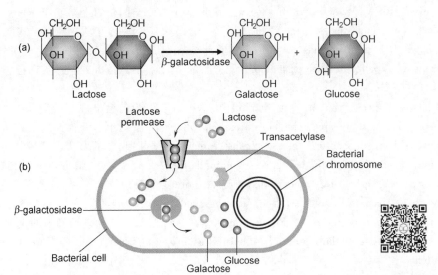

Figure 3.15 (a) Lactose is broken down into galactose and glucose; (b) Function and localization of proteins encoded by *lac* operon genes

图3.15 (a) 乳糖被打断产生半乳糖和葡萄糖；(b) *lac* 操纵子基因编码的蛋白质的功能及其在细胞中的位置

These three proteins only need to be made when lactose is present and glucose is not present. Making them under any other condition would be useless. If glucose is present (even if lactose is also present), there is no reason to metabolize lactose, because it is easier to metabolize glucose. If lactose is absent (whether or not glucose is present), there is no reason to make proteins that metabolize lactose. The production of the lactose-metabolizing proteins is tightly controlled because expressing proteins when they are not needed is wasteful; protein production is costly for the cell in terms of energy and nutrients.

How does E. coli ensure that lactose metabolizing proteins are only produced in the appropriate condition? Like the genes that make tryptophan, the three lactose metabolizing genes are organized into an operon. This is called the lac operon, and it has one regulatory region for all three genes.

(2) Negative Regulation—the Lac Repressor

When there is no lactose in the cell, a tetrameric protein called the **lac repressor** binds to the operator in the regulatory upstream region of the operon (Figure 3.16). The operator is located very near to the promoter on the DNA. Bound to the operator, the repressor protein physically prevents RNA polymerase from attaching to the promoter and transcribing the lac genes. This kind of regulation, in which a protein binds to the DNA to prevent transcription, is called **negative regulation**.

When lactose is present in the cell, it binds to the lac repressor. This binding causes the lac repressor protein to change shape, and accordingly to change function. In its new conformation, the lac repressor unbinds the DNA. Once the operator is unoccupied, RNA polymerase has space to bind to the promoter and transcribe the lac operon genes. Regulation of the lac repressor by lactose binding is an example of **allosteric regulation**.

In reality, the lac repressor does not bind directly to lactose, but to a derivative of it called allolactose; for

这三种蛋白质只有在存在乳糖并且不存在葡萄糖的情况下才产生出来。在任何其他条件下生产它们将是没有用的。当存在葡萄糖（即使乳糖也同时存在）时，没有理由去代谢乳糖，因为代谢葡萄糖更容易。当乳糖不存在时（不管有没有葡萄糖），就更没有理由产生这些用于代谢乳糖的蛋白质了。乳糖代谢蛋白质的生产受到很严格的控制，因为在不需要的时候表达这些蛋白质是很浪费的；从能量和营养成分的角度看，蛋白质生产对细胞来说代价是相当大的。

大肠杆菌是怎样确保只有在适宜的环境条件下才生产这些乳糖代谢蛋白的呢？与合成色氨酸的基因类似，三个乳糖代谢基因被组织在一个操纵子中，称为乳糖操纵子，它含有一个用来调控所有三个基因的调控区域。

(2) 负调控——lac 阻遏蛋白

当细胞中没有乳糖时，一种称为 **lac 阻遏蛋白**的四聚体蛋白质结合在操纵子上游调控区域的操纵基因上（图 3.16）。在 DNA 上，操纵基因的位置非常靠近启动子。结合在 DNA 上的阻遏蛋白从空间位置上防止了 RNA 聚合酶结合到启动子上去转录 lac 基因。这种当一种蛋白质结合到 DNA 上后阻止基因转录的调控方式叫作**负调控**。

当乳糖出现在细胞中时，乳糖会与 lac 阻遏蛋白结合。这一结合使 lac 阻遏蛋白改变了形状，相应地也改变了功能。处于新构象下，lac 阻遏蛋白不能与 DNA 结合。一旦操纵基因没有被结合，RNA 聚合酶就有了与启动子结合并转录 lac 操纵子基因的空间。通过乳糖结合而调控 lac 阻遏蛋白是**别构调节**的一个例子。

实际上，lac 阻遏蛋白并不是直接与乳糖结合，而是与一种称为异乳糖的衍生

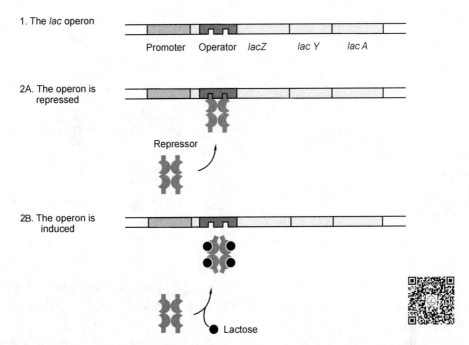

Figure 3.16 Negative regulation of the *lac* operon by lac repressor

图 3.16 lac 阻遏蛋白对 *lac* 操纵子的负调控

our purposes this is not an important detail.

物结合；对我们的学习目的来说，这并不是一个很重要的细节。

(3) Positive Regulation—CAP

Just removing repression by the lac repressor is not enough for transcription to actually occur. Another protein is required to actively recruit RNA polymerase to the promoter. This protein, **catabolite activator protein (CAP)**, takes into consideration the levels of glucose in the cell. It only activates transcription of *lac* genes if glucose levels are low (Figure 3.17).

Regulation by CAP protein is an example of **positive regulation**, in which a protein binds to the DNA to stimulate transcription. Under low glucose conditions, CAP binds to a site upstream of the operator. If the lac repressor is absent, CAP then strongly activates transcription and the *lac* genes are expressed. Under high glucose conditions, CAP does not bind to the DNA and no transcription occurs, whether or not lac repressor is bound. In this way, the cell can ensure that *lac* genes are only expressed when use of glucose is not an option.

(3) 正调控——CAP

仅仅去掉 lac 阻遏蛋白的阻遏还不足以使转录真正发生，还需要有另一种蛋白质主动地将 RNA 聚合酶召集到启动子的位置。这种蛋白，即**代谢物激活蛋白（CAP）**能感应细胞里的葡萄糖水平。只有在葡萄糖水平低的情况下，它才会激活 *lac* 基因的转录（图 3.17）。

CAP 蛋白的调控是一个**正调控**的例子，它以一种蛋白质结合到 DNA 上刺激转录发生的方式调控基因的表达。在低葡萄糖条件下，CAP 结合到操纵基因的上游。如果 lac 阻遏蛋白不存在，CAP 就会强烈地激活转录，*lac* 基因得到表达。在高葡萄糖条件下，CAP 不能结合到 DNA 上，转录不能发生，而不管 lac 阻遏蛋白有没有结合在 DNA 上。在这种调控方式下，细胞能够确保 *lac* 基因只有在没有葡萄糖可以利用的情况下才表达。

3.3 Regulation of Gene Expression in Prokaryotes | 63

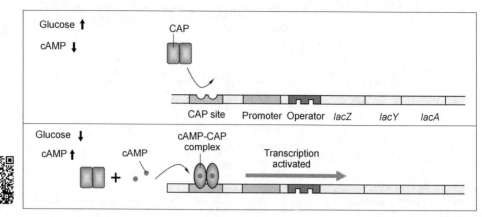

Figure 3.17 Positive regulation of the *lac* operon by CAP

图 3.17 CAP 对 *lac* 操纵子的正调控

CAP does not respond to glucose concentration by binding directly to glucose. Instead, it binds to cAMP, a small molecule whose concentration varies inversely with the concentration of glucose. When glucose level is low, cAMP level is high; when glucose level is high, cAMP level is low. CAP only attaches to the DNA and activates transcription when it binds cAMP, in other words, under the low glucose condition. The reason for this inverse relationship between glucose and cAMP is that transport of glucose into the cell inhibits the enzyme, called **adenylyl cyclase**, that produces cAMP.

We can compare the *lac* operon to a car with a brake and an accelerator. To drive the car, it is not enough to remove the brake; it is also necessary to press on the accelerator. Likewise in the *lac* operon, it is not enough to simply remove the lac repressor; it is also necessary to apply the CAP protein. That is to say, negative and positive regulations are both required to control expression of this operon (Figure 3.18).

3.3.3 The *Trp* Operon

The ***trp* operon** provides an interesting contrast to *lac* operon regulation. As mentioned above, the *trp* operon contains five genes for synthesizing the amino acid tryptophan (Figure 3.19). Therefore, these

CAP 不能通过与葡萄糖直接结合而对葡萄糖的浓度产生响应，它是通过与 cAMP 这样的小分子结合而发挥作用的。cAMP 的浓度与葡萄糖的浓度成反比，当葡萄糖浓度降低时，cAMP 浓度上升；当葡萄糖浓度上升时，cAMP 浓度变低。CAP 只有在与 cAMP 结合后才能结合到 DNA 上并激活转录，也就是说，在葡萄糖浓度低的情况下才激活转录。葡萄糖与 cAMP 之间存在这种颠倒关系的原因是葡萄糖运输进入细胞后能抑制**腺苷酸环化酶**的作用，该酶催化产生 cAMP。

我们可以把 *lac* 操纵子比作一辆有刹车和油门的小汽车。要开动这辆汽车，光松开刹车是没有用的，还应踩下油门。*lac* 操纵子中的情形是一样的：仅仅移开 lac 阻遏蛋白是不够的，还必须有 CAP 蛋白的作用。这就是说，需要负调控和正调控两者一起来控制 *lac* 操纵子的表达（图 3.18）。

3.3.3 色氨酸操纵子

trp 操纵子提供了一个有趣的与 *lac* 操纵子间的对比。正如上面已提到的那样，*trp* 操纵子含有 5 个合成色氨酸的基因（图 3.19）。因此，这些基因具有

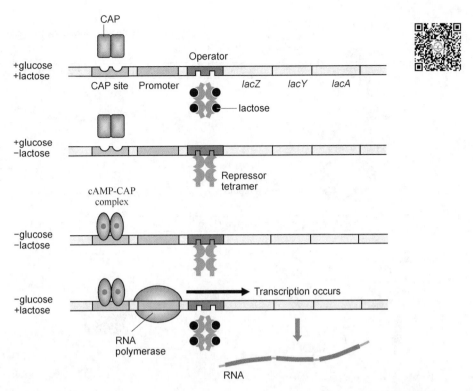

Figure 3.18　Summary of *lac* operon regulation

图 3.18　*lac* 操纵子调控小结

genes have an **anabolic** function (building molecules) as opposed to *lac* operon genes which have a **catabolic** function (breaking down molecules). This difference has an important effect on the regulation of the operon.

合成代谢的功能（即建造分子），*lac* 操纵子基因正相反，它们具有**分解代谢**的功能（即打断分子）。这一区别在操纵子的调控上具有重要的效应。

| Promoter | Operator | Leader-attenuator | trpE | trpD | trpC | trpB | trpA |

Figure 3.19　*Trp* operon structure

图 3.19　*trp* 操纵子的结构

(1) Negative Regulation—Trp Repressor

When the tryptophan level in the environment is high, the cell does not need to produce tryptophan, and the operon genes must not be transcribed. They are turned off by a dimerized protein called the **trp repressor** (Figure 3.20). The tryptophan in the cell binds to the trp repressor. This gives the trp repressor a structure which can bind to the operator and block transcription by RNA polymerase. As the lac repressor, this is an example of allosteric regulation.

(1) 负调控——trp 阻遏蛋白

当环境中的色氨酸水平高时，细胞不需要生产色氨酸，操纵子上的基因一定不能转录出来。它们被一种叫作 **trp 阻遏蛋白**的二聚体蛋白质关闭（图 3.20）。细胞中的色氨酸结合到了 trp 阻遏蛋白上。这给了 trp 阻遏蛋白一种可以结合到操纵基因上并阻止 RNA 聚合酶进行转录的结构。与 lac 阻遏蛋白一样，这也是一个别构调节的例子。

3.3　Regulation of Gene Expression in Prokaryotes　65

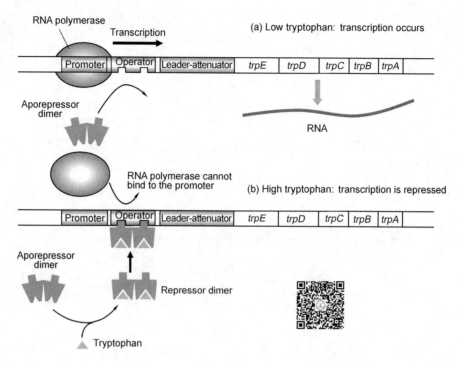

Figure 3.20 Negative regulation of the *trp* operon by trp repressor

图 3.20 trp 阻遏蛋白对 *trp* 操纵子的负调控

When tryptophan levels in the environment are low, the cell must produce tryptophan. In this case, there is no tryptophan present to bind to the trp repressor. The absence of tryptophan causes the repressor to adopt a structure that cannot attach to the DNA. As a result, the operon becomes unblocked and the *trp* operon genes can be expressed.

Notice the similarities and differences between the lac and trp repressor proteins. Both are involved in negative regulation, turning off transcription when they bind to DNA. But they bind to DNA under opposite circumstances. Lac repressor binds to DNA when lactose levels are low, whereas trp repressor binds to DNA when tryptophan levels are high.

The two repressors act differently because they have different responses to binding of small molecules. When lactose binds to lac repressor, the protein detaches from the DNA, making transcription of *lac* operon genes possible. By contrast, when tryptophan binds to

当环境中色氨酸水平低时，细胞必须生产色氨酸。在这种情况下，没有色氨酸与 trp 阻遏蛋白结合。没有了色氨酸，trp 阻遏蛋白采取的是一种不能结合到 DNA 上去的结构。这样一来，操纵子变得没有障碍了，*trp* 操纵子基因便可以得到表达。

请注意 lac 阻遏蛋白和 trp 阻遏蛋白之间的异同点。它们两者都起负调控作用，即当它们结合到 DNA 上时都关闭转录。但它们是在相反的情形下结合到 DNA 上的。lac 阻遏蛋白是在乳糖水平低的时候结合到 DNA 上，而 trp 阻遏蛋白是在色氨酸水平高的时候结合到 DNA 上。

这两种阻遏蛋白发生作用的方式不同，因为它们对小分子的结合有不同的响应。当乳糖结合到 lac 阻遏蛋白后，lac 阻遏蛋白从 DNA 上脱离，使得 *lac* 操纵子上基因的转录成为可能。相反，当

trp repressor, the protein attaches to the DNA, making transcription of *trp* operon genes impossible. This makes sense because when lactose is present lactose metabolizing gene may need to be produced, whereas when tryptophan is present tryptophan producing genes do not need to be produced.

Because of opposite effects of lactose and tryptophan on the proteins to which they bind, these two small molecules are assigned different terms. Lactose is called an **inducer**, because it induces transcription when it binds to the lac repressor. Tryptophan is called a **co-repressor** because it causes repression of transcription when it binds to the trp repressor. Any small molecule that affects the activity of a protein through binding is called a **ligand**.

Unlike the *lac* operon, the *trp* operon does not have positive regulation. Simply removing repression by trp repressor is enough to cause transcription of the genes. This is reasonable because the *trp* operon only responds to one condition, the concentration of tryptophan. By contrast, the *lac* operon has to respond to two conditions, the concentration of lactose and the concentration of glucose.

(2) Attenuation

Although the *trp* operon does not have positive regulation, it does have an extra mechanism for inhibition of transcription, called **attenuation**. This mechanism is important because repression by the trp repressor alone is not very strong, transcription still continues at a significant rate. Attenuation ensures that when tryptophan is present, transcription can be fully blocked.

Attenuation depends on two DNA elements that occur immediately downstream of the operator and promoter, but upstream of the operon genes. These elements are called the **leader** and the **attenuator** [Figure 3.21(a)]. RNA transcribed from the attenuator is capable of forming different hairpin structures. One of these hairpin loops is followed by a string of U's in the RNA, and

色氨酸结合到trp阻遏蛋白后，trp阻遏蛋白会结合到DNA上，使得*trp*操纵子上基因的转录变得不可能。这是符合情理的，因为当有乳糖时，就有可能需要表达用于乳糖代谢的基因；而当有色氨酸时，用于生产色氨酸的基因是不需要表达的。

由于乳糖和色氨酸结合相应的蛋白质后具有相反的效应，它们被给予了不同的名称。乳糖被称为**诱导物**，因为当它与lac阻遏蛋白结合后诱导了转录。色氨酸被称为**辅阻遏物**，因为当它与trp阻遏蛋白结合后阻止了转录的发生。任何通过结合到蛋白质上从而影响蛋白质活性的小分子都被叫作**配体**。

与*lac*操纵子不同的是，*trp*操纵子没有正调控功能。只要去掉trp阻遏蛋白的阻遏作用即足以导致基因发生转录。这是符合情理的，因为*trp*操纵子只对一种条件产生响应，即色氨酸的浓度。相反，*lac*操纵子需要对乳糖的浓度和葡萄糖的浓度两种条件产生响应。

(2) 衰减作用

虽然*trp*操纵子不具备正调控功能，但它确实具有一种额外的被称为**衰减作用**的转录抑制机制。这种机制相当重要，因为单一由trp阻遏蛋白发挥的阻遏作用不是非常强，转录仍旧能以一定的速率继续。衰减作用确保了在色氨酸存在的情况下转录能被彻底阻遏。

衰减作用通过紧随启动子和操纵基因之后但位于操纵子基因前的两个DNA元件发挥作用。这些元件叫作**前导子**和**衰减子** [图3.21(a)]。从衰减子转录出来的RNA能够形成不同的发夹结构。其中一个发夹结构的后面跟随着一串U，因此与能引起转录的内在型终止的结构很相

therefore closely resembles the structure formed during intrinsic termination [Figure 3.21(b)]. Indeed, when it forms it stops transcription before any *trp* operon genes can be transcribed. The other possible loop, however, has no string of U's adjacent to it and therefore no effect on transcription [Figure 3.21(c)].

像[图 3.21(b)]。实际的情形也确实是这样，当这种结构形成时转录便在任何 *trp* 操纵子基因被转录出来之前就停下来了。而另一可能的发夹结构由于没有紧随其后的一串 U，所以对转录不会有任何影响[图 3.21(c)]。

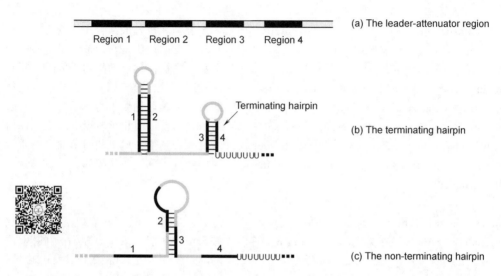

Figure 3.21 The leader-attenuator region and the folding of RNA transcribed from this region

图 3.21 前导子-衰减子区域以及从这一区域转录出来的 RNA 的折叠情况

In keeping with the purpose of the *trp* operon genes, it is important for the terminating loop to form when tryptophan concentration in the cell is high, and for the non-terminating loop to form when tryptophan concentration is low. To understand how this happens, we must jump ahead and briefly discuss translation. After transcription, mRNA is translated by ribosomes into protein. In prokaryotes, ribosomes are very hasty to translate the mRNA. They attach to the mRNA and begin translating while it is still being transcribed. This is called **coupled transcription-translation** (Figure 3.22).

为了与 *trp* 操纵子基因的目的保持一致，当细胞中色氨酸含量高时形成终止型发夹结构，而当细胞中色氨酸含量低时形成非终止型发夹结构是很重要的。为了理解这是如何发生的，我们需要越过此处，简单地讨论一下翻译作用。正如我们所知，转录以后，mRNA 将由核糖体翻译成蛋白质。在原核生物中，核糖体很能抓紧时间，在 mRNA 还处于转录过程中时它们就开始了翻译工作。这一现象被称为**偶联转录-翻译作用**（图 3.22）。

Ribosomes begin to translate the leader region of the *trp* operon soon after the region has been transcribed. The leader region codes for a polypeptide that requires two adjacent tryptophan amino acids. If tryptophan concentration is low in the cell, the ribosome stops temporarily at this point in translation and waits to find whatever small amount of tryptophan it can find. When the ribosome stops

核糖体在 *trp* 操纵子上的前导子序列被转录出来之后不久就会开始翻译前导区域的 mRNA。前导区域的 mRNA 含有两个相邻的色氨酸密码子。如果细胞中的色氨酸含量较低，则核糖体在翻译这一位置时会做临时的停顿，等待从细胞中找到为数不多的色氨酸。核糖体在这一地

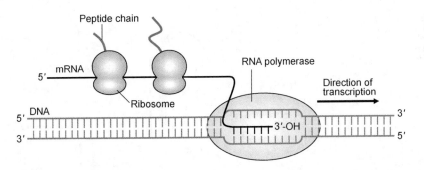

Figure 3.22 Coupled transcription-translation

图 3.22 偶联转录-翻译作用

in this place, it promotes formation of the non-terminating hairpin in the RNA being transcribed. This is the hairpin that forms between region 2 and 3 of the attenuator (Figure 3.23). RNA polymerase can move right through this hairpin to transcribe the operon genes.

方的停留促进了正在被转录出来的 RNA 形成非终止型发夹结构，即由衰减子区段 2 和 3 形成的发夹结构（图 3.23）。RNA 聚合酶能够毫无困难地通过这一发夹结构，继续转录操纵子上的基因。

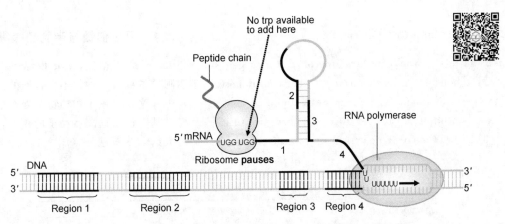

Figure 3.23 Non-terminating hairpin formed when tryptophan levels are low and ribosome pauses when tryptophan must be added into new peptide chain

图 3.23 当色氨酸处于低水平时形成非终止型发夹结构，在色氨酸必须被加到新合成肽链上的时候核糖体暂停

If tryptophan concentration is high in the cell, the ribosome does not stop when tryptophan must be added to the polypeptide (Figure 3.24). It continues, making a short stretch of protein, and then falls off of the mRNA. The ribosome is not there to influence folding of the mRNA transcript. As a result, two hairpins are formed, one between regions 1 and 2 of the attenuator, and another between regions 3 and 4. The latter is the termination signal. It causes RNA polymerase to fall off the DNA and transcription stops before even the first gene of the *trp* operon has been transcribed.

如果细胞中的色氨酸含量较高，核糖体不会在肽链合成中需要色氨酸的位置停顿（图 3.24）。它能继续向前，生产出一小段蛋白质，然后从 mRNA 上掉下来。核糖体并没有在那儿影响 mRNA 转录物的折叠。结果，形成了两个发夹结构，一个在衰减子区段 1 和 2 之间，另一个在区段 3 和 4 之间。后者是终止信号，它使 RNA 聚合酶从 DNA 上掉下来，这样在第一个基因被转录出来之前转录就终止了。

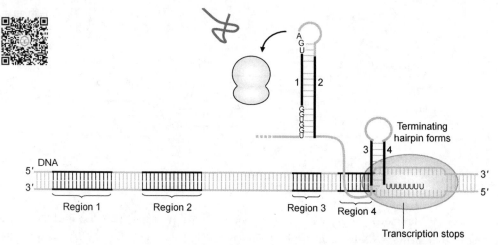

| Figure 3.24 When tryptophan levels are high, ribosome does not stall in protein production. It produces a short polypeptide and then falls off, allowing formation of a terminating hairpin | 图 3.24 当色氨酸处于高水平的时候，核糖体在蛋白质生产过程中不停顿。它产生出一条短的肽链，之后从 RNA 上掉下，使终止型发夹结构得以形成 |

3.3.4 *Ara* and *Gal* Operons

Like the *lac* operon, the **ara operon** contains genes for metabolizing a less-common sugar, **arabinose**. The operon is unique in that its operator is far away from the promoter, about 200 bases upstream (Figure 3.25). When arabinose is absent in the cell, *ara*

3.3.4 阿拉伯糖与半乳糖操纵子

与 *lac* 操纵子相似，***ara* 操纵子**含有代谢阿拉伯糖（一种并不很普通的糖）的基因。这一操纵子的独特之处在于它的操纵基因位于远离启动子的地方，在上游大约 200 碱基处（图 3.25）。当细胞

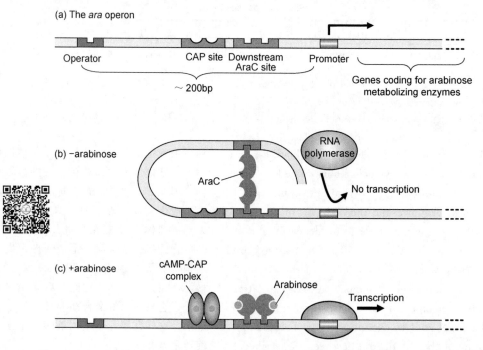

Figure 3.25　Control of the *ara* operon　　　　图 3.25　*ara* 操纵子的控制

operon genes must not be expressed. In this case, a protein called AraC is used as the repressor. One **AraC protein** binds at the operator, and another AraC protein binds near the promoter, far downstream. The two proteins also bind to each other. Because the operator and promoter are far away from each other, this causes a large loop in the DNA around the regulatory region of the operon. This DNA loop prevents transcription of operon genes by RNA polymerase.

When arabinose is present in the cell, it binds to the AraC proteins and allosterically changes their binding preferences. Instead of one protein binding to the operator, both proteins now bind near the promoter. This undoes the DNA loop, allowing transcription of *ara* operon genes. As the *lac* operon, however, *ara* operon genes must not be transcribed when glucose is available. As a result, the *ara* operon is also under positive regulation by CAP protein.

The **gal operon**, which contains the genes for metabolizing the sugar **galactose**, also uses DNA looping as a means of repression (Figure 3.26). Interestingly, the

中没有阿拉伯糖时，*ara* 操纵子基因不表达。在这种情况下，称为 AraC 的蛋白质被用来作为阻遏蛋白。一个 **AraC 蛋白**结合在操纵基因上，另一个 AraC 蛋白结合到远在下游的启动子附近。这两个蛋白质也互相结合在一起。由于操纵基因和启动子相距很远，这样的结合在操纵子 DNA 的调控区域形成了一个很大的环。这个 DNA 环防止了操纵子上的基因被 RNA 聚合酶转录。

当细胞中有阿拉伯糖时，阿拉伯糖与 AraC 蛋白结合，使 AraC 蛋白的结合倾向发生了别构化改变。原先是有一个 AraC 蛋白与操纵基因结合，现在是两个 AraC 蛋白都结合到了启动子附近。这样的结合方式消除了 DNA 环，使 *ara* 操纵子基因的转录可以进行。然而，与 *lac* 操纵子相似，*ara* 操纵子基因在能够使用葡萄糖的时候必须停止转录。因此，*ara* 操纵子也是处于 CAP 蛋白的正调控之下的。

gal 操纵子含有**半乳糖**代谢基因，该操纵子也采用使 DNA 形成环的方式实现阻遏作用（图 3.26）。有趣的是，*gal*

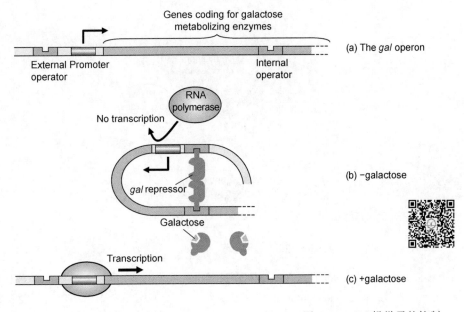

Figure 3.26　Control of the *gal* operon

图 3.26　*gal* 操纵子的控制

gal operon has two operators; one, called the **external operator**, is located near the promoter as expected. Another operator is located within one of the genes of the operon, and is called the internal operator. The *gal* operon is repressed by a protein called the **gal repressor**. When galactose is not bound to the repressor, the protein can bind to both operators of the operon. This causes a loop to form which creates a physical block to transcription.

3.4 Experiments

The concept of an operon was developed by the French scientists Jacob and Monod, who studied lactose metabolism. The first step was to determine which genes were involved in metabolizing lactose in *E.coli*. Hundreds of strains of *E. coli* were isolated that were unable to grow on medium with only lactose as the carbon source. This meant that in each strain one of the genes involved in using lactose was damaged. Next, many combinations were made between the strains, in which DNA from one mutant strain was put into another strain. This created cells called **partial diploid**s, with two sets of genes related to lactose (Figure 3.27).

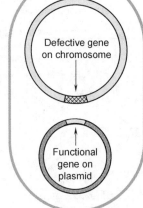

Figure 3.27 The partial diploid

In some cases, putting DNA from one strain into another strain created a cell that could grow on medium with only lactose as the carbon source. This meant that different genes were damaged in the two different strains. For example, imagine that two genes are involved in metabolizing lactose, A and B. When cell 1 and cell 2 are combined, the functional gene B from cell 1 could compensate for a damaged gene B in cell 2. Likewise, the functional gene A in cell 2 could compensate for a damaged gene A in cell 1.

In other cases, when DNA from one strain was put into another strain, the cell still could not grow on lactose. This meant that in both strains, the same gene was mutated. Even combining DNA from two cells could not overcome the lost function of an enzyme necessary for using lactose. These kinds of genetic experiments are called **complementation analysis**.

By making many such combinations, it was concluded that at least two genes were required for using lactose. They were called *lacZ* and *lacY*. Cells with mutant *lacZ* were found to be deficient for the enzyme β-galactosidase, so it was concluded that the gene normally makes β-galactosidase. Cells with mutant *lacY* could not absorb lactose. This indicated that it made a protein responsible for bringing lactose into the cell. The protein was later identified as lactose permease. Genetic mapping also showed that *lacZ* and *lacY* were present next to each other on the DNA.

In order to understand how the *lac* genes are regulated, *E. coli* strains were isolated that had abnormal regulation. Of particular interest were **constitutive mutants**, strains that expressed the *lac* genes whether or not lactose was present. Partial diploids were also made using these strains, combining DNA from one constitutive mutant with DNA from a normal *E. coli*. The mutants were found to fall into two groups. (1) When some constitutive mutant strains were combined with a normal strain, the partial diploid continued to have normal regulation. These constitutive mutants were called *lacI*⁻ mutants (Figure 3.28). (2) When other constitutive mutant strains were combined with a normal strain, the partial diploid created also transcribed *lac* genes constitutively (i. e., even in the absence of lactose). These mutant

味着两个不同菌株中的不同基因有损坏。例如，设想有两个基因（A 和 B）在代谢乳糖中发挥作用。当用细胞 1 和细胞 2 进行组合时，在细胞 1 中有功能的基因 B 会弥补细胞 2 中损坏了的基因 B。同样，在细胞 2 中有功能的基因 A 会弥补细胞 1 中损坏了的基因 A。

在另一些情形下，一个菌株的 DNA 放入另一个菌株中后，得到的细胞仍然不能在乳糖上生长。这意味着两个菌株中相同的基因发生了突变。即使将两个菌株的 DNA 进行合并也不足以克服那种乳糖代谢酶功能的丧失。这种类型的遗传实验叫作**互补分析**。

通过制备许多这样的组合，总结出使用乳糖至少需要两个基因。它们被称为 *lacZ* 和 *lacY*。发现具有 *lacZ* 突变的细胞中 β-半乳糖苷酶的功能有缺陷，因此认为该基因正常情况下编码的是 β-半乳糖苷酶。具有 *lacY* 突变的细胞不能吸收乳糖。这说明它产生的蛋白质是负责将乳糖带进细胞里的。这一蛋白质后来被鉴定出是乳糖渗透酶。遗传作图也发现 *lacZ* 和 *lacY* 在 DNA 上处于相邻的位置。

为了弄清楚 *lac* 基因是怎样被调控的，需要分离那些具有异常调控功能的大肠杆菌菌株。这当中引起特别关注的是一些**组成型突变体**，即那些在不管有没有乳糖的条件下都表达 *lac* 基因的菌株。这些菌株也用来制备部分二倍体，方法是用某一组成型突变体的 DNA 与正常的大肠杆菌 DNA 进行组合。发现这样的突变体可以分为两组：(1) 当有些组成型突变体与正常菌株组合时，得到的部分二倍体继续具有正常的调控功能。这些组成型突变体称为 *lacI*⁻ 突变体（图 3.28）。(2) 当其他组成型突变体与正常菌株组合时，得到的部分二倍体仍然继续以组成型转录 *lac* 基因（甚至在没有乳糖的情况下也转录）。这种突

strains (the haploid strain, not the partial diploid) were called $lacO^c$ mutants (Figure 3.29).

变体菌株（指单倍体菌株而不是部分二倍体）叫作 $lacO^c$ 突变体（图 3.29）。

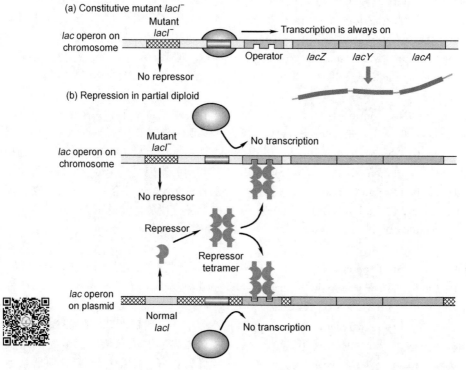

Figure 3.28　Constitutive mutant $lacI^-$　　图 3.28　组成型突变体 $lacI^-$

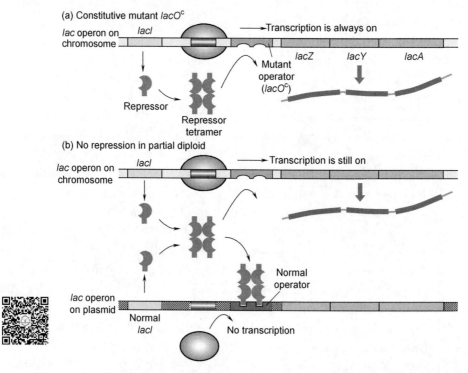

Figure 3.29　Constitutive mutant $lacO^c$　　图 3.29　组成型突变体 $lacO^c$

One more experiment was necessary before conclusions could be drawn. A *lacO^c* mutant was isolated that also had a mutation in the *lacZ* gene (*lacZ^-*). When this *lacO^c*/*lacZ^-* strain was combined with a normal strain, the result was different than when just *lacO^c* is used. The partial diploid no longer had constitutive expression of *lacZ* (Figure 3.30). Thus, it was concluded that *lacO^c* only controls the transcription of genes that lie on the same piece of DNA. It has no effect on the regulation of *lac* genes in the normal strain's DNA. This is not the case with the *lacI^-* mutants, however. As long as there is one normal *lacI* sequence in the partial diploid, the regulation of *lac* genes on both pieces of DNA in the cell will be normal.

在下结论之前还有必要做另一种实验，即还需要分离一个带有 *lacZ* 基因突变（*lacZ^-*）的 *lacO^c* 突变体。当这种 *lacO^c*/*lacZ^-* 菌株与正常菌株组合时，其结果与只使用 *lacO^c* 时不同。得到的部分二倍体不再能以组成型表达 *lacZ*（图3.30）。这样，可以认为 *lacO^c* 只控制位于相同 DNA 片段上基因的转录。它对位于正常菌株 DNA 上 *lac* 基因的转录没有调控作用。然而，在 *lacI^-* 突变体中情况是不同的。只要在部分二倍体中有正常的 *lacI* 序列存在，细胞中对两个位置 DNA 上的 *lac* 基因的调控便都能正常进行。

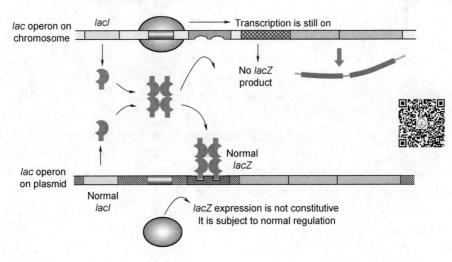

Figure 3.30 *lacO^c* functions only on the same DNA fragment

图 3.30 只有当 *lacO^c* 位于同一 DNA 片段上时才起作用

The results could be interpreted by this model: *lacI^-* strains are defective for a gene, call it *lacI*, that pro-

上述结果可以用这样的模型来解释：*lacI^-* 菌株是 *lacI* 基因有缺陷的菌株，*lacI* 基因

duces a protein. Only one normal gene is necessary in the diploid because the protein can diffuse around the cell. Even if the *lacI* gene is mutated on one piece of DNA, the protein produced on the normal piece of DNA can regulate the transcription of genes on the mutant piece of DNA.

By contrast, *lacO^c* mutants are defective for a regulatory DNA element, call it *lacO*. This element is normally involved in repressing transcription of the *lac* genes near it on the DNA. If *lacO* is damaged, the genes downstream are always transcribed, producing a constitutive mutation. Even if the *lacO^c* strain is combined with a normal strain, the *lac* genes will be constitutively transcribed, because the normal *lacO* element in the normal strain DNA has no effect on the regulation of genes in the *lacO^c* strain DNA.

The *lacO* element was later named the operator, and the *lacI* gene product was called the *lac* repressor. It was found that the operator was directly upstream of *lacZ* and *lacY*, and that these two genes were always transcribed and translated together. Also, the *lac* repressor was found to function by binding directly to the operator. Thus, the operon model was formed.

Summary

Before proteins are made from information in the genome, the DNA base sequence is first copied into RNA. This helps to protect the DNA, and offers an added step in which protein production can be regulated. Promoters play an important part in regulating transcription. Promoters help to mark the start site of transcription. RNA polymerase, the enzyme responsible for transcription, can only recognize the promoter if it is accompanied by the σ subunit. The holoenzyme then separates DNA strands at the promoter, creating a transcription bubble that moves down the length of the gene. Along the way, complementary ribonucleotides are bound to the DNA and linked together to form an RNA strand. As transcription proceeds, the new RNA strand is gradually separated from the DNA template. At the end of the gene, termination sequences cause full dissociation of the new RNA strand from the RNA polymerase and the DNA. Cells are able to control which genes are made into proteins by controlling transcription of the gene. Genes with related functions are often regulated together in prokaryotes. The collection of genes under one regulatory region is called an operon. The *lac* operon contains genes for metabolizing lactose and is under both positive and negative regulation by proteins. The *ara* operon is similar to the *lac* operon, except that it is repressed by a different mechanism, involving a loop in the DNA. The *trp* operon is only under negative regulation, but uses an extra mechanism, attenuation, to ensure that genes are transcribed only under appropriate conditions.

被调控。处于一个调控区域控制下的所有基因被称为操纵子。*lac* 操纵子含有代谢乳糖的基因，它处于由蛋白质实施的正调控和负调控之下。*ara* 操纵子与 *lac* 操纵子类似，不同之处在于它采用不同的阻遏机理（在 DNA 上形成环）。*trp* 操纵子只有负调控，但采用衰减作用这样的特殊机理来确保基因只在适当的条件下才进行转录。

Vocabulary 词汇

allolactose [æləʊ'læktəʊs]	异乳糖	inducer [in'dju:sə]	诱导物
anabolic [ə'næbəlik]	合成代谢的	initiation [iˌniʃi'eiʃən]	起始
anabolism [ə'næbəlizəm]	合成代谢	intrinsic termination [in'trinsik]	内在型终止
antitermination ['æntiˌtə:mi'neiʃən]	抗终止作用	lactose ['læktəus]	乳糖
associate [ə'səuʃieit]	联合，结合	leader ['li:də]	前导子
attenuation [əˌtenju'eiʃən]	衰减作用	metabolic [ˌmetə'bɔlik]	代谢的
attenuator [ə'tenjueitə]	衰减子	metabolism [me'tæbəlizəm]	代谢
catabolic [ˌkætə'bɔlik]	分解代谢的	negative regulation	负调控
catabolism [kə'tæbəlizəm]	分解代谢	open promoter complex	开放启动子复合体
catabolite activator protein (CAP) [kə'tæbəlait]	代谢物激活蛋白	operator ['ɔpəreitə]	操纵基因
chorismate ['kɔrismeit]	分支酸	operon ['ɔpərɔn]	操纵子
cistron ['sistrɔn]	顺反子	permease ['pə:mieis]	渗透酶
close promoter complex	闭合启动子复合体	polycistronic mRNA [pɔlisis'trɔnik]	多顺反子 mRNA
consensus sequence [kən'sensəs] ['si:kwəns]	共有序列	polymerization ['pɔliməraiˈzeiʃən]	聚合（反应）
		positive regulation	正调控
coordinate regulation [kəu'ɔ:dinit]	协同调控	primer ['praimə]	引物
		progressively [prə'gresivli]	逐渐地
core enzyme ['enzaim]	核心酶	promoter [prə'məutə]	启动子
co-repressor [ˌkəuri'presə]	辅阻遏物	proofreading [ˌpru:f'ri:diŋ]	校正（功能）
coupled transcription-translation	偶联转录-翻译作用	pyrophosphate ion [pairəu'fɔsfeit] ['aiən]	焦磷酸离子
ρ-dependent termination	ρ 依赖型终止	regulatory gene	调控基因
dissociate [di'səuʃieit]	脱离，解离	repressor [ri'presə]	阻遏蛋白
DNA duplex ['dju:pleks]	DNA 双链	rewind [ri:'waind]	重旋（重新形成螺旋）
elongation [ˌi:lɔŋ'geiʃən]	延伸		
galactose [gə'læktəus]	半乳糖	ribosome ['raibəsəum]	核糖体
galactosidase [gəːlæktəu'saideis]	半乳糖苷酶	RNA polymerase ['pɔliməreis]	RNA 聚合酶
gene expression	基因表达	roadblock ['rəudblɔk]	路障
genome ['dʒi:nəum]	基因组	scarce [skɛəs]	稀少的
holoenzyme [ˌhɔləu'enzaim]	全酶	structural gene	结构基因

template ['templit]	模板	transcription bubble [træns'krip∫ən] ['bʌbl]	转录泡
terminating hairpin ['hɛəpin]	终止型发夹	trp operon	色氨酸操纵子
termination [ˌtə:mi'nei∫ən]	终止	untranslated region (UTR)	非转译区域
transacetylase [ˌtrænzə'setileis]	转乙酰基酶	unwind ['ʌn'waind]	解旋
transcript ['trænskript]	转录本，转录产物		

Review Questions / 习题

Ⅰ. True/False Questions（判断题）

1. All genes in *E. coli* have the same promoter.
2. Sigma subunit is necessary for RNA polymerase to make contact with DNA.
3. Transcription occurs rarely in cells.
4. RNA folding is important for termination of transcription.
5. Messenger RNAs are used to bring amino acids to the ribosome for translation.
6. β-Galactosidase and lactose permease proteins are found in the cell at the same concentration.
7. Transcription of the *lac* operon is weak when cAMP levels are low.
8. β-Galactosidase and lactose permease proteins are joined together.
9. Ribosomes are involved in regulation of the *trp* operon.
10. *E. coli* is able to produce tryptophan when it is not available in the environment.

Ⅱ. Multiple Choice Questions（选择题）

1. Which of the following does not occur during initiation of transcription?
    a. formation of closed promoter complex
    b. promoter clearance
    c. polymerization of ribonucleotides
    d. formation of a hairpin loop
    e. formation of open promoter complex

2. What does not play a role in termination of transcription?
    a. −10 box
    b. hairpin loop
    c. rho protein
    d. string of A's and U's
    e. RNA polymerase

3. What kind of bond is most important for binding of RNA to DNA?
    a. hydrophobic interactions
    b. ionic bonds
    c. van der Waals forces
    d. covalent bonds
    e. none of the above

4. You find an *E. coli* cell that transcribes β-galactosidase even when there is no lactose present in the cell. What might be the reason?
    a. There is a mutation in the beta-galactosidase gene.
    b. The *lac* operon has no promoter in this cell.
    c. There is a mutation in the gene for the *lac* repressor.
    d. Glucose levels in the cell are always low.
    e. All of the above.

5. When tryptophan levels in the cell are low, _____.
    a. the ribosome slows down while translating the leader region
    b. the terminating hairpin loop forms
    c. the ribosome never attaches to the leader region
    d. the ribosome continues to transcribe the *trp* operon without stopping
    e. the *lac* repressor does not bind to the *lac* operator

Exploration Questions 思考题

1. Why do you think the cell transcribes DNA into RNA, instead of transcribing DNA into more DNA?
2. Do you think that every gene is transcribed in the same way? What are some of the things you expect might control how strongly each gene is transcribed?
3. What are the main differences in the function of the repressor protein in the *trp* operon vs. the *lac* operon?
4. What are the advantages to organizing genes in operons? What might be some disadvantages?

Chapter 4 Transcription in Eukaryotes: Mechanism and Regulation

第 4 章 真核生物转录：机理与调控

Although the basic scheme of transcription is similar between eukaryotes and prokaryotes, the detailed mechanisms are quite different. Eukaryotes have many more genes and require much more complex regulation of gene expression. This increased complexity is reflected in the complexity of the eukaryotic transcriptional mechanisms.

Eukaryotic transcription depends on RNA polymerase and proteins called transcription factors. Transcription factors can be divided into two major categories: general transcription factors and specific transcription factors. General transcription factors are required for transcription of all genes. Specific transcription factors are proteins that can increase or decrease the rate of transcription of specific genes. They provide eukaryotic cells with a very fine control over transcription. We will begin by describing RNA polymerase and the general transcription factors, and then explain how specific transcription factors control the work of this basic machinery.

虽然真核生物和原核生物转录的基本框架是相似的，但是其详细机理有很大不同。真核生物拥有更多的基因，需要对基因表达进行更复杂的调控。这种增加的复杂性反映在真核生物转录机理的复杂性上。

真核生物转录依赖于 RNA 聚合酶和称为转录因子的蛋白质。转录因子可被分为两大类：通用转录因子和特异转录因子。通用转录因子是所有基因转录都需要的。特异转录因子是那些能提高或降低特殊基因转录速率的蛋白质。它们为真核细胞提供了非常精确的转录调控。我们将从描述 RNA 聚合酶和通用转录因子开始，然后解释特异转录因子怎样对基本转录装置进行调控。

4.1 Eukaryotic RNA Polymerases

4.1 真核生物 RNA 聚合酶

Recall from chapter 3 that all transcription in *E. coli* is performed by one kind of RNA polymerase. By contrast, eukaryotes employ three different RNA polymerases, which each transcribe different classes of genes. The three polymerases are termed RNA polymerase Ⅰ, Ⅱ and Ⅲ (Table 4.1). The genes they transcribe are called Class Ⅰ, Ⅱ, and Ⅲ genes, respectively.

回忆第 3 章中所讲，大肠杆菌中的所有转录都由一种 RNA 聚合酶完成。与此相反，真核生物采用了三种不同 RNA 聚合酶，每一种聚合酶转录不同类型的基因。这三种聚合酶称为 RNA 聚合酶 Ⅰ、Ⅱ 和 Ⅲ（表 4.1）。它们转录的基因也分别称为 Ⅰ 类、Ⅱ 类和 Ⅲ 类基因。

Table 4.1 Eukaryotic RNA polymerases and their roles

表 4.1 真核生物 RNA 聚合酶和它们的作用

Polymerase	Location	Products
Polymerase Ⅰ	Nucleolus	A large rRNA precursor for 5.8S, 18S, and 28S rRNAs
Polymerase Ⅱ	Nucleoplasm	hnRNAs (precursors for mRNAs) and most of the snRNAs
Polymerase Ⅲ	Nucleoplasm	Precursors for 5S rRNA and 4.5S tRNAs, and U6 snRNA, etc

RNA polymerase Ⅰ is used to transcribe only one gene, the large rRNA precursor. This large transcript is later used to produce three mature rRNAs. There are many copies of the large rRNA precursor gene in each cell. Thus, although RNA polymerase Ⅰ only transcribes one kind of gene, it is quite active. **RNA polymerase** Ⅲ also transcribes a relatively limited set of genes, notably the tRNA genes. Class Ⅲ genes comprise some other small RNAs as well, including the 5S rRNA and the U6 RNA, which are described in later chapters. **RNA polymerase** Ⅱ is the most prolific RNA polymerase. It transcribes all genes that produce mRNA and thus code for proteins. It also transcribes small RNAs called snRNAs, explained in chapter 5. We will mainly discuss transcription by RNA polymerase Ⅱ. Later we will return to briefly discuss transcription by RNA polymerase Ⅰ and Ⅲ.

RNA polymerase Ⅱ is composed of 12 subunits, named Rpb1-12 (Figure 4.1). 'Rp' stands for RNA polymerase. 'b', the second letter of the alphabet, indicates that these are the subunits for RNA polymerase Ⅱ. Rpb1, Rpb2, and Rpb3 are called the 'core subunits' because they are fundamentally required for enzyme activity. These subunits are related to the three subunits of the E. coli RNA polymerase core: β, β' and α. They bind to the DNA and catalyze the reaction that joins ribonucleotides.

RNA 聚合酶Ⅰ只用来转录一种基因——大的 rRNA 前体基因。这一大的转录物在之后用来产生三种成熟的 rRNA。在每个细胞中有许多这一大的 rRNA 前体基因的拷贝。因此，虽然 rRNA 聚合酶Ⅰ只用来转录一种基因，它却是相当活跃的。**RNA 聚合酶**Ⅲ也转录相对来说较少的基因，它主要转录 tRNA 基因。第Ⅲ类基因也包含一些小 RNA 基因，如 5S rRNA 和 U6 RNA 基因，这两个基因在后面的章节中还会述及。**RNA 聚合酶**Ⅱ是最多产的 RNA 聚合酶。它转录所有产生 mRNA 并编码了蛋白质的基因。它也转录一些称为 snRNA 的小 RNA 基因，这部分内容将在第 5 章中讲述。本章我们主要讨论 RNA 聚合酶Ⅱ的转录情况，之后回过头来简要讨论 RNA 聚合酶Ⅰ和Ⅲ的转录情况。

RNA 聚合酶Ⅱ由 12 个亚基组成，称为 Rpb1-12（图 4.1）。"Rp" 代表 RNA 聚合酶。"b"，字母表中的第二个字母，指这些亚基属于 RNA 聚合酶Ⅱ。Rpb1、Rpb2 和 Rpb3 称为"核心亚基"，因为它们是产生酶活性的最基本组分。这几种亚基与大肠杆菌 RNA 聚合酶核心的 β、β'和 α 亚基有一定的关系。它们与 DNA 结合并催化连接核糖核苷酸的反应。

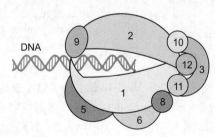

Figure 4.1 Schematic drawing of RNA polymerase Ⅱ showing all 12 subunits

图 4.1 RNA 聚合酶Ⅱ结构示意图（显示所有 12 个亚基）

The function of the other nine RNA polymerase Ⅱ subunits is not clear, but most of them are also very important for the function of the polymerase. Like E. coli RNA polymerase, the overall structure of the 12

RNA 聚合酶Ⅱ其他九个亚基的功能还不清楚，但多数亚基对聚合酶的功能也很重要。像大肠杆菌 RNA 聚合酶一样，由 12 个亚基组成的聚合酶总体结

subunit polymerase resembles a crab claw. The space within the claw can accommodate a double-stranded DNA molecule. The clamp then closes during transcription to stabilize the association between the DNA and the newly polymerized RNA.

4.2 Eukaryotic Promoters

In order to initiate transcription, RNA polymerase Ⅱ and the general transcription factors form a complex around the promoter. Recall that *E. coli* promoters usually have two elements, a −10 box and a −35 box. Eukaryotic promoters are more varied (Figure 4.2). The core promoter, closest to the start site of transcription, may have four elements: the TATA box, the initiator, the TFⅡB recognition element, and the downstream promoter element. Beyond the core promoter, upstream promoter elements may contain a variety of conserved sequences including the GC box and the CCAAT box. A eukaryotic promoter almost never includes all of the elements just listed. Different promoters contain different combinations of these elements, providing a diversity that allows tens of thousands of individual genes to be clearly distinguished from each other.

4.2 真核生物启动子

为了启动转录，RNA 聚合酶Ⅱ与通用转录因子在启动子周围形成一个复合体。回忆一下，大肠杆菌启动子通常含有两个元件，即−10 框和−35 框。而真核生物启动子有很大不同（图 4.2）。最靠近转录起始位置的核心启动子可以包含四个元件：TATA 框、起始子、TFⅡB 识别元件和下游启动子元件。在核心启动子以外，上游启动元件可能含有一些不同的保守序列，包括 GC 框和 CCAAT 框。一个真核启动子几乎从来不会包括所有上述元件。不同启动子含有这些元件的不同组合，这种组合的多样性使得数以万计的单个基因可以清楚地互相区别开来。

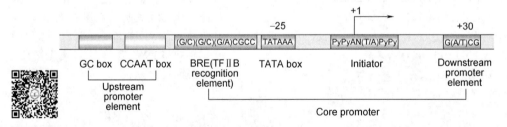

Figure 4.2 Common promoter elements in eukaryotes

图 4.2 真核生物共有的启动子元件

The **TATA box** is a particularly common and well-studied eukaryotic promoter element. Its name derives from its consensus sequence, TATAAA in the non-template strand. Notice that this sequence is quite similar to the −10 box sequence in prokaryotes (TATAAT). However, these two elements should not be confused. The TATA box only refers to an element in eukaryotes. The TATA box is usually centered at position −25 relative to the transcription start site, not −10. It is most often present in specialized genes that

TATA 框是特别常见并且研究得很清楚的真核启动子元件。其名称来源于它的共有序列，即在非模板链上的 TATA-AA。注意这一序列与原核生物的−10 框序列（TATAAT）非常相似。然而，这两个元件不应该被混淆。TATA 框只是指真核生物中的一个元件，它通常以转录起始位点上游−25 位为中心，而不是−10 位。它最常出现在特殊的基因中，这些基因只在特殊的细胞里产生

make proteins only found in particular cells. It tends to be absent from genes that are always expressed, called **housekeeping genes**, and genes that are important for organismal development.

The **downstream promoter element (DPE)** is another very common promoter element. Its name derives from the fact that, unlike the other elements, it begins downstream of the transcription start site, at position +30, approximately. The consensus sequence of the DPE is G(A/T)CG, meaning that the consensus sequence may have A or T at the second position.

The consensus sequence of the **initiator (Inr)** promoter element is even more generalized. In mammals, it is PyPyAN(T/A)PyPy. 'Py' refers to Pyrimidine- T or C. 'N' stands for any nucleotide. This sequence is positioned with A at the transcription start site. The initiator itself can drive basic transcription, but its function is greatly amplified by the presence of other promoter elements.

The **TFⅡB recognition element (BRE)** is located upstream of the TATA box and has the consensus sequence (G/C)(G/C)(G/A)CGCC. As the name suggests, the BRE sequence binds to TFⅡB, one of the general transcription factors, which is introduced in the next section. The BRE is not the only promoter element that binds to a transcription factor. Most promoter elements function by binding to a transcription factor or directly to RNA polymerase.

4.3 General Transcription Factors and Initiation

Recall that in prokaryotes RNA polymerase core has to associate with the σ subunit in order to recognize promoters and begin transcription at the correct site. In eukaryotes, recognizing promoters and initiating transcription requires at least six proteins in addition to RNA polymerase Ⅱ. These proteins are the general transcription factors, TFⅡA, B, D, E, F, and H. The acronym TFⅡ stands for Transcription Factor of RNA polymerase Ⅱ.

蛋白质。在那些总是需要表达的基因里一般没有它,那样的基因叫**持家基因**;在一些对生物发育起重要作用的基因中也没有。

下游启动子元件(DPE)是另一个很常见的启动子元件。其名称来源于这样一个事实:不同于其他元件,它开始于转录起始位点下游约+30位置。DPE的共有序列是G(A/T)CG,意思是共有序列在第二个位置是A或T。

起始子(Inr)启动子元件的共有序列甚至更具有普遍性。在哺乳动物中,它是PyPyAN(T/A)PyPy。"Py"指嘧啶,即T或C。"N"代表任何核苷酸。这个序列的A被放在转录起始位点上。起始子本身能够驱动基础转录,但在其他启动子元件存在时它的功能会变得更强。

TFⅡB识别元件(BRE)位于TATA框上游,具有共有序列(G/C)(G/C)(G/A)CGCC。顾名思义,BRE序列结合到TFⅡB上,TFⅡB是通用转录因子中的一种,我们将在下一节中进行介绍。BRE并不是仅有的与转录因子结合的启动子元件。大多数启动子元件通过结合到转录因子上或直接结合到RNA聚合酶上而发挥作用。

4.3 通用转录因子与转录起始

回忆一下,在原核生物中RNA聚合酶核心必须与σ亚基结合以识别启动子而后在正确位置开始转录。在真核生物中,识别启动子和起始转录需要除RNA聚合酶Ⅱ以外至少六种蛋白质。这些蛋白质是通用转录因子,即TFⅡA、B、D、E、F和H。缩略词TFⅡ表示RNA聚合酶Ⅱ的转录因子。

General transcription factors are required for transcription of all class Ⅱ genes. They are particularly important for initiation, at which point they help to determine the start site and direction of transcription. The collection of general transcription factors with RNA polymerase Ⅱ at the promoter is called the **pre-initiation complex** (Figure 4.3). Many of the proteins in the complex are themselves composed of multiple subunits, making this a very formidable collection of peptides.

通用转录因子在转录所有Ⅱ类基因中都需要。它们对转录起始来说特别重要，在这一环节它们帮助确定转录起始位点和方向。通用转录因子与RNA聚合酶Ⅱ在启动子位置形成的组合称为**前起始复合体**（图4.3）。复合体中的许多蛋白质自身也是由许多亚基组成的，这使得复合体成为非常惊人的肽的集结地。

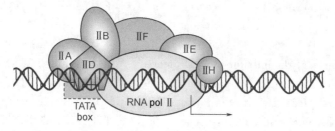

Figure 4.3　The pre-initiation complex

图4.3　前起始复合体

4.3.1　TFⅡD

The first TFⅡ to bind the promoter is **TFⅡD** (Figure 4.4). This is a large protein composed of 8~10 subunits. One of these subunits is a protein called **TATA-binding protein (TBP)** that binds to the TATA box. TBP is found in most eukaryotes and its amino acid sequence is quite well conserved between very different organisms. This fact indicates the importance of TBP and its design in eukaryotic transcription.

4.3.1　TFⅡD

第一个结合到启动子上的TFⅡ是**TFⅡD**（图4.4）。这是一个由8~10个亚基组成的大蛋白。这些亚基中的一个称为**TATA结合蛋白（TBP）**，它与TATA框结合。TBP在大多数真核生物中都有，其氨基酸序列在差异很大的生物中也相当保守。这一事实表明了TBP的重要性和它在真核生物转录中的用途。

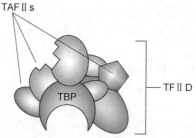

Figure 4.4　Schematic drawing of TFⅡD showing TBP and TAFⅡs

图4.4　TFⅡD结构示意图（显示TBP和所有TAFⅡ）

The shape of TBP has been described as resembling a saddle. In the 'stirrups' of the protein are phenylalanine residues. The phenylalanine R group contains an aromatic ring that resembles a nitrogenous base in size and shape. When TBP binds to the minor groove of the

TBP的形状被描述成类似于马鞍。在蛋白质的"马镫"处是苯丙氨酸残基。苯丙氨酸的R基团含有一个芳香环，它在大小和形状上类似于含氮碱基。当TBP结合到DNA小沟上时，这个环可

DNA, this ring can insert between the DNA bases and cause it to bend sharply, to an angle of almost 80 degrees (Figure 4.5). The bend allows easy separation of the DNA double strands so that a single-stranded template can be exposed for transcription. The weak binding between T-A base pairs allows the TATA box to be denatured relatively easily.

以插入 DNA 的碱基之间并引起 DNA 链的明显弯曲，形成一个将近 80°的角（图 4.5）。这种弯曲使 DNA 双链很容易分开，从而暴露出单链模板用于转录。T-A 碱基对之间弱的结合力使 TATA 框相对较为容易发生变性。

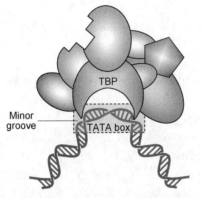

Figure 4.5　TBP binds to TATA box and causes sharp bend in DNA

图 4.5　TBP 结合到 TATA 框引起 DNA 链的明显弯曲

The other subunits of TFⅡD are called **TBP-Associated Factors Ⅱ (TAFⅡ)**. Be sure not to confuse TAFⅡs with TFⅡs. TFⅡs refers to the whole set of general transcription factors for RNA polymerase Ⅱ, while TAFⅡs only refer to subunits of one general transcription factor, TFⅡD. Within TFⅡD, TAFⅡs bind to TBP and to each other. TAFⅡs have two major functions in transcription. First, they bind to core promoter elements other than the TATA box, such as DPE and Inr, to stimulate transcription (Figure 4.6). Second, they bind to specific transcription factors to allow enhancement or inhibition of transcription (Figure 4.7).

TFⅡD 的其他亚基称为 **TBP 相关因子 Ⅱ (TAFⅡ)**。请一定不要将 TAFⅡs 与 TFⅡs 相混淆。TFⅡs 指 RNA 聚合酶Ⅱ的所有通用转录因子，而 TAFⅡs 只是指一种通用转录因子（TFⅡD）的亚基。在 TFⅡD 当中，TAFⅡs 结合到 TBP 上并互相结合。TAFⅡs 在转录中具有两种主要功能。第一，它们结合到核心启动子元件（如 DPE 和 Inr）上而不是结合到 TATA 框上去促进转录（图 4.6）。第二，它们与特异转录因子结合以增强或抑制转录（图 4.7）。

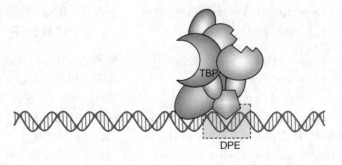

Figure 4.6　TAFⅡs are used to recognize promoter elements

图 4.6　TAFⅡs 用于识别启动子元件

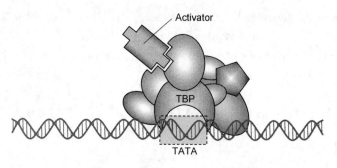

Figure 4.7 TAFⅡs are able to bind to specific transcription factors

图 4.7 TAFⅡs 能与特异转录因子结合

Not all TAFⅡs in a cell are required for transcription of every class Ⅱ gene. However, TBP is required for all transcription of class Ⅱ genes, even for genes that don't have a TATA box (Figure 4.6). This is because, in addition to recognizing the TATA box, TBP serves as the organizational center of TFⅡD, holding the TAFⅡs in the right orientation to perform their functions.

在细胞中并不是转录每种Ⅱ类基因都需要所有 TAFⅡs 的。然而，TBP 是转录所有Ⅱ类基因都需要的，甚至那些不含有 TATA 框的基因也需要它（图4.6）。这是因为除了识别 TATA 框以外 TBP 还作为 TFⅡD 的组织中心，使 TAFⅡs 保持在正确的方向来行使它们的功能。

4.3.2　Other TFⅡs

TFⅡD begins the assembly of the pre-initiation complex. It binds first to the promoter with help from TFⅡA. Next, TFⅡB joins them, followed by TFⅡF and RNA polymerase Ⅱ. TFⅡE and TFⅡH bind last. We will not detail the individualized functions of these general transcription factors here. To give some idea of their functions: TFⅡB serves as a link between TFⅡD and RNA polymerase Ⅱ. TFⅡH participates in opening the DNA helix to allow transcription. Even without going into more detail, the reader should by now appreciate the complexity of initiation of transcription in eukaryotes.

4.3.2　其他 TFⅡs

前起始复合体的装配开始于 TFⅡD。它在 TFⅡA 的帮助下首先结合到启动子上。之后，TFⅡB 加入进来，随后是 TFⅡF 和 RNA 聚合酶Ⅱ。TFⅡE 和 TFⅡH 最后结合上来。在此我们不详细介绍这些通用转录因子各自的功能。对它们功能方面的一些认识是：TFⅡB 起到连接 TFⅡD 和 RNA 聚合酶Ⅱ的作用，TFⅡH 参与打开 DNA 螺旋以便转录发生。即使没有深入更多的细节中，到现在为止读者应该能体会到真核生物转录起始的复杂性。

4.3.3　General Transcription Factors for RNA Polymerase Ⅰ and Ⅲ

Although still more complicated than prokaryotic initiation, transcription initiation for class Ⅰ genes is much simpler than for class Ⅱ genes (Figure 4.8). Only two general transcription factors are required, **SL1** (sometimes called TIF-IB) and **UBF** (sometimes called UAF). The class Ⅰ promoter contains two elements, a core element centered at the transcription start site and an upstream promoter element. SL1 binds

4.3.3　RNA 聚合酶Ⅰ和Ⅲ的通用转录因子

虽然仍旧比原核生物的起始更复杂，Ⅰ类基因的转录起始要比Ⅱ类基因简单得多（图4.8）。它只需要两个通用转录因子，即 **SL1**（有时称为 TIF-IB）和 **UBF**（有时称为 UAF）。Ⅰ类启动子含有两个元件，即一个以转录起始位点为中心的核心元件和一个上游启动子元件。SL1 结合到核心元件上并帮助召集

to the core element and helps recruit RNA polymerase Ⅰ to the promoter. UBF binds to the upstream promoter element to help SL1 to bind to the core element.

RNA 聚合酶Ⅰ到启动子位置。UBF 结合到上游启动子元件帮助 SL1 结合到核心元件上。

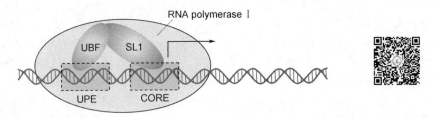

Figure 4.8　The transcription factors of class Ⅰ genes

图 4.8　Ⅰ类基因的转录因子

Transcription initiation of class Ⅲ genes is more varied. For tRNA genes, the promoter contains two elements, **box A** and **box B**, both of which are downstream of the transcription start site (Figure 4.9). A

Ⅲ类基因的转录起始有更多不同之处。对 tRNA 基因来说，启动子含有两个元件，即 **A 框**和 **B 框**，两者都位于转录起始位点下游（图 4.9）。一种称为 **TF**

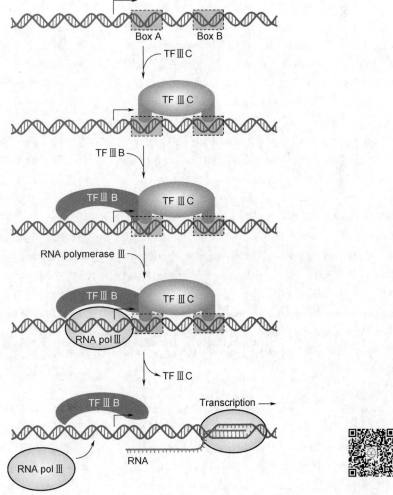

Figure 4.9　Transcription factors of class Ⅲ (tRNA) genes

图 4.9　Ⅲ类基因（tRNA 基因）的转录因子

class Ⅲ general transcription factor called **TF Ⅲ C** binds to these elements. TF Ⅲ C helps another protein **TF Ⅲ B** to bind to the DNA near the transcription start site, with help from TATA-binding protein (TBP). TF Ⅲ B helps RNA polymerase Ⅲ to bind at the gene. Even though TF Ⅲ C detaches after transcription initiation, TF Ⅲ B stays bound and can be used for several rounds of transcription. Note that this mechanism does not apply to transcription initiation of all class Ⅲ genes.

4.4 Specific Transcription Factors and Transcriptional Regulation

The general transcription factors and RNA polymerase Ⅱ rarely join together spontaneously and begin transcription. When only these proteins are present, transcription occurs at a low rate called **basal transcription**. In order to modify the rate of transcription, eukaryotic cells use another class of proteins called **specific transcription factors**. These may also be called gene-specific transcription factors.

Specific transcription factors are a huge class of proteins, providing very fine regulation of transcription in eukaryotes. We already encountered the concept of transcriptional regulation in the last chapter. We saw that it is advantageous for *E. coli* to transcribe different genes depending which nutrients are available in the environment. In eukaryotes, the need for transcriptional regulation is much greater even. Consider this fact: essentially every cell in your body has exactly the same set of genes. But think of how many different kinds of cells exist in the body, from neurons to skin cells to blood cells. How can cells that are so different be made from the same set of genes?

The answer is that within each cell type, genes are transcribed differently. Red blood cells are red because they transcribe massive amounts of hemoglobin, which carries oxygen. Other cells, like skin cells, have no need for hemoglobin, and do not transcribe its gene. Multicellularity is only made possible by very precise regulation of transcription. This regulation has

ⅢC的Ⅲ类通用转录因子结合到这些元件上。在TATA结合蛋白（TBP）的帮助下，TFⅢC帮助另一种蛋白质（即**TFⅢB**）结合到靠近转录起始位点的DNA上。TFⅢB又帮助RNA聚合酶Ⅲ结合到该基因上。即使在转录起始后TFⅢC已经从DNA上脱离，TFⅢB仍然与DNA结合并在后面几轮转录中仍被利用。请注意，这一机理并不适用所有Ⅲ类基因的转录起始。

4.4 特异转录因子与转录调控

通用转录因子和RNA聚合酶Ⅱ很少自发地结合到一起去启动转录。当只有这些蛋白质存在时，转录以很低的速率进行，这称为**基础转录**。为了调整转录速率，真核生物使用另一类称为**特异转录因子**的蛋白质。它们也可以被称为基因特异性转录因子。

特异转录因子是一大类蛋白质，它们对真核生物转录实施非常精确的调控。我们在上一章中已经遇到了转录调控的概念。我们看到，对大肠杆菌来说根据环境中可以获得哪些营养成分而转录不同的基因是非常有利的。在真核生物中，对转录过程进行调控的要求甚至更高。考虑这样一个事实：从本质上讲在你身体中的每个细胞都具有完全相同的一整套基因。再想一下在你的身体中存在多少种不同种类的细胞，从神经元到皮肤细胞到血细胞。具有相同基因的细胞怎么会变得如此不同呢？

答案是：在各种类型的细胞里，基因转录情况是不同的。红细胞是红的，因为它转录了很多起输送氧气作用的血红蛋白。其他细胞，如皮肤细胞，不需要血红蛋白，也就不需要转录血红蛋白基因。细胞的多样性只有通过非常精确的转录调控才有可能实现。这种调控在通

its foundation in the complex world of general transcription factors and specific transcription factors, and their interaction with regulatory sequences in DNA.

4.4.1 Activators

Specific transcription factors that enhance the rate of transcription are called **activators**. These proteins work through a variety of mechanisms. The most basic way is to enhance binding between the general transcription factors, RNA polymerase Ⅱ and the promoter (Figures 4.10 and Figures 4.11). Increased interaction between these components causes the pre-initiation complex to form more often, and to stay together more stably once it forms. As a result, transcription initiates more frequently at the promoter. This mechanism is somewhat similar to that of CAP protein in *E. coli*, which attracts the RNA polymerase holoenzyme to activate the *lac* operon.

4.4.1 激活蛋白

提高转录速率的特异转录因子称为**激活蛋白**。这些蛋白通过多种不同机理发挥作用。最基本的方式是增强通用转录因子、RNA 聚合酶 Ⅱ 和启动子之间的结合（图 4.10 和图 4.11）。这些成员之间的更多互相作用使得前起始复合体更频繁地形成，并且一旦形成后又能更稳定地结合在一起。结果，转录在启动子的位置更频繁地发生。这种机理有点类似于大肠杆菌中的 CAP 蛋白，CAP 蛋白通过吸引 RNA 聚合酶全酶来激活 *lac* 操纵子。

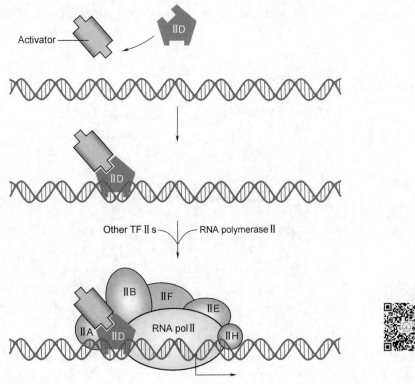

Figure 4.10 Activators can help recruit the pre-initiation complex

图 4.10 激活蛋白能帮助召集前起始复合体

Activators also enhance transcription by altering the higher-level packaging of a gene's DNA. In eukaryotes, DNA is tightly wound around proteins called histones.

激活蛋白也通过改变基因中 DNA 的高级包装情况来促进转录。在真核生物中，DNA 紧紧缠绕在组蛋白外面，这

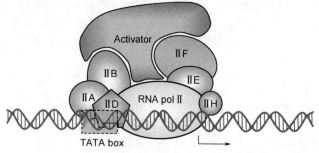

Figure 4.11 Activators can help organize the subunits of the pre-initiation complex, and maintain cohesion of the complex

图 4.11 激活蛋白能帮助组织前起始复合体的亚基并保持复合体的结合力

This helps to organize the genome, but it also makes genes difficult to transcribe. The pre-initiation complex alone does not have the necessary proteins to free the DNA for transcription. Activators recruit enzymes that dislodge the histones associated with DNA, or alter them so that they associate less tightly with DNA (Figure 4.12). We will discuss this kind of activator and the association of DNA with protein in the next section.

种缠绕帮助组织了基因组，但也使基因转录更困难。前起始复合体中并没有必需的蛋白质可以把转录所需要的 DNA 释放出来。激活蛋白能够召集一些酶来挪开与 DNA 结合的组蛋白，或者改变它们使它们与 DNA 结合得不那么紧密（图 4.12）。我们将在下一节中讨论这一类型的激活蛋白以及它们与 DNA 之间的相互作用。

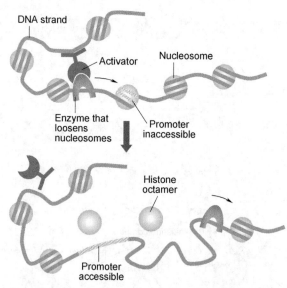

Figure 4.12 Activators can attract enzymes that loosen DNA from protein, freeing important regions like the promoter

图 4.12 激活蛋白能够吸引酶来从蛋白质上松开 DNA，让启动子这样的重要区域游离出来

Activators, as well as repressors discussed in the next section, are frequently modular. This means that one protein contains separate regions, or modules, that have separate functions. Most activators have a DNA

激活蛋白以及下一节将讨论的阻遏蛋白通常是模块化的。这意味着一种蛋白质包含具有不同功能的区域或模块。大多数激活蛋白具有一个识别特异调控序列

90 | Chapter 4 Transcription in Eukaryotes: Mechanism and Regulation

binding module that recognizes specific regulatory sequences, and a separate activation module that regulates transcription.

4.4.2 Repressors

Cells can also use specific transcription factors to prevent the transcription of a gene. In this case, the transcription factor is called a **repressor**. Some repressors act by simply blocking the function of an activator (Figure 4.13). Because transcription occurs very slowly in the absence of activators, this alone can cause strong repression of a gene. Other repressors act more directly on the transcriptional machinery. They might, for example, bind to the DNA and prevent association of general transcription factors with the promoter (Figure 4.14). This type of mechanism is similar to that of the lac repressor. Finally, some repressors act by attracting enzymes that tighten the association of a gene with protein, making the DNA inaccessible for transcription (Figure 4.15).

4.4.2 阻遏蛋白

细胞也可以使用特异转录因子来阻止一个基因的转录。在这种情形下，转录因子叫作**阻遏蛋白**。一些阻遏蛋白通过简单地阻断激活蛋白的功能而发挥作用（图4.13）。因为在缺少激活蛋白的情形下转录发生得很慢，单是这样就已经能对基因起到相当强的阻遏作用。其他阻遏蛋白则更直接地在转录装置上发挥作用。例如它可以结合到DNA上防止通用转录因子与启动子的结合（图4.14）。这种机理与lac阻遏蛋白的机理类似。最后，有些阻遏蛋白通过吸引一些能够使基因与蛋白质结合得更紧密的酶来使DNA不易接近而抑制转录（图4.15）。

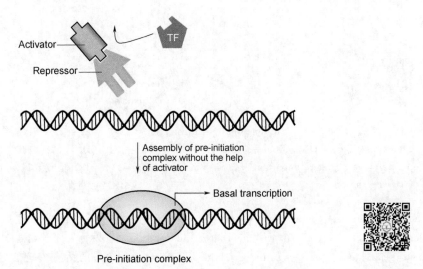

Figure 4.13 Repressors can function by blocking the activity of activators

图4.13 阻遏蛋白能通过阻碍激活蛋白的活性而发挥作用

4.4.3 Enhancers and Silencers

Specific transcription factors rely on DNA sequences to tell them how each gene should be regulated. These sequences are called **enhancers** or **silencers**, depending on whether they bind activators or repressors, respectively. Enhancers and silencers are sometimes located

4.4.3 增强子和沉默子

特异转录因子依靠DNA序列来告诉它们应该怎样调控每一种基因。这些序列称为**增强子**或**沉默子**，分别依赖于它们结合的是激活蛋白还是阻遏蛋白。增强子和沉默子有时位于距离基因几千个碱

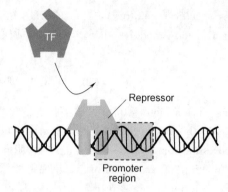

Figure 4.14 Repressors can function by blocking access of the general transcription factors to the promoter

图 4.14 阻遏蛋白能通过阻止通用转录因子接近启动子而发挥作用

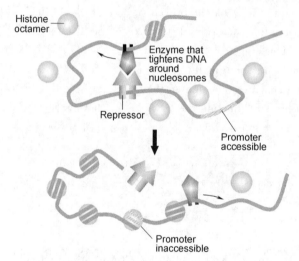

Figure 4.15 Repressors can function by recruiting enzymes that tighten DNA around proteins, making the promoter and other important regions inaccessible for transcription

图 4.15 阻遏蛋白能通过召集酶使 DNA 紧密地缠绕在蛋白上，让启动子和其他重要区域不可接近，从而阻止转录

thousands of base pairs away from the rest of the gene. It might seem surprising that transcription factors can have any effect if they are bound to a site so distant from the promoter. Actually, DNA can loop around, so that these distant regulatory regions come quite near to the promoter (Figure 4.16). As a result, the specific transcription factors are able to interact directly with the pre-initiation complex.

The regulatory regions of eukaryotic genes can become extremely complex. The *Endo 16* gene in sea urchins, for example, contains six regulatory enhancers and silencers upstream of the promoter which can respond to

基对的位置。这似乎有点奇怪，因为如果转录因子结合在离启动子那么远的地方的话它们怎么还能起作用呢。实际上，DNA 可以弯过来，这样这些距离远的调控区域就可以来到启动子的附近（图 4.16）。结果，特异转录因子就能够与前起始复合体发生直接相互作用。

真核基因的调控区域可以变得极其复杂。例如海胆的 *Endo 16* 基因在启动子上游含有六个起调控作用的增强子和沉默子，它们能够响应十几种不同蛋白的

92　Chapter 4　Transcription in Eukaryotes: Mechanism and Regulation

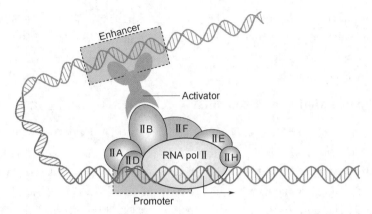

Figure 4.16 Enhancers and silencers may be located far from the promoter. DNA looping allows the enhancer and the attached activator to come into contact with the pre-initiation complex

图 4.16 增强子和沉默子可以位于远离启动子的位置。DNA 绕成环状使增强子及其结合的激活蛋白与前起始复合体发生接触

over a dozen different proteins. Complex regulation is necessary in this instance because *Endo 16* is a gene that guides the development of sea urchin embryos. It must be expressed in only a subset of cells, but the pattern of when and where it is expressed changes as the embryos develops. Whether the *Endo 16* gene is transcribed in a cell or not ultimately depends on which combinations of activators are present in the cell at a particular time and are able to bind the enhancers.

4.5 Structures of Specific Transcription Factors

We have seen that the binding of proteins to DNA is required for transcription and its regulation, in both prokaryotes and eukaryotes. DNA-binding proteins are usually specific for a DNA sequence. TFⅡB, for instance, specifically binds to the sequence of the BRE promoter element. How can a protein read a DNA sequence? Surprisingly, DNA-binding domains usually recognize DNA sequence without even opening the DNA helix. All four bases must therefore be distinguishable from the outside of the helix.

A, G, T, and C can only be fully distinguished from the surface of major grooves in the DNA. From the minor groove, G is indistinguishable from C, and A is indistinguishable from T. Therefore, most DNA-binding proteins that recognize specific sequences bind

调控作用。在这个例子中复杂调控是必需的，因为 *Endo 16* 是指导海胆胚胎发育的基因。它必须在一部分细胞中进行表达，但什么时间表达、在什么位置表达则随胚胎发育时期的不同而改变。在细胞中 *Endo 16* 基因是否转录最终依赖于在一个特定时间里哪些蛋白组合出现在细胞中以及能否结合到增强子上。

4.5 特异转录因子的结构

我们已经看到，在原核生物和真核生物中转录和转录调控都需要有蛋白质与 DNA 发生结合。DNA 结合蛋白对 DNA 序列来说通常是特异性的。例如，TFⅡB 特异性地结合到 BRE 启动子元件序列上。蛋白质是怎样解读 DNA 序列的呢？令人惊奇的是，DNA 结合域通常在不打开 DNA 螺旋的情况下就能识别 DNA 序列。因此，所有四种碱基从螺旋外面看肯定是能区分开的。

A、G、T 和 C 只能从 DNA 大沟的表面才能完全加以区别。从小沟上看，G 与 C 没有区别，A 与 T 没有区别。因此，绝大多数识别特异序列的 DNA 结合蛋白都结合到 DNA 的大沟上。成功

in the major groove of the DNA. Successful binding requires interaction between the amino acids of the protein and the chemical groups of the bases that are accessible from the major groove.

4.5.1 DNA-Binding Motifs in Prokaryotes

In prokaryotes, one type of DNA-binding structure appears in many different DNA-binding proteins. This is called the **helix-turn-helix (HTH)** motif (Figure 4.17). A **motif** is part of a domain with a highly conserved shape. The HTH motif is made of two α helices connected by a short peptide turn. One of these helices, called the recognition helix, is oriented so that it can fit tightly into major grooves, without actually opening the double helix.

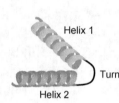

Figure 4.17 The helix-turn-helix motif

Many proteins have the HTH motif but recognize completely different sequences in the DNA. This is possible because although the overall structure of the motif is conserved, the individual amino acids that compose it can vary. Different arrangements of amino acids on the recognition helix recognize different DNA sequences. This principle applies to all DNA-binding proteins that recognize specific sequences.

The trp repressor is an example of a prokaryotic DNA-binding protein with a HTH motif. The HTH motif of the trp repressor is only able to bind *trp* operator DNA in the presence of tryptophan. Binding of tryptophan in a regulatory site on the repressor causes a structural shift that allows the recognition helix to contact the major groove. The trp repressor is actually a **dimer** of two proteins, both of which contain an HTH motif. The use of dimerization in DNA-binding proteins is common because it greatly

4.5.1 原核生物 DNA 结合基序

在原核生物中，有一种类型的 DNA 结合结构出现在很多种不同的 DNA 结合蛋白里，它被称为**螺旋-转角-螺旋（HTH）**基序（图 4.17）。**基序**是具有高度保守形状的结构域的组成部分。HTH 基序由通过一个短的肽链转角连接起来的两个 α 螺旋组成。两个螺旋中的一个称为识别螺旋，具有一定的方向性，它的大小正好适合紧密地插入大沟中，在识别过程中实际上不用打开 DNA 双螺旋。

图 4.17 螺旋-转角-螺旋基序

许多蛋白质具有 HTH 基序却识别完全不同的 DNA 序列。这是有可能的，因为虽然基序的总体结构是保守的，但是组成它的氨基酸可以不同。在识别螺旋上氨基酸的不同排列会导致识别不同的 DNA 序列。这一原理适用于所有识别特定序列的 DNA 结合蛋白。

trp 阻遏蛋白是原核生物中一种具有 HTH 基序的 DNA 结合蛋白。trp 阻遏蛋白的 HTH 基序只有在存在色氨酸的情况下才能结合到 *trp* 操纵基因 DNA 上。色氨酸结合到阻遏蛋白的调控位点上能引起结构的改变，从而允许识别螺旋与大沟的接触。trp 阻遏蛋白实际上是两个蛋白的**二聚体**，两个蛋白各含有一个 HTH 基序。DNA 结合蛋白中发生二聚化作用是很普遍的，因为它极大

increases the strength of binding to the DNA, roughly squaring the affinity between DNA and the DNA-binding protein.

4.5.2 DNA-Binding Motifs in Eukaryotes

DNA binding motifs are more diverse in eukaryotes than in prokaryotes. The **homeodomain** is a common motif in proteins involved in organismal development. The homeodomain contains three α helices connected by short peptide turns (Figure 4.18). Two of these helices together appear quite similar to the HTH motif, with one providing support, while the other makes specific contacts in the major groove. The third helix of the homeodomain has a peptide extension that fits into the minor groove of the DNA.

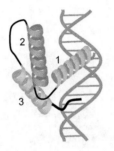

Figure 4.18   The homeodomain

Another important class of DNA-binding motifs in eukaryotes is **zinc finger**. In this motif, an α helix and an anti-parallel β strand are held together by a zinc ion [Figure 4.19(a)]. The structure is elongated, some-

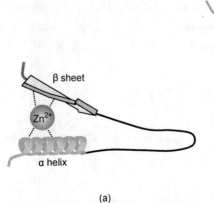

(a)

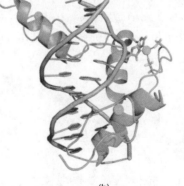

(b)

Figure 4.19   (a) A zinc finger motif; (b) Representation of the DNA binding domain of the transcription factor Egr1, showing multiple zinc finger domains (Source: http://www.wikipedia.org)

what resembling a finger. The finger shape can easily insert into the major groove, allowing the α helix to make specific contacts with bases. DNA-binding domains often have several zinc finger motifs adjacent to each other, each recognizing a piece of a target DNA sequence [Figure 4.19(b)]. It should be noted that zinc fingers are a diverse group of motifs, and not all are structurally similar.

The **leucine zipper** has a quite different structure from the motifs above (Figure 4.20). To make a leucine zipper, two α helices from different proteins come together at the DNA and hold the major groove like a pair of pinchers. Each helix has a line of (non-polar) leucine amino acids, which hold the helices together by hydrophobic forces. The helices coil slightly around each other, perhaps giving the impression of a zipper.

亮氨酸拉链具有与上述基序很不相同的结构（图4.20）。要形成一个亮氨酸拉链，需要不同蛋白质上的两个α螺旋一起靠近DNA链，形成类似钳子那样的结构来夹住DNA的大沟。每个螺旋具有一排（非极性的）亮氨酸，通过疏水作用力把螺旋保持在一起。两个螺旋互相靠近并发生轻微卷曲，或许正是这给我们留下了拉链的印象。

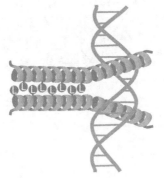

Figure 4.20 Leucine zipper

图4.20 亮氨酸拉链

An interesting feature of leucine zippers is that they can form from the combination of two different proteins (Figure 4.21). This is called **heterodimerization**. The heterodimers can recognize different DNA

亮氨酸拉链一个有趣的特征是两种不同的蛋白质组合在一起也能形成这样的结构（图4.21）。这叫作**异源二聚化作用**。与从只有一种蛋白质产生的亮氨酸拉链

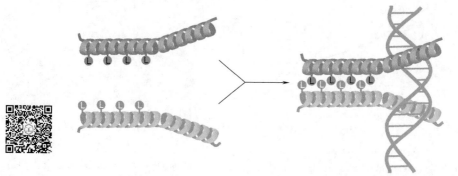

Figure 4.21 Leucine zippers formed by heterodimerization

图4.21 通过异源二聚化作用形成的亮氨酸拉链

sequences than the leucine zipper made from only one kind of protein. This property can be used to create more DNA-binding domains from fewer parts.

A motif called the **helix-loop-helix (HLH)** is quite similar to the leucine zipper. Like the leucine zipper, the HLH domain is formed from two separate proteins, and grips the DNA like a pair of pinchers. HLH domains can also form from protein heterodimerization. Unlike the leucine zipper, however, each half of the motif is composed of two α helices connected by a loop (Figure 4.22). The helix-loop-helix structure allows the two halves of the motif to dimerize. The helix-loop-helix should not be confused with the helix-turn-helix.

相比，异源二聚体能够识别不同的 DNA 序列。这一特性可以被用来从较少的蛋白结构部件产生更多种类的 DNA 结合域。

一种称为**螺旋-环-螺旋（HLH）**的基序与亮氨酸拉链很相似。像亮氨酸拉链一样，HLH 结构域是由两个蛋白质形成的，它也像一把钳子一样抓住 DNA。HLH 结构域也可以通过蛋白质异源二聚化而形成。然而，与亮氨酸拉链不同，这个基序的每一半由通过一个环连接在一起的两个 α 螺旋组成（图 4.22）。螺旋-环-螺旋结构使基序的两半发生二聚化。注意不要将螺旋-环-螺旋结构与螺旋-转角-螺旋相混淆。

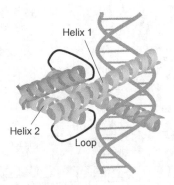

Figure 4.22　Helix-loop-helix motif

图 4.22　螺旋-环-螺旋基序

4.6　Experiments

4.6.1　RNA Polymerase Targets

We saw in this chapter that eukaryotes have three different RNA polymerases that transcribe different genes. The activity of the different polymerases was determined using a toxin called **α-amanitin**, which is produced by a mushroom that is deadly to humans. α-amanitin kills by binding to RNA polymerases and preventing them from functioning. The use of the toxin in experiments is that it inhibits the three different eukaryotic RNA polymerases to very different extent. Low concentrations completely inhibit RNA polymerase Ⅱ, but have no effect on the activity of the other polymerases. Much higher concentrations can also inhibit RNA polymerase Ⅲ, but still have no effect on RNA polymerase Ⅰ.

4.6　实验研究

4.6.1　RNA 聚合酶的目标

本章我们看到真核生物拥有三种不同的 RNA 聚合酶，用于转录不同的基因。不同聚合酶的活性是使用了一种叫作 **α-鹅膏蕈碱**的毒素来测定的，这种毒素是一种蘑菇产生的，对人具有致死作用。α-鹅膏蕈碱通过结合到 RNA 聚合酶上阻止它们发挥功能而杀死细胞。在实验中使用这一毒素是因为它对三种不同的真核 RNA 聚合酶的抑制程度非常不同。低浓度毒素完全抑制 RNA 聚合酶Ⅱ的活性，而对其他聚合酶的活性没有任何影响。更高浓度的毒素也能抑制 RNA 聚合酶Ⅲ，但仍然对 RNA 聚合酶Ⅰ没有影响。

To see which genes are transcribed by which polymerase, cells are exposed to α-amanitin. Cells exposed to very high concentrations of α-amanitin only produce the rRNA precursor, suggesting that this is the only gene transcribed by RNA polymerase Ⅰ. Cells exposed to low concentrations of the toxin also produce tRNA and other small RNAs like the 5S rRNA, suggesting that these are the genes transcribed by RNA polymerase Ⅲ. mRNA is not produced even at low concentrations of α-amanitin, suggesting that RNA polymerase Ⅱ is responsible for transcription of mRNA.

为了弄清哪些基因由哪种聚合酶转录，细胞被暴露在α-鹅膏蕈碱中。暴露在很高浓度α-鹅膏蕈碱中的细胞只产生rRNA前体，意味着这正是RNA聚合酶Ⅰ所转录的基因。暴露在低浓度毒素中的细胞还能产生tRNA和其他小RNA如5S rRNA，意味着这些基因是由RNA聚合酶Ⅲ转录的。即使在低浓度α鹅膏蕈碱下mRNA也不能产生，说明RNA聚合酶Ⅱ负责转录mRNA。

4.6.2 Modularity of Specific Transcription Factors

Specific transcription factors are usually modular, containing a DNA binding module and a distinct activation module. **Modularity** was most notably shown using the activator Gal4. Gal4 binds to a sequence called UAS to activate transcription (Figure 4.23).

4.6.2 特异转录因子的模块化

特异转录因子通常是模块化的，含有一个DNA结合模块和一个分开的激活模块。**模块化**最明显的例子是Gal4。Gal4能结合到一段称为UAS的序列去激活转录（图4.23）。

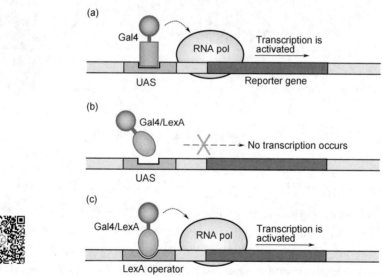

Figure 4.23 The modularity of Gal4

图4.23 Gal4结构的模块化

First, a UAS sequence was placed in front a reporter gene. A **reporter gene** is a gene whose transcription can be easily detected and measured. Thus, it is convenient for testing how well an activator protein works. When the **UAS** was present in front of the reporter gene, **GAL4** activated transcription of the gene.

首先，把一段UAS序列置于报告基因之前。**报告基因**是一种转录情况可以很容易地被检测并测量的基因。这样，就能方便地试验一种激活蛋白工作情况的好坏。当**UAS**出现在报告基因前面的时候，**GAL4**激活了这个基因的转录。

Next, the GAL4 protein was replaced by an artificially

然后，用一种经过人工改造过的蛋白质

engineered protein. This protein contained the activation domain from GAL4, but the DNA-binding domain from a different protein, **LexA**. LexA normally binds to a DNA sequence called the LexA-operator and represses transcription. This Gal4-LexA protein was unable to bind to the UAS and therefore unable to activate transcription of the reporter gene.

Finally, a LexA-operator sequence was placed in front of a reporter gene. Now, the Gal4-LexA protein was able to bind to the DNA. Most importantly, although the Gal4 domain was connected to a different DNA-binding domain, it could still activate transcription of the reporter gene. This proved that Gal4 was modular. As long as the activation domain of Gal4 could be brought near the target gene, its function was not affected by the identity of the DNA-binding domain. The same is true of LexA DNA binding domain. It was able to bind the LexA-operator regardless of which activation or repression domain was attached to it.

Summary

Transcription of mRNA in eukaryotes requires RNA polymerase Ⅱ, as well as two groups of proteins: general transcription factors and specific transcription factors. TFⅡs are required for recognition of promoters and initiation of transcription in class Ⅱ genes. Along with RNA polymerase Ⅱ they form a complex of proteins at the promoter called the pre-initiation complex. The most important TFⅡ is TFⅡD. It is composed of TBP and a variety of proteins named TAFⅡs, which recognize promoter elements and make contacts with specific transcription factors. Transcription occurs very rarely when only the proteins of the pre-initiation complex are present. Specific transcription factors modify the rate of transcription through a variety of mechanisms, most often by recruiting or blocking the pre-initiation complex, or altering the association of DNA with protein at the promoter. Specific transcription factors can be activators or repressors, and bind to sequences called enhancers or silencers, respectively. Specific transcription factors function by binding to DNA. Binding may occur via a number of motifs. Each motif is conserved in overall structure,

替换GAL4蛋白。这种蛋白含有来源于GAL4的激活域，但是它的DNA结合域来自一种不同的蛋白**LexA**。LexA通常结合到一段叫作LexA-操纵基因的DNA序列上并阻遏转录。这种Gal4-LexA蛋白不能结合到UAS上，因此也就不能激活报告基因的转录。

最后，一段LexA-操纵基因序列被放到报告基因之前。现在，这种Gal4-LexA蛋白能够结合到DNA上。更重要的是，虽然Gal4域被连接到一个不同的DNA结合域上，它仍然能够激活报告基因的转录。这证明了Gal4是模块化的。只要Gal4的激活域能够被吸引到靠近目标基因的位置，它的功能就不会因为DNA结合域的不同而受影响。这一情况对LexA DNA结合域也是如此。它能够结合到LexA-操纵基因上而不管它上面带有什么激活域或阻遏域。

小结

真核生物mRNA的转录需要RNA聚合酶Ⅱ，同时还需要两组其他蛋白质：通用转录因子和特异转录因子。Ⅱ类基因的转录需要TFⅡs来识别启动子和启动转录。它们与RNA聚合酶Ⅱ一起在启动子的位置形成了一种叫作前起始复合体的蛋白质复合体。TFⅡD是最重要的TFⅡ。它由TBP和一些称为TAFⅡs的蛋白质组成，它识别启动子元件并与特异转录因子接触。当只有前起始复合体蛋白存在的时候转录很少发生。特异转录因子通过多种机理改变转录的速率，大多数通过召集或阻止前起始复合体，或改变在启动子位置DNA与蛋白质的结合情况来实施其影响。特异转录因子可以是激活蛋白或阻遏蛋白，它们分别结合到称为增强子或沉默子的序列上。特异转录因子通过与DNA结合而发挥作用。这种结合可以通过一些基序来实现。每种基序在总体结构中是保守的，但在特殊的氨基酸组成上会有不同，这就使得一种基序可以

but varies in specific amino acid composition, allowing one motif to recognize many different DNA sequences.

识别多种不同的 DNA 序列。

Vocabulary 词汇

activator ['æktiveitə]	激活蛋白	initiator [i'niʃieitə]	起始子
core promoter	核心启动子	leucine zipper ['luːsiːn] ['zipə]	亮氨酸拉链
crab claw [kræb] [klɔː]	蟹钳	modularity [ˌmɔdju'læriti]	模块化
dimer ['daimə]	二聚体	motif [məu'tiːf]	基序（基本序列）
dimerization [ˌdaiməri'zeiʃən]	二聚化（作用）	nucleolus [njuː'kliːələs]	核仁
domain [dəu'mein]	域	nucleoplasm ['njuːkliəplæzm]	核质
downstream promoter element	下游启动子元件（DNA）元件	precursor [pri'kəːsə]	前体
element ['elimənt]		pre-initiation complex	前起始复合体
enhancer [in'hɑːnsə]	增强子	recognition helix [ˌrekəg'niʃən]	识别螺旋
general transcription factor	通用转录因子	recruit [ri'kruːt]	召集
helix-loop-helix (HLH)	螺旋-环-螺旋	reporter gene [ri'pɔːtə]	报告基因
helix-turn-helix (HTH)	螺旋-转角-螺旋	silencer ['sailənsə]	沉默子
heterodimerization ['hetərəudaiməriˌzeiʃən]	异源二聚化（作用）	specific transcription factor	特异性转录因子
		TATA box	TATA 框
homeodomain ['həumiəuˌdəumein]	同源异形域	TF Ⅱ B recognition unit	TF Ⅱ B 识别单元
		transcription factor	转录因子
housekeeping gene	持家基因	zinc finger	锌指

Review Questions 习题

Ⅰ. True/False Questions（判断题）

1. TF Ⅱ D is not involved in the transcription of tRNA genes.
2. Enhancers are proteins that activate transcription.
3. Most class Ⅱ genes require TATA-binding protein (TBP) for transcription.
4. Most class Ⅱ genes require the TATA box for transcription.
5. TAF Ⅱ s frequently bind to activator proteins.
6. RNA polymerase Ⅲ transcribes tRNA genes in prokaryotes.
7. When both are present, the initiator promoter element is always downstream of the GC box.
8. Helix-loop-helix motifs are similar to helix-turn-helix motifs except for the shape of the loop.
9. The helix-turn-helix motif is common in prokaryotes.
10. One protein may have more than one zinc finger motif.

Ⅱ. Multiple Choice Questions（选择题）

1. If the protein UBF is absent from the cell, which kind of RNA may not be transcribed?
   a. mRNA   b. rRNA
   c. tRNA   d. snRNA
   e. all of the above

2. You find a gene that is lacking a downstream promoter element. You conclude that this gene _____ .
   a. cannot be a eukaryotic gene
   b. must have a serious mutation
   c. probably came from a prokaryote
   d. is frequently transcribed
   e. none of the above

3. A friend claims that the same attenuation mechanism used in the *trp* operon (chapter 3) also is used for

some eukaryotic genes. You respond:
  a. That is not surprising because eukaryotes also produce tryptophan.
  b. That is not surprising because eukaryotic RNA also forms hairpins.
  c. That is impossible because eukaryotes and prokaryotes are so different.
  d. That is impossible because there are no ribosomes present during transcription in eukaryotes.
  e. That is impossible because prokaryotes only have one RNA polymerase, while eukaryotes have three.

4. Which statement is not accurate?
  a. Some activators can contact the DNA near a promoter.
  b. Some activators can contact enhancer DNA.
  c. Some activators can make contact with general transcription factors.
  d. Some activators can make contact with more than two different proteins.
  e. None of the above.

5. Which of the following motifs contains a beta-sheet?
  a. zinc finger
  b. homeodomain
  c. helix-turn-helix
  d. helix-loop-helix
  e. leucine zipper

Exploration Questions   思考题

1. Why is it important that eukaryotic genes may have so many different elements in their regulatory DNA regions?
2. It is known that many enhancers can be moved hundreds or thousands of base pairs away from their normal location and still function normally. Is this surprising? Why or why not?
3. How does a protein recognize a specific DNA base sequence? What kinds of amino acids would you expect are most important for recognition and binding?
4. Would you expect proteins that bind to RNA to have similar structure as proteins that bind to DNA? Why or why not?

Chapter 5　mRNA Modifications in Eukaryotes

第 5 章　真核生物 mRNA 的修饰

In prokaryotes, transcription produces a nearly exact mRNA copy of the DNA, and the transcript is immediately translated into protein. In eukaryotes, a series of modifications occur to mRNA during and after transcription. Compared with the gene that served as its template, a final eukaryotic mRNA may have some extra bases, some missing bases, and some covalent modifications. In this chapter we introduce these mRNA modifications and their functions.

在原核生物中，转录产生的 mRNA 几乎是 DNA 的准确拷贝，并且这一转录产物会立即被翻译成蛋白质。在真核生物中，转录时以及转录后会对 mRNA 进行一系列修饰。与作为模板的基因相比，最终得到的 mRNA 也许会含有一些额外的碱基或少了一些碱基并经过了一些共价修饰。本章我们介绍这些 mRNA 修饰作用以及它们的功能。

5.1　Capping

5.1　加帽

An mRNA that has been transcribed but is not yet ready for translation is called a **pre-mRNA**, or a **primary transcript**. The first and simplest modification that is made to pre-mRNAs is called **capping**. This term denotes the addition of a modified guanosine (7-methylguanosine) to the 5′ end of the pre-mRNA (Figure 5.1).

一条已经转录出来但还没有准备好用于翻译的 mRNA 称为**前体 mRNA** 或**初级转录本**。对前体 mRNA 所做的最早也是最简单的修饰是**加帽**。这一术语表示在 mRNA 前体的 5′ 末端加上一个修饰过的鸟嘌呤核苷（7-甲基鸟嘌呤核苷）（图 5.1）。

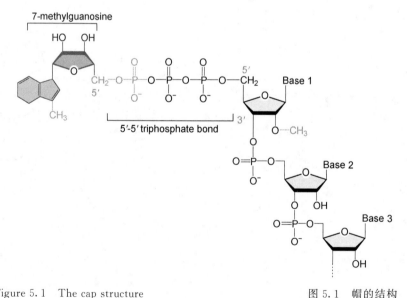

Figure 5.1　The cap structure

图 5.1　帽的结构

The guanosine derivative is attached to the 5′ end of the pre-mRNA by a 5′-5′ linkage. This is an unusual

这一鸟嘌呤核苷的衍生物通过 5′-5′ 键连接到前体 mRNA 的 5′ 末端上。对核酸

bond for nucleic acids; nucleotides are usually joined by $5'$-$3'$ bonds. The result is that the cap nucleotide is kinked relative to the rest of the strand. Also, it is joined to the RNA strand by a triphosphate bond, not the normal phosphodiester bond.

Recall that transcription of an mRNA occurs in the $5'\rightarrow 3'$ direction. Thus, the $5'$ end of the mRNA is the first to be transcribed. As a result, capping on the $5'$ end can be completed before the rest of the gene has even been transcribed. Three enzymes are involved in the capping process (Figure 5.2). RNA triphosphatase first removes one phosphate from the $5'$ end of the pre-mRNA. Guanylyl transferase then adds a guanosine nucleotide to make a $5'$-$5'$ linkage. Finally, methyl transferase adds a methyl group to the guanosine nucleotide. These enzymes aggregate around the C-terminal domain of the Rpb1 subunit of RNA polymerase Ⅱ as it begins transcription of the pre-mRNA. Because of the importance of this domain, it is often just called the **'CTD'**.

来说这是一个不寻常的键，因为核苷酸通常是由 $5'$-$3'$ 键连接的。结果，帽结构的核苷酸相对于前体mRNA链的其他部分来说发生了扭折。还有，它是通过三磷酸键而不是正常的磷酸二酯键与RNA链相连的。

回想一下，mRNA的转录是以 $5'\rightarrow 3'$ 的方向进行的。因此，mRNA的 $5'$ 端最先被转录出来。这样，对 $5'$ 端的加帽工作甚至在基因的其他部分还没有被转录出来的时候就完成了。加帽过程需要有三种酶参与（图5.2）。RNA 三磷酸酶首先从前体mRNA的 $5'$ 端移去一个磷酸；之后鸟苷酸转移酶加上一个鸟嘌呤核苷酸以形成 $5'$-$5'$ 键；最后，甲基转移酶在鸟嘌呤核苷酸上加上一个甲基。这些酶在 RNA 聚合酶Ⅱ开始转录前体mRNA的时候聚集在其 Rpb1 亚基的 C-末端功能域中。由于这一功能域的重要性，它常常就被称为"**CTD**"。

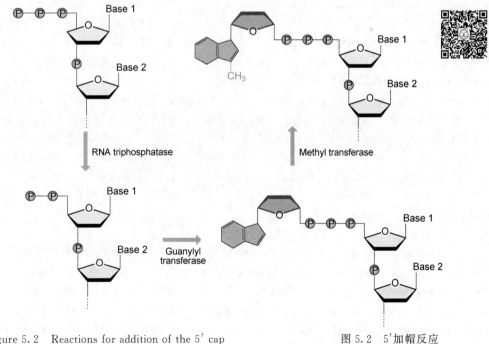

Figure 5.2 Reactions for addition of the $5'$ cap

图 5.2 $5'$ 加帽反应

The cap structure has four principal functions. First, it helps to prevent degradation of the mRNA at the $5'$ end. RNAs in the cell can be rapidly degraded by

帽结构具有四种主要功能。首先，它对防止mRNA在 $5'$ 端发生降解有帮助。细胞中的 RNA 会被称为**核糖核酸酶**的酶

enzymes called **ribonucleases**. However, these enzymes are generally not capable of degrading the triphosphate bond introduced by the cap structure. Second, the cap structure helps the RNA transcript to pass through the selective pores of the nuclear membrane and into the cytoplasm. This is important because transcription occurs in the nucleus, but the ribosomes that translate mRNA are located in the cytoplasm.

Third, the cap structure enhances translation. In order to bind the ribosome, mRNA requires help from cap-binding protein. As the name implies, this protein requires the presence of the cap structure in order to bind the mRNA. Finally, the cap is required for another modification of mRNA, called splicing, to occur completely. Specifically, its presence is required for splicing of the first intron. We will explain this concept shortly.

5.2　Polyadenylation

In addition to the modification on the 5′ end of the pre-mRNA, a string of approximately 250 adenosine nucleotides is added to the 3′ end of the transcript. The process that produces this **poly(A) tail** is called **polyadenylation**. Polyadenylation does not occur at the natural end of transcribed pre-mRNA. In mammals, an AAUAAA sequence in the pre-mRNA designates that the transcript must be cut approximately 20 nucleotides downstream, near a GU-rich sequence. This cleavage creates a new 3′ end to the pre-mRNA. The string of adenosine nucleotides is added at this new terminus (Figure 5.3). The original 3′ end of the transcript is degraded after cleavage.

Various proteins are required for the polyadenylation steps. For cleavage, a protein called **cleavage and polyadenylation specificity factor (CPSF)** binds to the AAUAAA sequence, while a **cleavage stimulation factor protein (CstF)** binds to the GU rich sequence downstream. With the aid of **cleavage factors Ⅰ and Ⅱ (CF Ⅰ and CF Ⅱ)**, these proteins cleave the mRNA transcript between the AAUAAA sequence and the GU rich sequence to produce a terminus for

迅速降解掉。然而，这些酶一般不能降解帽结构中的三磷酸键。第二，帽结构帮助 RNA 转录产物穿过核膜的选择性孔道而进入细胞质。这一点很重要，因为转录发生在细胞核，而翻译 mRNA 的核糖体位于细胞质中。

第三，帽结构能增强翻译。为了与核糖体结合，mRNA 需要帽结合蛋白的帮助。正如它的名称提示的那样，这种蛋白需要 mRNA 上有帽结构才能与 mRNA 结合。最后，要完成全部剪接过程（另一种 mRNA 修饰作用），帽结构也是必需的。准确地说，是第一个内含子的剪接需要帽结构的存在。我们稍后将对这一概念进行解释。

5.2　聚腺苷酸化

除了在前体 mRNA 的 5′端进行修饰外，转录产物的 3′端也会被加上一串约 250 个腺嘌呤核苷酸。产生这种 **poly(A) 尾**的过程称为**聚腺苷酸化**。聚腺苷酸化并不是在转录的前体 mRNA 本身的尾部发生的。在哺乳动物中，前体 mRNA 上的 AAUAAA 序列提示它应在下游约 20 个核苷酸处靠近富含 GU 的地方被切断。这一切割在前体 mRNA 上产生出一个新的 3′端。聚腺苷酸尾就是加在了这一新的末端上（图 5.3）。转录产物原先的 3′端序列在切下后会被降解掉。

在聚腺苷酸化步骤中需要有多种蛋白质的参与。对切割来说，一种称为**切割与聚腺苷酸化特异因子（CPSF）**的蛋白质结合到 AAUAAA 序列上，与此同时一种称为**切割激活因子（CstF）**的蛋白质结合到富含 GU 的下游序列上。在**切割因子Ⅰ和Ⅱ（CFⅠ和CFⅡ）**的帮助下，这些蛋白质在 AAUAAA 序列和富含 GU 的序列之间将 mRNA 转录产物切断，

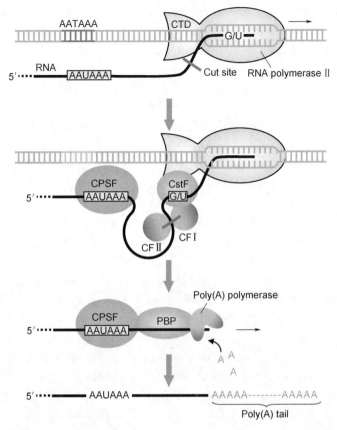

Figure 5.3 The polyadenylation mechanism

图 5.3 聚腺苷酸化机理

polyadenylation.

After cleavage of the transcript, CPSF remains bound at the AAUAAA site. A specialized RNA polymerase called **poly(A) polymerase** is then required to add adenosine nucleotides. The enzyme begins slowly at first, and is eventually accompanied by a protein called poly(A)-binding protein (PBP). As with capping, many of the proteins involved in polyadenylation aggregate around the CTD of RNA polymerase Ⅱ. Increased phosphorylation of the CTD causes capping proteins that were originally present to leave, and allows polyadenylation proteins to take their place.

The main function of the poly(A) tail is to protect the mRNA from degradation by ribonucleases. Unlike the cap structure, the tail does not provide protection by introducing a triphosphate bond; it is the large number of adenosine nucleotides separating the coding region

从而为聚腺苷酸化产生出所需要的末端。

在转录产物被切断之后，CPSF 仍然结合在 AAUAAA 位点。需要一种称为 **poly(A) 聚合酶**的特殊 RNA 聚合酶来将腺嘌呤核苷酸加上去。这种酶刚开始的时候动作很慢，最后会与一种称为聚腺苷酸结合蛋白（PBP）的蛋白质一起发挥作用。跟加帽中的情况相似，与聚腺苷酸化有关的多种蛋白质都聚集在 RNA 聚合酶Ⅱ的 CTD 周围。原先存在于 CTD 周围的加帽蛋白随着 CTD 磷酸化程度的提高而逐渐离去，以便让聚腺苷酸化蛋白取代它们的位置。

poly(A) 尾的主要功能是保护 mRNA 免受核糖核酸酶的降解。与帽结构不同的是，poly(A) 尾不是通过引入一个三磷酸键来提供保护；它是通过用较多数量的腺嘌呤核苷酸将 mRNA 末端的编码区

from the end of the mRNA that provides protection. There is also some evidence that the poly(A) tail is involved in splicing and enhances translation of mRNAs.

5.3 Splicing

One surprising feature of eukaryotic genes is that they are peppered with a lot of mysterious sequences. The finely evolved sequence of nucleotides that codes for amino acids is periodically interrupted by long stretches of DNA that have no apparent meaning. These are called **introns**. The short coding regions produced when introns interrupt a gene are called **exons**.

During transcription, the entire eukaryotic gene is transcribed as one big piece, and the pre-mRNA includes introns. However, these introns must be carefully removed from the pre-mRNA after transcription. If they are not removed, the protein translated from the mRNA will be completely nonfunctional, and perhaps dangerous to the cell. The process of removing introns from a pre-mRNA is called **splicing** (Figure 5.4).

5.3 剪接

域隔开而提供保护的。也有一些证据证明 poly(A) 尾与剪接以及增强 mRNA 的翻译有关。

真核基因的一个令人吃惊的特性是它们充斥着许许多多难解的序列。那些精确进化而来的编码氨基酸的核苷酸被一些没有明显含义的 DNA 长片段有规律地隔开。这些 DNA 片段称为**内含子**。当内含子隔开一个基因时产生的短编码区域称为**外显子**。

在转录过程中，整个真核基因会被转录成包括内含子的一条前体mRNA长链。然而，在转录后这些内含子必须被小心地去除。如果它们没有被去除，那么从mRNA翻译出来的蛋白质会是彻底没有功能的，也可能对细胞有危害。将内含子从前体mRNA中去除的过程称为**剪接**（图 5.4）。

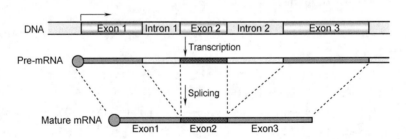

Figure 5.4 Gene structure showing introns and exons. Removal of introns by splicing is necessary to form a mature mRNA

图 5.4 显示了内含子与外显子的基因结构。形成成熟的mRNA需要进行剪接以去除内含子

5.3.1 The Basic Splicing Reaction

Sequences within the DNA tell the cell exactly where an intron begins and ends. This signal must be very precise, because if so much as one nucleotide of an exon is removed, or one nucleotide of an intron is left in the mRNA, a completely nonfunctional protein may be produced. The first two bases on the 5′ side of an intron are almost always GU, and the last two bases on the 3′ side are almost always AG. Depending on the

5.3.1 基本的剪接反应

DNA 中的序列能告诉细胞一个内含子开始与结束的确切位置。这个信号必须非常准确，因为如果mRNA中就算只有一个外显子核苷酸被去掉，或只有一个内含子核苷酸被留了下来，那么得到的蛋白质就可能是完全没有功能的。内含子 5′端的头两个碱基几乎全部都是 GU，3′端的最后两个碱基几乎全部都是 AG。

organism, other consensus sequences are also involved in marking introns. For example, the exon-intron boundaries in yeast often look like:

5′ AG/GUAUGU... body of intron... UACUA<u>A</u>C-YAG/3′

where slashes represent the intron/exon boundary. These boundaries are also called **splice sites**. The leftward boundary in the sequence above would be considered the 5′ splice site of that intron, the rightward boundary would be considered the 3′ splice site.

Notice also in the sequence above that one of the adenine bases is underlined. This base is marked because it is central to the mechanism of splicing (Figure 5.5). In normal splicing, a hydroxyl group on this A nucleotide 'attacks' the 5′ splice site. The result is that the bond between the G of the exon and the G of the intron at the boundary is severed, and the G of the intron forms a new bond with the attacking A.

不同生物有一些其他共有序列来标识它们的内含子。例如，酵母的外显子-内含子交界处一般是这样的：

其中斜线代表内含子/外显子的交界处。这些交界处也叫作**剪接位点**。上述序列的左边界被认为是那一内含子的5′剪接位点，右边界被认为是3′剪接位点。

也请注意上述序列中的一个腺嘌呤碱基被加了下划线。对它作标记是因为它在剪接机理中起重要的作用（图5.5）。在正常的剪接中，在这一核苷酸A上的羟基会"进攻"5′剪接位点。结果是在交界处外显子的G和内含子的G之间的键受到威胁，导致内含子中的G与来进攻的A形成新的键。

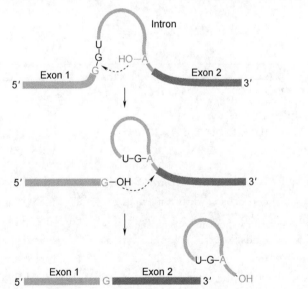

Figure 5.5 The basic splicing reaction

图5.5 基本的剪接反应

The intron now has become a loop, called a **lariat**. The exon at 5′ splice site dangles freely, but not for long, the hydroxyl group of the exon's terminal G attacks the 3′ splice site. This time, the connection between the lariat and the exon at the 3′ splice site is severed. A new bond is formed between the 5′ exon and the 3′ exon, and removal of the intron is complete (Figure 5.5).

内含子现在变成了一个环，称为**套索**。5′剪接位点的外显子自由悬挂在那里，过一会儿之后，外显子末端G上的羟基就会开始进攻3′剪接位点。这次，3′剪接位点处套索和外显子之间的连接被切断。在5′外显子和3′外显子之间形成新的键之后，去除内含子的整个过程就算完成了（图5.5）。

5.3 Splicing | 107

5.3.2 Proteins Involved in Splicing

Although often considered a 'post-transcriptional' modification, like capping and polyadenylation splicing also takes place mostly during transcription, as the pre-mRNA is synthesized. The set of molecules that carry out splicing is collectively termed the **spliceosome**. The central agents of the spliceosome are **small nuclear ribonucleoproteins**, or **snRNPs** (pronounced as 'snurps'). As the name indicates, snRNPs are small particles found in the nucleus, and they are formed from protein and RNA. The RNA within snRNPs, called snRNA, can form specific base pairs with the pre-mRNA. This pairing correctly positions the snRNPs on the transcript so that they can mediate the splicing reactions (Figure 5.6).

5.3.2 在剪接中发挥作用的蛋白质

虽然通常被认为是"转录后"修饰作用，但像加帽和聚腺苷酸化一样，剪接也主要是在转录过程中进行的，这时前体mRNA还处于被合成的状态。执行剪接作用的分子集群一起被称为**剪接体**。剪接体的中心成员是**核内小核糖核蛋白或snRNPs**（读音参照单词"snurps"）。正如这一名称所表明的那样，snRNPs是在细胞核里发现的小颗粒，是由蛋白质和RNA组成的。在snRNPs中的RNA（称为snRNA）能够与前体mRNA中的碱基配对。这样的配对使snRNPs正确定位到前体mRNA上，之后它们才能开始引导剪接反应（图5.6）。

Figure 5.6 A snRNP, bound to pre-mRNA by hybridization between the snRNA of the snRNP and the pre-mRNA

图5.6 一个snRNP，其中的snRNA与前体mRNA杂交使snRNP结合到前体mRNA上

The action of the spliceosome on the pre-mRNA is dynamic. Throughout the process, different factors bind and unbind to each other and to the transcript. First, the U1 snRNP binds at the 5′ splice site, committing the intron to splicing (Figure 5.7). Next, the U2 snRNP binds at the attacking adenosine nucleotide. Two snRNPs which are bound to each other, U4 and U6, then join the 5′ splice site. Meanwhile, the U5 snRNP holds the 5′ and 3′ splice sites near each other. The reaction does not begin until the U4 unbinds U6. This activates U6, and once U6 is activated it removes U1 from the 5′ splice site. Finally, U6 and U2 are able to carry out the major reactions of splicing-attack of the 5′ splice site by the activated adenosine, and joining of the 5′ splice site to the 3′ splice site.

剪接体在前体mRNA上的作用是动态的。在整个过程中，不同因子与前体mRNA之间会发生互相结合和解离。首先，snRNP U1结合到5′剪接位点，将内含子送交剪接（图5.7）。接着，snRNP U2结合到具有进攻作用的腺嘌呤核苷酸上。然后，U4和U6这两种互相结合在一起的snRNPs来到5′剪接位点。同时，snRNP U5将5′和3′剪接位点保持在互相靠近的位置。在U4从U6解离之前反应是不会发生的。U4的解离活化了U6，处于活化状态的U6会从5′剪接位点移走U1。最后，U6与U2才能够去执行剪接过程中的主要反应，也就是使活化的腺嘌呤核苷对5′剪接位点发起进攻并将5′和3′剪接位点连接在一起的反应。

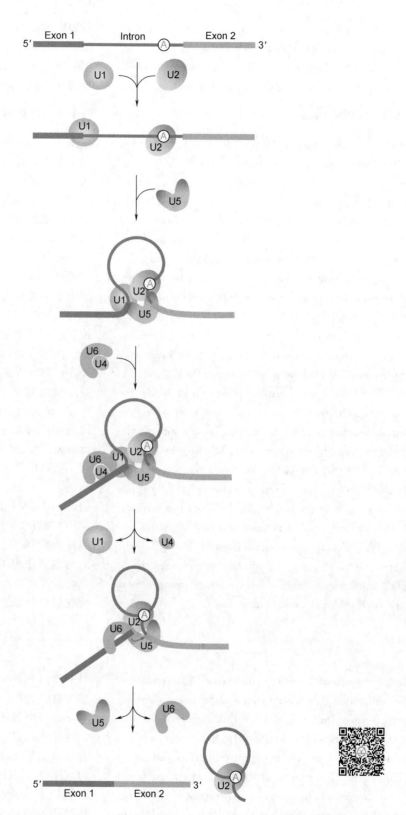

Figure 5.7 Splicing reaction showing involvement of spliceosome

图 5.7 剪接反应（显示剪接体的参与）

5.3 Splicing

As with capping and polyadenylation, the CTD of RNA polymerase Ⅱ binds to snRNPs and helps to organize their activity. The CTD also seems to play an important role in marking exon-intron boundaries.

5.3.3 Self-Splicing

Considering how complex the normal process of splicing is, it is amazing that sometimes it can occur without any outside factors at all! Splicing that occurs without the help of proteins or snRNPs, called **self-splicing**, was first discovered in the rRNAs of a small eukaryote called tetrahymena. These introns, called **group Ⅰ introns**, differ in the fundamental splicing mechanism in that the attacking residue does not come from within the intron, but rather from a free guanosine molecule. As a result, no lariat structure is formed during the reaction because the end of the 5′ exon is never covalently bound to the intron. Group Ⅰ introns have two principal helical domains termed P3-P9 and P4-P6. The P3-P9 helix forms a pocket that binds the guanosine substrate. Thus, even though the attacking guanosine is not part of the intron, in self-splicing the initiating event in cleavage is coordinated by the intron. The P4-P6 domain provides structural support to the P3-P9 helix, and aids in selection of the 5′ splice site by stabilizing the interaction between the terminal U nucleotide of the 5′ exon and a G residue within the intron. As in typical splicing, following cleavage of the 5′ splice site, the free end of the exon attacks the 3′ splice site, displacing the intron and forming a new bond with the next exon (Figure 5.8).

Another group of introns, called **group Ⅱ introns**, can also self-splice. The basic reaction for splicing these introns is very similar to the reaction mediated by spliceosome proteins, but, again, it can occur without catalytic contributions from proteins. Here, attack by adenosine in the intron results in the formation of a lariat intermediate. The adenosine is brought into position for attack by a complex tertiary structure caused by the formation of stem loops within the intron. Added structural organization is created by base pairing between se-

与加帽和聚腺苷酸化相似，RNA聚合酶Ⅱ的CTD也与snRNP结合并帮助组织它们的活动。CTD似乎也在标明外显子与内含子的边界上起着一定的作用。

5.3.3 自我剪接

细想一下，正常的剪接过程是相当复杂的，令人惊奇的是有时剪接可以在没有任何外界因子的帮助下得以完成！在没有蛋白质或snRNP的帮助下发生的剪接称为**自我剪接**，它是在一种称为四膜虫的小真核生物中首次发现的。这种内含子被称为**Ⅰ类内含子**，具有与基本剪接机理不同的剪接方式：起进攻作用的基团不是来自内含子自身，而是来自游离的鸟嘌呤核苷分子。其结果是，在剪接反应中不产生套索状结构，因为5′外显子的末端并不与内含子通过共价键连接在一起。Ⅰ类内含子具有两个重要的螺旋形结构域，称之为P3-P9结构域和P4-P6结构域。P3-P9结构域形成一个口袋形状用来结合鸟嘌呤核苷底物。如此看来，即使起进攻作用的鸟嘌呤核苷并不来自内含子，在自我剪接中最初的切割事件也还是由内含子来进行协调的。P4-P6结构域为P3-P9结构域提供结构支撑，同时通过稳定5′外显子末尾的核苷酸U与内含子中的核苷酸G之间的相互作用而协助选择5′剪接位点。之后，像经典剪接一样，在5′剪接位点断开后，外显子的游离末端进攻3′剪接位点，代替内含子与后面的外显子形成新的化学键（图5.8）。

另一类内含子（称为Ⅱ类内含子）也能发生自我剪接。剪接这一类内含子的基本反应与由剪接体蛋白协调完成的剪接反应非常相似；但是，它同样可以在没有蛋白质催化作用下完成剪接。其中，内含子中腺嘌呤核苷的攻击导致形成套索状中间产物。内含子中形成的茎环结构导致产生复杂的三级结构而将腺嘌呤核苷置于攻击位置。内含子中的外显子结合序列（EBS）与外显子

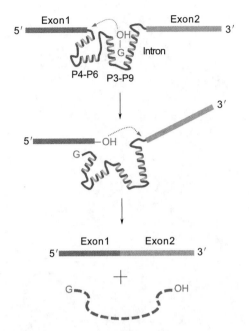

Figure 5.8 Self-splicing of group Ⅰ intron

图 5.8 Ⅰ类内含子的自我剪接

quences in the intron (exon binding sequences, EBS) and sequences in the exons (intron binding sequences, IBS). While the sequences of introns may vary across genes and species, there are typically six domains (D1-D6). It is the 2′ hydroxyl group of an unpaired adenosine in D6, held in proper conformation by the other domains, that attacks the 5′ splice site, followed by attack of the 3′ splice site from the free 3′ hydroxyl group (Figure 5.9).

Self-splicing is possible because, as noted in chapter 2, RNA can fold into complex structures that have enzymatic activity, much like a protein. Group Ⅰ and Ⅱ introns are most often found in mitochondria and chloroplast mRNAs. We should emphasize that the vast majority of genes in the vast majority of eukaryotes are not self-splicing and require involvement of the spliceosome.

5.3.4 Trans-Splicing

An interesting variation on splicing occurs in a variety of organisms, including trypanosomes and the roundworm Caenorhabditis elegans. Many genes in these organisms have the same short sequence of nucleotides

中的内含子结合序列（IBS）发生碱基配对又提供了额外的结构支撑。尽管内含子序列在不同基因及物种中有所不同，但它们均具有六个典型的结构域（D1~D6）。D6结构域中一个未配对腺嘌呤核苷的2′羟基在其他结构域的帮助下维持攻击5′剪接位点的适当构象；之后，游离出来的3′羟基再攻击3′剪接位点（图5.9）。

自我剪接是可能的，因为正如在第2章中讲到的那样，RNA能够折叠成复杂的结构从而像蛋白质一样具有催化功能。Ⅰ类和Ⅱ类内含子大多存在于线粒体和叶绿体mRNA中。应该强调的是，在绝大多数真核生物中，大多数基因并不是采用自我剪接方式进行剪接的，它们的剪接过程都需要有剪接体参与。

5.3.4 反式剪接

在锥虫和秀丽隐杆线虫等生物中存在一种有趣的剪接方式。这些生物中许多基因成熟RNA的5′端都有一段相同的短序列。这是因为在细胞中有一段短

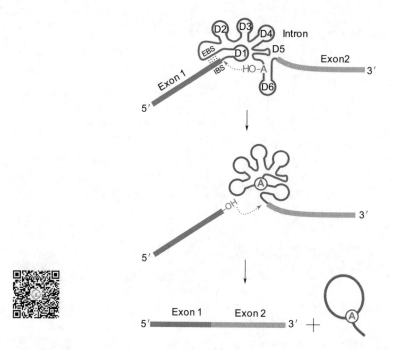

Figure 5.9　Self-splicing of group Ⅱ intron

图 5.9　Ⅱ类内含子的自我剪接

on the 5′ end of the mature RNA. This is because a short RNA is transcribed in large quantities in the cell, and attached by a splicing-like reaction to the 5′ end of many other transcribed genes (Figure 5.10). This attached piece of RNA is called the **spliced leader (SL)**.

RNA 转录的量较大，转录后又在类似于剪接的反应中被连接到许多其他基因转录产物的 5′端上（图 5.10）。这一被连接上去的 RNA 片段称为**剪接前导序列（SL）**。剪接前导序列原本是作为单

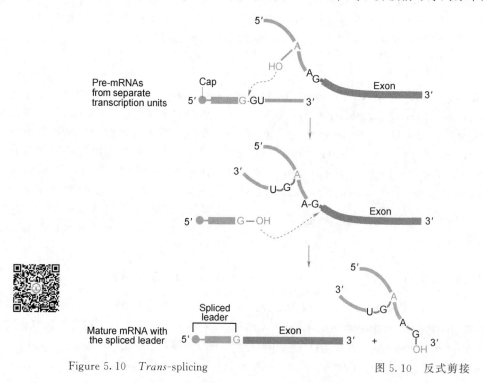

Figure 5.10　*Trans*-splicing

图 5.10　反式剪接

The spliced leader is originally transcribed as a single transcript that contains what might be considered the 5′ end of an intron, and a 5′ splice site. Much as in a normal splicing reaction, a hydroxyl group from an adenylate nucleotide in a pre-mRNA can attack the 5′ splice site of this SL-containing RNA, removing the intron portion, and allowing attachment of the SL to the pre-mRNA. This is known as *trans*-splicing, because in contrast to normal splicing, the spliced RNAs are derived from different genes.

5.3.5 Reasons for Introns

The presence of introns in eukaryotic genes clearly adds a significant complication to gene expression. So why do eukaryotic genomes have introns? And why don't prokaryotes have them? This question is still being debated, and there currently exist two broad theories, the **introns-early theory** and the **introns-late theory**.

The introns-early theory holds that very early life forms contained introns. Prokaryotes lost them because these organisms are under selective pressure to be small. Also, the frequent reproduction of prokaryotes has allowed more opportunities to evolve genomes free of introns. Eukaryotes, which have more space in the nucleus and have had fewer generations during which to evolve, have kept the introns present in early life forms.

The introns-late theory holds that the original life forms did not have introns, and that eukaryotes acquired these extra pieces of DNA after their lineage had separated from prokaryotes. Introns would have been introduced perhaps by viruses or by pieces of the genome switching locations. The presence of introns in eukaryotes might have been favored because they actually serve some sort of function.

Indeed, there are indications that introns are more than just junk sequences. Some introns contain important regulatory elements, like promoters and enhancers. By providing interruptions in the coding region, introns give these elements the unique opportunity to be present after the start site of transcription.

一产物被转录出来的，它含有内含子的5′端与5′剪接位点那样的序列。就像是在正常的剪接反应中那样，来自前体mRNA中腺嘌呤核苷酸上的羟基能够进攻SL-RNA的5′剪接位点、去除内含子部分并帮助SL连接到前体mRNA上。与正常的剪接过程相反，这一过程称为**反式剪接**，因为被连接到一起的RNA片段来自不同的基因。

5.3.5 内含子存在的原因

很明显，真核基因中出现内含子大大增加了基因表达的复杂性。那么真核生物的基因组为什么会带有内含子呢？还有，为什么原核生物中没有呢？这一问题仍然处于争论之中，现在存在着两种广泛的理论，即**内含子早期论**和**内含子晚期论**。

内含子早期论认为很早期的生命中具有内含子。原核生物失去了它们是因为原核生物在选择压力的作用下朝更微小的方向进化。还有，原核生物频繁的繁殖行为使它们具有更多的机会进化出没有内含子的基因组。而真核生物在细胞核中具有更多的空间以及较少的进化世代，所以保留了在早期生命形式中存在的内含子。

内含子晚期论认为，原先的生命形式中没有内含子，真核生物是在它们的谱系与原核生物分开后获得这些额外DNA片段的。内含子有可能是由病毒引入或由基因组中的片段发生位置交换而产生的。在真核生物中出现内含子可能是有利的，因为它们也确实具有某些功能。

实际上，已经有一些征兆说明内含子并不仅仅是无用的序列。一些内含子具有重要的调控元件比如启动子和增强子。通过在编码区域提供间隔地带，内含子为这些元件提供了独特的出现在转录起始位点之后的机会。

The clearest advantage of introns is that they can allow one gene to be used to make several different proteins. This occurs through a process called **alternative splicing**. The pre-mRNA from most human genes can be spliced in more than one way (Figure 5.11). Sometimes an 'intron' is kept in the transcript and will code for protein, acting therefore as an exon. Sometimes an exon is removed from the pre-mRNA so that the region it codes for will not be present in the final protein. Using alternative splicing, the variety of proteins made from one gene can be enormous; the *Dscam* gene in *Drosophila*, for example, contains 116 exons. By choosing various combinations of 17 exons from among the 116 total, 8 000 different proteins can be produced from this one gene.

内含子最清楚的优点是它们可以使一个基因产生不同的蛋白质。这是通过称为**可变剪接**的过程实现的。绝大多数人类基因的前体mRNA都可以按不止一种方式进行剪接（图5.11）。有时一个"内含子"会被保留在转录产物中并为蛋白质编码，因此起到了外显子的作用。有时一个外显子会从前体mRNA上去除，它编码的区域在最后的蛋白质中并不出现。采用可变剪接可使来源于一个基因的蛋白质数量大幅度上升；例如，果蝇的*Dscam*基因具有116个外显子，通过从总数116中选择17个外显子进行组合，从这一个基因就可以产生出8 000种不同的蛋白质。

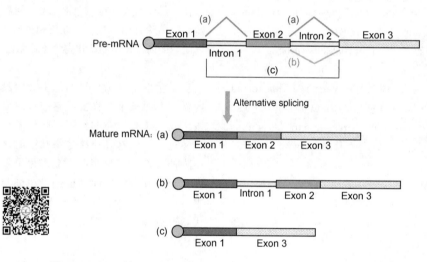

Figure 5.11 Alternative splicing

图5.11 可变剪接

Another example of alternative splicing occurs in immune cells, which contain proteins called antibodies that bind to foreign particles. Sometimes it is useful for antibodies to be mounted on the surface of cells; other times antibodies must be released into the blood to perform their function. By alternatively splicing the transcript of one gene, immune cells can determine whether antibodies are stuck in the membrane or secreted (Figure 5.12). If the protein must be mounted in the membrane, exons for the transmembrane region are kept in the final transcript. If the protein is to be secreted, these exons are spliced out, and a shorter protein is produced that cannot stick in the membrane.

可变剪接的另一个例子发生在免疫细胞中，免疫细胞中含有能与外来颗粒结合的称为抗体的蛋白质。有时需要让抗体固定在细胞的表面；其他时候则必须将它们释放到血液中去行使功能。通过对其基因转录产物采用可变剪接，免疫细胞可以决定抗体是被固定在细胞膜上还是被分泌到细胞外（图5.12）。如果抗体必须被固定在细胞膜上，那么为跨膜区编码的外显子就会被保留在最后的转录产物中。如果抗体需要被分泌出去，则这些外显子就会被剪接掉，产生的是不能附着到细胞膜上的较短的蛋白质。

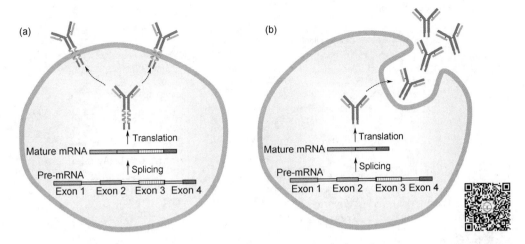

Figure 5.12　Example of alternative splicing. Antibodies can be produced in a membrane-bound form (a) or a secreted form (b), depending on splicing of the gene

图 5.12　可变剪接实例。根据基因剪接方式的不同，抗体生产分为"与膜结合型"(a) 和"细胞外分泌型"(b)

5.4　mRNA Editing

So far, the RNA modifications discussed have entailed adding or removing nucleotides from the pre-mRNA. Another form of modification, called **RNA editing**, results in covalent modification of nucleotides within the transcript.

In a kind of RNA editing called **C→U editing**, enzymes remove an amino group from a cytosine in the RNA, converting the base to uracil (Figure 5.13). This can have important effects on the translation of the mRNA. A classic case is **apolipoprotein B (APOB)**, which carries cholesterol in the body. In the liver, a large version of this APOB is needed, and the gene is transcribed and translated as normal. However, in the small intestine, a short version of APOB is needed. In this case, C→U editing is performed on one nucleotide of the mRNA transcript. The change of base causes a translation stop signal in the middle of the transcript. When the transcript is translated, the short version of the protein is produced (Translation stop signals, called stop codons, are discussed in next chapter).

A more common kind of editing is **A→I editing**. In this

5.4　mRNA编辑

到现在为止，我们讨论的 RNA 修饰作用还只限于向前体mRNA增加或去除核苷酸。另一种称为 **RNA 编辑**的修饰形式能对转录产物中的核苷酸进行共价修饰。

在一种称为 **C→U 编辑**的 RNA 编辑中，酶能去掉 RNA 中胞嘧啶上的氨基从而将它转变成尿嘧啶（图 5.13）。这会对 mRNA的翻译产生重要的影响。一个经典的实例是**载脂蛋白 B (APOB)**，它在体内起着运送胆固醇的作用。在肝脏中，需要这种 APOB 以完整的形式出现，这时基因以正常方式转录和翻译。而在小肠中，需要 APOB 以较短的形式出现，这时会对mRNA转录产物上的一个核苷酸进行 C→U 编辑。这一碱基变化在转录产物的中部产生了一个翻译终止信号。当这样的 mRNA 被翻译的时候，产生的就是较短的蛋白质（翻译终止信号称为终止密码，在下一章中讨论）。

一种更普遍的编辑修饰是 **A→I 编辑**。

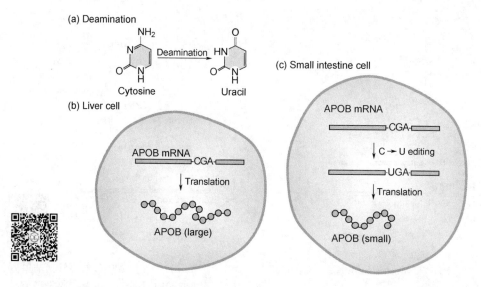

Figure 5.13 C→U editing. (a) Deamination leads to a C→U change; (b) In the liver, translation of a normal APOB mRNA produces a large protein; (c) In the small intestine, C→U editing creates a stop codon and results in a small protein being produced

图 5.13 C→U 编辑。(a) 脱氨基导致 C→U 变化；(b) 在肝脏中，正常mRNA的翻译产生大蛋白；(c) 在小肠中，C→U编辑产生一个终止密码，导致生产出小蛋白

process, removal of an amino group from an adenine base creates an inosine base. Inosine is an uncommon base that is read by the ribosome as though it were guanine. Thus, the modification essentially produces an A→G change. This kind of editing is most common in proteins of the nervous system. One case is the receptor GluR-B, which requires A→I editing at a specific site (Figure 5.14). The modification causes an arginine to replace glutamine in the protein. The change of amino acid is required for the protein to function normally.

在这一过程中，从腺嘌呤碱基中去掉氨基得到的是次黄苷碱基。次黄苷是一种不常见的碱基，核糖体会把它当成鸟嘌呤来读取。因此，这样的修饰实际上产生了一个 A→G 变化。这种类型的编辑在神经系统的蛋白质中最为普遍。比如，受体 GluR-B 需要在特定的位置进行 A→I 编辑（图 5.14）。这一修饰导致了蛋白质中的一个谷氨酰胺被精氨酸取代。这样的改变是这种蛋白质发挥正常功能所需要的。

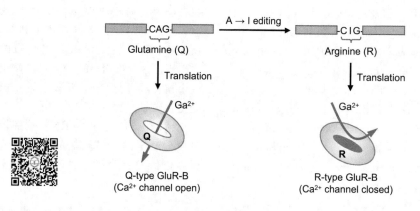

Figure 5.14 A→I editing in GluR-B mRNA

图 5.14 GluR-B mRNA 中的 A→I 编辑

In A→I editing, eligible adenosines are not edited exactly the same in each transcript. One *Drosophila* membrane protein (a voltage-gated calcium channel) has ten possible sites of A→I editing. As different combinations of these adenosines are modified in different transcripts of the same gene, over 1000 distinct proteins can be made from one gene (Figure 5.15).

进行 A→I 编辑时，在每一条转录产物中被选中的腺嘌呤核苷并不是完全按照同样的方式进行编辑的。一种果蝇的膜蛋白（电压控制的钙通道蛋白）含有十个可能的 A→I 编辑位点。在对同一个基因的不同转录产物中不同组合的腺嘌呤核苷进行编辑后，可以从一个基因产生 1000 多种各不相同的蛋白质（图 5.15）。

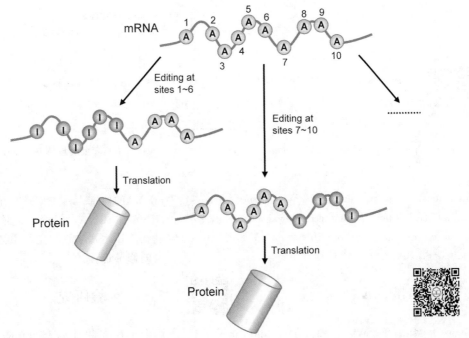

Figure 5.15　Potential A→I editing sites in mRNA of a membrane protein

图 5.15　一种膜蛋白 mRNA 中的潜在 A→I 编辑位点

The term RNA editing is also sometimes used to refer to the addition of nucleotides into the RNA transcript. This is less common than the kind of editing we have just described. The classic example of this kind of editing occurs in trypanosomes, which are protozoa that cause African sleeping sickness. Trypanosome mitochondrial genes produce unfinished mRNA transcripts into which uridylate (U) nucleotides must be added or removed. Specialized RNA molecules called **guide RNAs (gRNAs)** hybridize to the transcript (Figure 5.16). At nucleotides where hybridization is not complete, cellular proteins either add or remove uridylate nucleotides, depending on context.

RNA 编辑这一术语有时也用来指向 RNA 转录产物中添加核苷酸。这种编辑方式与我们刚才讨论过的编辑方式相比较而言较为少见。这种编辑方式的经典实例发生在锥虫中。锥虫是一种能引起非洲昏睡病的原生生物。锥虫线粒体基因产生的是未完成的 mRNA 转录产物，还需要向其中添加或去除尿苷酸（U）。称为**指导 RNA（gRNA）**的特殊 RNA 分子与转录产物杂交（图 5.16）。在杂交不完全的核苷酸位置，细胞中的蛋白质会根据序列的上下文添加或去除尿苷酸。

Note: This chapter has mainly focused on the modifi-

注：本章把焦点主要集中在了真核生物

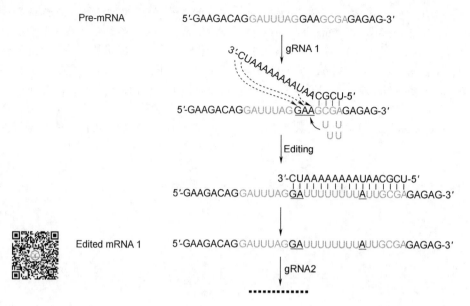

Figure 5.16 RNA editing of mitochondrial genes in trypanosome

图 5.16 锥虫线粒体基因的 RNA 编辑

cations to mRNA in eukaryotes. We do not discuss rRNA and tRNA here, but the reader should be aware that these also undergo modifications in both prokaryotes and eukaryotes.

mRNA 的修饰上。我们没有在这儿讨论 rRNA 和 tRNA，但读者应该意识到它们在原核生物和真核生物中都会经历一定的修饰过程。

5.5 Experiments

5.5 实验研究

Evidence for introns came from experiments in which mRNA was hybridized to its gene in the DNA. Because the mRNA does not contain introns, it is shorter than the DNA. The exons in the DNA hybridize to the mRNA, but the introns in the DNA form large loops because they cannot hybridize (Figure 5.17). These loops could be seen by electron microscopy.

内含子存在的证据来自 mRNA 与基因 DNA 的杂交实验。因为 mRNA 没有内含子，它比相应的 DNA 短。DNA 中的外显子会与 mRNA 杂交，DNA 中的内含子由于不能杂交而形成很大的环（图 5.17）。这些环可以用电子显微镜观察到。

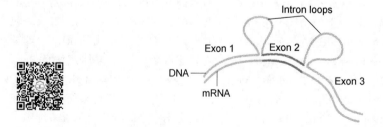

Figure 5.17 Hybridization between mRNA and DNA

图 5.17 mRNA 与 DNA 之间的杂交

Another way to detect the presence of loops was by exposing the hybrid molecule to nucleases that only cut

另一种检测环是否存在的方式是将这一杂种分子用只能切割单链 DNA 的核酸

single-stranded DNA. In this case, only the DNA in the loops was cut into little pieces. The exon DNA, which hybridized to the mRNA, was protected (Figure 5.18). Thus, by examining the DNA that was not cut, it was possible to determine which parts of the gene were exons.

酶进行处理。这时，只有位于环上的DNA才会被切成短的片段。与mRNA杂交的外显子DNA得到了保护（图5.18）。之后，通过检查未被切断的DNA就有可能确定基因中的哪些部分是外显子。

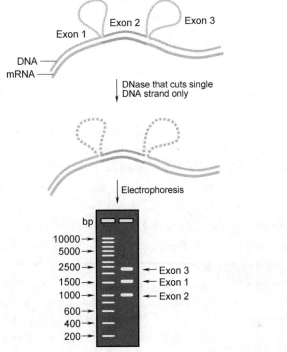

Figure 5.18 Detection of introns with DNase that cuts single DNA strand only

图5.18 用只切割DNA单链的DNase检测内含子

These experiments showed the presence of introns, but left some questions. Were the introns even transcribed? Or were they transcribed and then removed? In order to verify the existence of RNA splicing, the pre-mRNA for a gene was isolated and hybridized to DNA (Figure 5.19). Unlike the mRNA, pre-mRNA hybridized perfectly to the DNA without leaving any loops. This meant that introns were present in the pre-mRNA, and that introns are indeed transcribed and then removed later.

这些实验显示了内含子确实存在，但也留下了其他问题。内含子是否也被转录了呢？或者，它们是否先被转录出来然后才被去除呢？为了证实RNA剪接的存在，需要将基因的前体mRNA分离出来与DNA杂交（图5.19）。与mRNA不同，前体mRNA与DNA能完全杂交而不留下任何环。这意味着在前体mRNA中是存在内含子的，因此内含子确实是先被转录出来之后才被去除的。

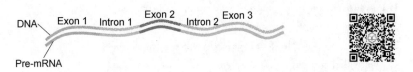

Figure 5.19 Hybridization between pre-mRNA and DNA

图5.19 前体mRNA与DNA之间的杂交

Summary

RNA transcripts in eukaryotes destined to be translated into protein must undergo a series of modifications. The most significant of these modifications are capping, polyadenylation, and splicing. All three occur during or immediately after transcription, and are coordinated by the CTD of RNA polymerase Ⅱ. Capping and polyadenylation have a range of functions, from increasing stability of the mRNA to increasing its ability to be translated. Splicing is necessary to remove introns, which interrupt the coding regions of eukaryotic genes. The reason why introns exist is still being debated, but introns do seem to serve some purpose. In particular, they permit alternative splicing of genes, which causes more than one protein product to be produced from one coding region. RNA editing, which often involves substitution of a base for another base, is another important modification and can also result in the production of more than one protein from one gene.

小结

真核生物中用于翻译产生蛋白质的 RNA 转录产物必须经历一系列修饰作用。最显著的修饰作用是加帽、聚腺苷酸化和剪接。三种修饰作用均在转录过程中进行或在转录结束后立即开始,它们由 RNA 聚合酶 Ⅱ 的 CTD 来协调。加帽和聚腺苷酸化具有一些不同的功能,它们能增加 mRNA 稳定性或增强 mRNA 的翻译能力。为去除隔开真核基因编码区域的内含子有必要进行剪接修饰。内含子存在的原因仍然处于争论之中,但内含子看起来确实也有一些作用。特别是它们允许基因的可变剪接,从而使一个编码区产生不止一种蛋白质。RNA 编辑通常牵涉到一个碱基被另一个碱基替换,它是另一种重要的修饰作用,也能导致从一个基因产生不止一种蛋白质。

Vocabulary 词汇

African sleeping sickness	非洲昏睡病	modification [ˌmɔdifi'keiʃən]	修饰(作用)
alternative splicing [ɔːl'təːnətiv]	可变剪接	polyadenylation ['pɔliˌædenilei ʃən]	聚腺苷酸化(作用)
antibody ['æntiˌbɔdi]	抗体	poly(A) tail	聚腺苷酸尾
antigen ['æntidʒən]	抗原	protozoa [ˌprəutəu'zəuə]	原生生物
apolipoprotein ['æpəˌlaipəu'prəutiːn]	载脂蛋白	RNA editing	RNA 编辑
capping ['kæpiŋ]	加帽(过程)	secrete [si'kriːt]	分泌
degradation [ˌdegrə'deiʃən]	降解	self-splicing	自我剪接
degrade [di'greid]	降解	snRNPs [snəːps]	核内小核糖核蛋白
exon ['iksən]	外显子	spliceosome ['splaisiəusəm]	剪接体
guide RNA (gRNA)	指导 RNA	splice site	剪接位点
inosine ['inəsiːn]	次黄苷	splicing ['splaisiŋ]	剪接(作用)
intron ['intrən]	内含子	tetrahymena [ˌtetrə'haimənə]	四膜虫
lariat ['læriət]	套索	trypanosome ['tripənəsəum]	锥虫

Review Questions   习题

Ⅰ. **True/False Questions** (判断题)

1. Polyadenylation occurs on the *N*-terminus of proteins.
2. The cap structure is joined to the pre-mRNA by a 5′-5′ link.
3. Capping occurs while the pre-mRNA is being transcribed.
4. Most splicing occurs with the help of snRNPs.
5. Most genes only have one intron.
6. RNA editing can change the length of a protein translated from an mRNA.

7. Most RNA modifications occur in the nucleus.
8. After receiving a cap and a poly(A) tail, mRNA cannot be degraded.
9. Each snRNP is made by one gene.
10. RNA polymerase Ⅱ adds the poly(A) tail only after the transcript is cleaved near an AAUAAA sequence.

Ⅱ. **Multiple Choice Questions**（选择题）
1. Which of the following might occur if an mRNA does not receive a 5′ cap?
   a. It will not be translated.
   b. It will not be exported to the cytoplasm.
   c. It will be quickly degraded.
   d. It will not have proper splicing.
   e. All of the above.
2. Which of the following proteins is not involved in polyadenylation?
   a. CPSF   b. APOB   c. PBP   d. CstF   e. CF Ⅰ
3. Which part of intron is cleaved first?
   a. the 5′ splice site
   b. the 3′ splice site
   c. the attacking adenylate nucleotide
   d. the lariat
   e. the 5′ cap
4. Which of the following statements is not true about introns?
   a. They are shorter than exons.
   b. They are not associated with histones.
   c. They are present in E. coli.
   d. They never serve a purpose.
   e. All of the above.
5. Which of the following is true of self-splicing of group Ⅱ introns?
   a. A lariat is formed.
   b. A free guanylate nucleotide is used.
   c. U1 binds to the 5′ splice site.
   d. It is assisted by gRNAs.
   e. None of the above.

Exploration Questions 思考题

1. Why do you think that so many modifications to RNA are made in eukaryotes?
2. What is the 'CTD' and what does it do during transcription?
3. Why do you think snRNPs contain RNA, instead of being exclusively made of protein, like most enzymes?
4. Give some examples of how post-transcriptional modifications allow one gene to make more than one kind of protein.

Chapter 6　Translation

The last several chapters have focused on nucleic acids, as DNA is transcribed to mRNA, and mRNA is modified. In this chapter we see how the information from these nucleic acids is transformed into life itself, as the sequence of bases becomes the basis for production of proteins. Yet even as we discuss the production of proteins, our focus remains largely on nucleic acids, because two of the most important components of the translation mechanism are in fact tRNA and rRNA. We begin by discussing how a base sequence can hold information for making an amino acid sequence, and then discuss the mechanism by which a sequence is actually translated.

6.1　The Genetic Code

In some ways, translation is less intuitive than transcription. The concept of transcription is simple: open the DNA helix, and allow complementary ribonucleotides to bind deoxyribonucleotides. Because of the binding properties of nucleic acids, DNA can be used directly as a template for making RNA (Figure 6.1).

第 6 章　翻译

在前面几章中我们把目光集中在了核酸上，因为 DNA 被转录成 mRNA，mRNA 又被进一步修饰。本章我们将要看到，在碱基序列成为蛋白质生产的基础时，核酸中的信息如何转化成生命本身。而即使在讨论蛋白质生产的时候，我们的焦点多多少少还会放在核酸上，因为翻译机理中两种最重要的成分实际上是 tRNA 和 rRNA。我们先来讨论碱基序列如何储存生产氨基酸序列的信息，之后再讨论实际翻译生成氨基酸序列的机理。

6.1　遗传密码

某种程度上，翻译并不像转录那样简单明了。转录的概念很简单：打开 DNA 螺旋，使互补的核糖核苷酸结合到脱氧核糖核苷酸上。由于核酸的结合特性，DNA 可以直接用作生产 RNA 的模板（图 6.1）。

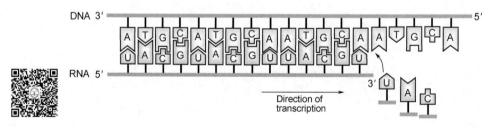

Figure 6.1　In transcription, DNA can be used directly as a template for making RNA because ribonucleotides bind to deoxyribonucleotides

图 6.1　转录时 DNA 可以直接作为生产 RNA 的模板，因为核糖核苷酸能与脱氧核糖核苷酸结合

But amino acids are completely different from nucleotides. The two kinds of molecules cannot bind directly to each other in a consistent way. In order to use RNA as a template in translation, a converter must be used. This converter is tRNA. tRNA has two parts: one side is RNA and hybridizes to an mRNA transcript; the other side can be connected to an amino acid (Figure 6.2).

但是氨基酸与核苷酸完全不同。这两种分子不能以相互一致的方式直接结合。为了在翻译中使用 RNA 作为模板，必须采用一种转换器。这种转换器就是 tRNA。tRNA 含有两部分：一端是 RNA，它能与 mRNA 转录产物杂交；另一端能够与氨基酸连接（图 6.2）。

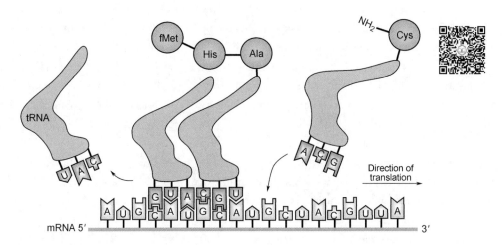

Figure 6.2 In translation, RNA cannot be used directly as template for making amino acid polymers. tRNA must be used as a converter between bases and amino acids

图 6.2 翻译时 RNA 不能直接作为生产氨基酸聚合物的模板。必须使用 tRNA 作为碱基与氨基酸之间的转换器

To produce a functional protein, the base sequence of an mRNA must precisely specify the order of amino acids in the protein. However, there are four different bases in RNA, and 20 different amino acids in a protein. How can such a seemingly simple base sequence represent a much more complicated amino acid sequence? If one kind of base was connected to one kind of amino acid, then only four different amino acids could be coded for in RNA (Table 6.1). Thus, the mRNA base sequence is clearly not read one base at a time.

为了生产有功能的蛋白质，mRNA 中的碱基序列必须精确地决定蛋白质中氨基酸的顺序。然而，RNA 中只有四种不同的碱基，蛋白质中却有 20 种不同的氨基酸。看起来如此简单的碱基序列是如何表示复杂得多的氨基酸序列的呢？假如一种碱基与一种氨基酸相关联，那么 RNA 只能编码四种不同的氨基酸（表 6.1）。这样，很显然 mRNA 的碱基序列不是一次只被读取一个碱基。

Table 6.1 The number of amino acids that can be coded for by combining bases in groups of 1, 2, or 3

表 6.1 以 1、2 或 3 个碱基进行组合所能编码的氨基酸数目

Number of bases in each codon	1	2	3			
Possible codons	U C A G	UU UC UA UG CU CC CA CG AU AC AA AG GU GC GA GG	UUU UUC UUA UUG UCU UCC UCA UCG UAU UAC UAA UAG UGU UGC UGA UGG	CUU CUC CUA CUG CCU CCC CCA CCG CAU CAC CAA CAG CGU CGC CGA CGG	AUU AUC AUA AUG ACU ACC ACA ACG AAU AAC AAA AAG AGU AGC AGA AGG	GUU GUC GUA GUG GCU GCC GCA GCG GAU GAC GAA GAG GGU GGC GGA GGG
Number of amino acids that can be coded for	4	16	64			

6.1 The Genetic Code

If two bases were read at one time, 16 different amino acids could be coded for. This is because from among four different bases, you can make sixteen different combinations of two bases (assuming that the combinations are considered different if the bases are in a different order). There are more than 16 amino acids, however, so this is still not sufficient.

If three bases are read at a time, there are 64 possible combinations. This is more than enough to code for all of the amino acids. Indeed, during translation groups of three bases correspond to one amino acid. A group of three bases is called a **codon**. Each represents one, and only one, amino acid to be placed in the growing protein. tRNA binds to codons and joins them to their corresponding amino acid.

The list of which codon corresponds to which amino acid is called the **genetic code** (Figure 6.3). Notice that several different codons may correspond to one amino acid. Because of this, the genetic code is said to be **redundant**. Redundancy in the code is not surprising because there are 64 codons, and only 20 different amino acids. Notice also that three of the codons do not code for amino acids; instead, they signal the end of translation and are therefore called **stop codons**. The AUG codon, in addition to coding for methionine, also signals the beginning of translation. It is called the **start codon**.

如果一次读取两个碱基，那么总共可以编码16种不同的氨基酸。这是因为从四种不同的碱基中，你能够得到由两个碱基组成的十六种不同组合（假如碱基以不同顺序出现时得到的组合被认为是不同的）。然而氨基酸不止16种，所以这仍然是不够的。

如果一次读取三个碱基，就会有64种可能的组合，足够用来编码所有的氨基酸。实际上，在翻译中确实是三个碱基一组与一个氨基酸对应。三个碱基一组称为一个**密码子**。每个密码子代表了一种且仅仅是一种将被置于生长中的蛋白质链上的氨基酸。tRNA结合到密码子上并将它们与相应的氨基酸连接。

密码子对应于氨基酸的列表称为**遗传密码**（图6.3）。请注意，几种不同的密码子可以对应于一种氨基酸。正因如此，遗传密码是**冗余的**。密码的冗余性并不令人惊讶，因为有64个密码子，而只有20种不同的氨基酸。也请注意，密码子中有三个并不编码氨基酸；相反，它们是翻译结束的信号，因此被称为**终止密码子**。密码子AUG除了编码甲硫氨酸以外还起到翻译起始信号的作用。它被称为**起始密码子**。

		Second position				
		U	C	A	G	
First position (5'end)	U	UUU, UUC } Phe UUA, UUG } Leu	UCU, UCC, UCA, UCG } Ser	UAU, UAC } Tyr UAA=Stop UAG=Stop	UGU, UGC } Cys UGA=Stop UGG=Trp	U C A G	Third position (3'end)
	C	CUU, CUC, CUA, CUG } Leu	CCU, CCC, CCA, CCG } Pro	CAU, CAC } His CAA, CAG } Gln	CGU, CGC, CGA, CGG } Arg	U C A G	
	A	AUU, AUC, AUA } Ile AUG=Met	ACU, ACC, ACA, ACG } Thr	AAU, AAC } Asn AAA, AAG } Lys	AGU, AGC } Ser AGA, AGG } Arg	U C A G	
	G	GUU, GUC, GUA, GUG } Val	GCU, GCC, GCA, GCG } Ala	GAU, GAC } Asp GAA, GAG } Glu	GGU, GGC, GGA, GGG } Gly	U C A G	

Figure 6.3　The genetic code　　　　　　　　图6.3　遗传密码

6.2 Mechanism of Translation in Prokaryotes

Translation occurs in a large structure called a **ribosome**. As the prefix ribo-suggests, ribosomes are made of ribonucleic acid as well as protein. They consist of two large subunits (Figure 6.4). In prokaryotes, the subunits are called 30S (small subunit) and 50S (large subunit). The **30S subunit** is made of 21 proteins and one large rRNA, 16S. The **50S subunit** is made of 32 proteins and two rRNAs, 5S and 23S. See Figure 6.4 for the structural composition of eukaryotic ribosomes.

6.2 原核生物翻译机理

翻译发生的场所是一种称为**核糖体**的大型结构。正如前缀 ribo-所提示的那样，核糖体由核糖核酸和蛋白质共同组成。它们含有两个亚基（图 6.4）。在原核生物中，这两个亚基称为 30S（小亚基）和 50S（大亚基）。**30S 亚基**由 21 种蛋白质和一条 16S 的 rRNA 组成。**50S 亚基**由 32 种蛋白质和两条 rRNA（5S 和 23S）组成。真核生物核糖体的结构组成情况请参照图 6.4。

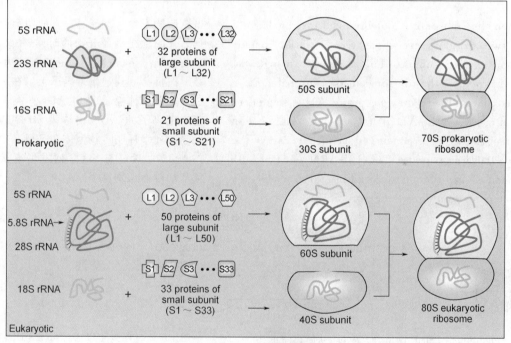

Figure 6.4　Composition of prokaryotic and eukaryotic ribosomes

图 6.4　原核生物与真核生物核糖体的组成

Translation at the ribosome can be divided, like transcription, into three phases: initiation, elongation, and termination. We begin by discussing these steps in prokaryotes.

像转录一样，在核糖体上进行的翻译过程也可以被分为三个阶段：起始、延伸和终止。我们首先讨论原核生物中的这些步骤。

6.2.1　Initiation

In prokaryotes, translation of an mRNA often begins while the mRNA is still being transcribed. Coupled

6.2.1　起始

在原核生物中，mRNA 的翻译常常在 mRNA 还在被转录的时候就已经开始

transcription-translation allows the cell to produce protein as quickly as possible. This is important because mRNAs in prokaryotes are very short-lived. To add to this efficiency, multiple ribosomes can translate one mRNA at the same time. One mRNA with multiple ribosomes attached is called a **polysome** (Figure 3.2).

Not all of an mRNA molecule is used to make proteins. At the beginning and end of the transcript are **untranslated regions (UTRs)**. The coding region, the part which is translated, begins with a start codon and ends with a stop codon. Prokaryotic mRNAs often have several coding regions; each one begins and ends with its own start and stop codons.

When the ribosome is not involved in translation, its two subunits are separated. Initiation begins when the 30S subunit alone binds to an mRNA. At this point, the ribosome must center itself at a start codon. There may be many AUG sequences in an mRNA; the AUG that acts as a start codon is designated by a conserved sequence called the **Shine-Dalgarno sequence**. The 16S rRNA in the 30S subunit of the ribosome recognizes this sequence by base pairing to it directly (Figure 6.5).

了。偶联转录-翻译使细胞能够尽可能快地生产蛋白质。这一点很重要，因为在原核生物中 mRNA 寿命是很短的。为了提高生产效率，会有很多核糖体同时去转录一条 mRNA。一条带有很多核糖体的 mRNA 称为**多聚核糖体**（图 3.2）。

并不是 mRNA 分子的所有部分都用来生产蛋白质的。转录产物的开头和末尾部分是**非翻译区（UTR）**。需要翻译的编码区域开始于一个起始密码子而结束于一个终止密码子。原核 mRNA 常常含有多个编码区域；每一个编码区域有它自己的起始密码子和终止密码子。

当核糖体不参与翻译时，它的两个亚基是分开的。在翻译开始的时候，30S 亚基结合到一条 mRNA 上。这时，核糖体必须把自己放到起始密码子上。在一条 mRNA 上可能会有许多 AUG；能作为起始密码子的 AUG 由称为 **SD 序列**的保守序列来标明。核糖体 30S 亚基中的 16S rRNA 通过与 SD 序列发生直接的碱基配对而识别这一序列（图 6.5）。

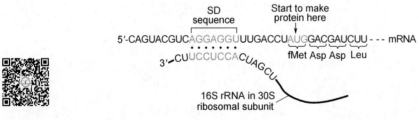

Figure 6.5 Shine-Dalgarno sequence is recognized by hybridization with 16S rRNA. The Shine-Dalgarno sequence designates which AUG will be used as a start codon

图 6.5 SD 序列可以通过与 16S rRNA 杂交而被识别。SD 序列标明了哪一个 AUG 应该作为起始密码子

Once the 30S subunit has attached near the start codon, a specialized tRNA called **initiator tRNA** binds to the start codon. The initiator tRNA carries **N-formylmethionine**, an uncommon molecule derived from the amino acid methionine (Figure 6.6). The collection of the 30S subunit, the mRNA, and the initiator tRNA is called the **30S initiation complex** (Figure 6.7).

一旦 30S 亚基附着到了靠近起始密码子的地方，一种称为**起始 tRNA** 的特殊 tRNA 就会结合到起始密码子上。起始 tRNA 带有 **N-甲酰甲硫氨酸**，它是一种从甲硫氨酸衍生而来的不寻常的分子（图 6.6）。30S 亚基、mRNA 和起始 tRNA 的集合称为 **30S 起始复合体**（图 6.7）。

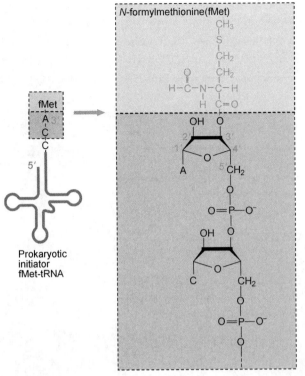

Figure 6.6　Initiator tRNA in prokaryotes, bound to *N*-formylmethionine

图 6.6　原核生物中与 *N*-甲酰甲硫氨酸连接的起始 tRNA

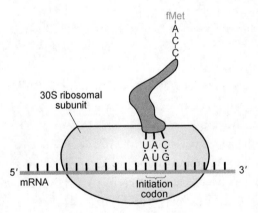

Figure 6.7　The 30S initiation complex

图 6.7　30S 起始复合体

Three proteins, called translation **initiation factors** (**IF**), are involved in assembling the initiation complex. IF1 and IF3 help to dissociate ribosomal subunits and help the 30S subunit bind to the mRNA at the Shine-Dalgarno sequence. IF2 is mainly involved in bringing initiator tRNA to the 30S.

The next step of initiation is the binding of the 50S subunit to the 30S complex to form the 70S initiation

称为**翻译起始因子**（**IF**）的三种蛋白质也参与到了组装起始复合体的过程中。IF1 和 IF3 帮助解离核糖体的亚基，同时帮助 30S 亚基在 SD 序列的位置结合到 mRNA 上。IF2 主要参与将起始 tRNA 带到 30S 亚基上来。

起始的下一个步骤是 50S 亚基与 30S 复合体结合形成 70S 起始复合体（图

complex (Figure 6.8). Note that subunit names do not add arithmetically. The 70S complex contains three tRNA binding sites: the **A (aminoacyl) site**, the **P (peptidyl) site**, and the **E (exit) site**. These sites allow tRNA to join the ribosome and bind directly to mRNA codons.

6.8)。注意亚基的名称不是按算术加法得来的。70S 复合体含有三个 tRNA 结合位点：**A（氨酰基）位、P（肽基）位和 E（退出）位**。这些位点使 tRNA 进入到核糖体中并直接结合到 mRNA 密码子上。

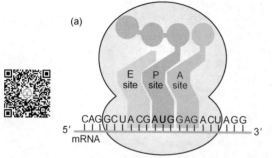 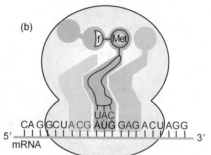

Figure 6.8 (a) The three tRNA binding sites of the 70S initiation complex; (b) Initiator tRNA binds initially at the P site

图 6.8 (a) 70S 起始复合体上的三个 tRNA 结合位点；(b) 起始 tRNA 最初结合在 P 位

6.2.2 Elongation

We will now walk through the mechanism of translation (Figure 6.9). For simplicity, various components will be assigned abbreviated names. These are not official names.

At the start of translation, the initiator tRNA is automatically positioned in the P site, and the A site is vacant. The P site is aligned with the start codon, and the A site is aligned with the adjacent codon in the 3′ direction. Next, a tRNA (tRNA1) binds into the A site. This tRNA will carry the amino acid (aa_1) corresponding to the codon at the A site. Next, the bond between the *N*-formylmethionine (Nmeth) and the initiator tRNA is broken, and the Nmeth is attached to aa_1. The first peptide bond has been formed. The molecule at the A site is now Nmeth-aa_1-tRNA1. The enzymatic function within the ribosome that forms the peptide bond is called **peptidyl transferase** (Figure 6.10).

Next, Nmeth-aa_1-tRNA1 moves with its codon into the P site. This movement is called **translocation**. The initiator tRNA is knocked into the E site and eventually leaves the ribosome. The A site is now empty, and

6.2.2 延伸

现在让我们来初步了解一下翻译的机理（图 6.9）。为简明起见，许多成分采用了缩写名称。请注意它们并不是正式的名称。

在翻译开始时，起始 tRNA 自动进入 P 位，那时 A 位是空的。P 位是与起始密码子对齐的，A 位与 3′方向相邻的密码子对齐。然后，一个 tRNA（tRNA1）进来结合到 A 位。这个 tRNA 携带着对应于 A 位密码子的氨基酸（aa_1）。接着，*N*-甲酰甲硫氨酸（Nmeth）和起始 tRNA 之间的键断开，Nmeth 连接到 aa_1 上。这就形成了第一个肽键。A 位上的分子现在变成了 Nmeth-aa_1-tRNA1。在核糖体内催化形成肽键的酶称为**肽基转移酶**（图 6.10）。

下一步，Nmeth-aa_1-tRNA1 与它的密码子一起移到了 P 位上。这种移动称为**移位**。起始 tRNA 被挪到了 E 位，并最终离开核糖体。A 位现在空了出来，它含有一

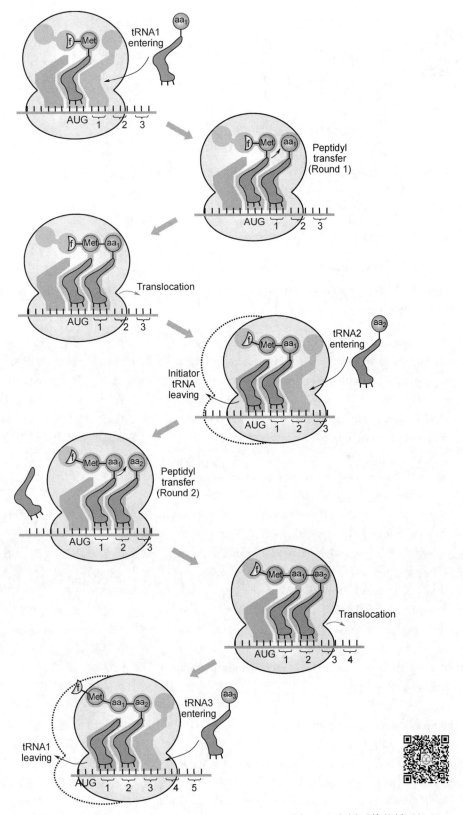

Figure 6.9　Cyclical mechanism of translation elongation

图 6.9　翻译延伸的循环机理

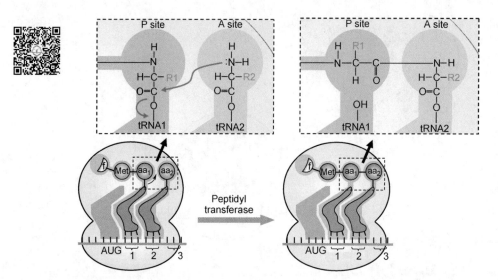

Figure 6.10　Peptide bond formation, catalyzed by peptidyl transferase (not shown)

图 6.10　肽键形成，由肽基转移酶催化（酶未显示）

contains a new codon. A tRNA（tRNA2）carrying an amino acid（aa_2）corresponding to this new codon binds at the A site. Next, Nmeth-aa_1 releases from tRNA1 and reattaches to aa_2-tRNA2. The result is Nmeth-aa_1-aa_2-tRNA2, located in the A-site. The amino acids are all connected by peptide bonds.

Let us go through one more cycle. Nmeth-aa_1-aa_2-tRNA2 translocates to the P site. tRNA1 is no longer attached to an amino acid; it is knocked into the E site and eventually leaves the ribosome. The A site is again empty, and has a new codon. tRNA3 carrying aa_3 corresponding to this codon binds at the A site. Now, Nmeth-aa_1-aa_2 releases from tRNA2 in the P site, and reattaches to aa_3-tRNA3 in the A site. The result is Nmeth-aa_1-aa_2-aa_3-tRNA3, with all the amino acids connected by peptide bonds. This cycle continues hundreds or thousands of times, until a long peptide has been assembled.

Elongation is aided by proteins called translation **elongation factors (EF)**. One such factor, EF-Tu, helps to bring new tRNAs to the A-site. In doing so, it requires the hydrolysis of GTP (Figure 6.11). Another factor, EF-G, is necessary for translocation, the process of moving a tRNA from the A site to the P site and placing a new codon at the A site (Figure 6.12).

个新密码子。对应于这一新密码子并携带了氨基酸（aa_2）的 tRNA（tRNA2）结合到 A 位。之后，Nmeth-aa_1 从 tRNA1 上释放又与 aa_2-tRNA2 连接。得到的结果是位于 A 位的 Nmeth-aa_1-aa_2-tRNA2。这些氨基酸都是由肽键连接的。

让我们再看一个循环。Nmeth-aa_1-aa_2-tRNA2 移位到 P 位。tRNA1 上面已经没有氨基酸；它被挪到了 E 位并最终离开核糖体。A 位又空了并且又有了新密码子。tRNA3 携带着与这个密码子对应的 aa_3 结合到 A 位。现在，Nmeth-aa_1-aa_2 在 P 位从 tRNA2 上释放，又与 A 位上的 aa_3-tRNA3 连接。得到的结果是 Nmeth-aa_1-aa_2-aa_3-tRNA3，所有的氨基酸都由肽键连接。这样的循环可以继续几百次或几千次，直到一条长的肽链被组装完成。

延伸需要称为翻译**延伸因子（EF）**的蛋白质帮助。其中一个因子称为 EF-Tu，它帮助新的 tRNA 进入 A 位。它需要水解 GTP 来完成上述任务（图 6.11）。在移位中需要另一个因子 EF-G，移位过程就是将 tRNA 从 A 位移到 P 位并将新的密码子放到 A 位（图 6.12）。

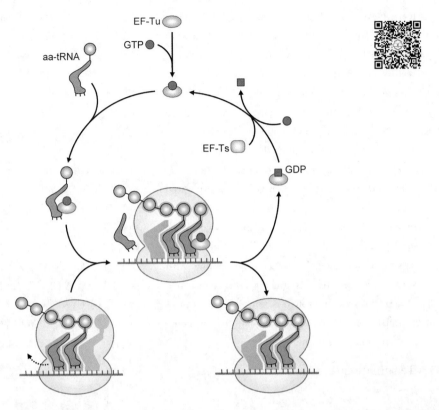

Figure 6.11 Function of elongation factor EF-Tu   图 6.11 延伸因子 EF-Tu 的功能

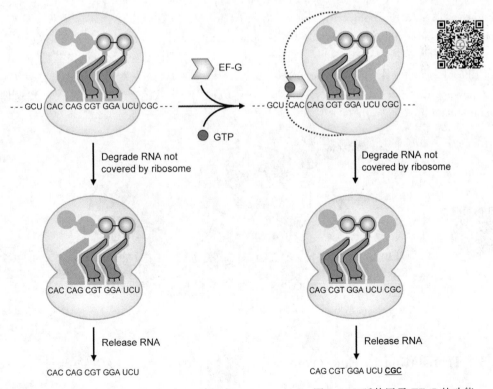

Figure 6.12 Function of elongation factor EF-G   图 6.12 延伸因子 EF-G 的功能

6.2 Mechanism of Translation in Prokaryotes

In summary, the ribosome begins to read an mRNA coding region at a start codon near its 5′ end, and proceeds toward the 3′ end. As translation moves down the mRNA, new codons are read, and for each new codon, a corresponding amino acid is added to the growing polypeptide. Peptide growth also occurs in a consistent direction. As explained in chapter 1, the two ends of a polypeptide are distinct: one end has an amino group, and is called the N-terminus; The other end has a carboxyl group, and is called the C-terminus. The amino group of the first amino acid incorporated into a peptide is never reacted. This amino group provides the N-terminus of the protein. Thus, the amino end is established at the beginning of translation. Every time a new amino acid is incorporated, it is attached by reacting with the carboxyl group of the previous amino acid to be incorporated. The C-terminus of the protein is not established until the end of translation. Therefore, it is said that proteins grow from N-terminus to C-terminus.

6.2.3 Termination

Translation terminates when the ribosome reaches a stop codon. No tRNA is able to bind a stop codon in the A site. Instead, a protein called a **release factor** binds (Figure 6.13). At this point, the recently created polypeptide is held at the ribosome only by its connection with tRNA in the P site. The release factor stimulates the ribosome to break this connection, releasing the new polypeptide. Finally, the tRNA and the two ribosomal subunits all separate.

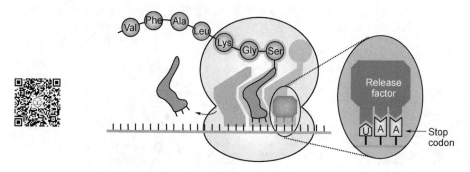

Figure 6.13　Mechanism of translation termination

6.3　Translation in Eukaryotes

Translation is quite similar in eukaryotes as in prokaryotes. The most important differences pertain to initiation. In eukaryotes, the initiator tRNA carries normal methionine, not N-formylmethionine. However, eukaryotic initiator tRNA is still unique in that it only binds to start codons; a different tRNA is used to add methionine within polypeptides.

Initiator tRNA joins with the 40S subunit and then with mRNA to form the 48S initiation complex. Once the start codon has been determined, the 60S ribosomal subunit joins to form an 80S complex, analogous to the 70S complex in prokaryotes.

To determine the site of the start codon, the 48S complex slides across the mRNA beginning at the 5′ cap structure, a process called **scanning**. In around 90% of mRNAs, the first AUG encountered is then used as a start codon. However, sometimes sequences within the mRNA can cause the ribosome to continue and use an AUG further downstream. In some rare cases translation begins at a site determined by a sequence called an **internal ribosome entry sequence (IRES)**. IRESs are usually several hundred bases long, and fold into simple structures (Figure 6.14).

最主要的区别在起始阶段。在真核生物中，起始 tRNA 携带的是正常的甲硫氨酸，而不是 N-甲酰甲硫氨酸。然而，真核生物的起始 tRNA 仍然是独特的，它只能与起始密码子结合；多肽链内部的甲硫氨酸由另一种 tRNA 负责添加。

起始 tRNA 先与 40S 亚基结合，之后再与 mRNA 结合形成 48S 起始复合体。一旦确定了起始密码子，60S 亚基便会加入形成 80S 复合体，它类似于在原核生物中形成的 70S 复合体。

为了确定起始密码子的位置，48S 复合体会从 mRNA 的 5′ 帽结构开始沿 mRNA 滑行，这一过程称为**扫描**。在大约 90% 的 mRNA 中，遇到的第一个 AUG 会被作为起始密码子。然而，有时 mRNA 中的序列能够使核糖体继续前进从而使用下游的 AUG。在一些并不常见的情形中，翻译的起始位点是由**内部核糖体进入序列（IRES）**决定的。IRES 一般长度在几百个碱基对并折叠成较为简单的结构（图 6.14）。

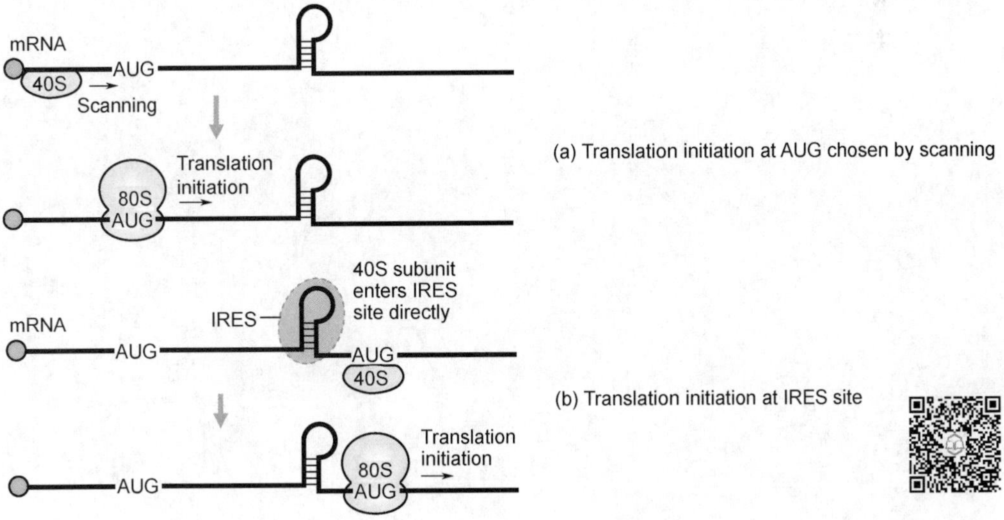

(a) Translation initiation at AUG chosen by scanning

(b) Translation initiation at IRES site

Figure 6.14　Translation initiated by scanning for AUG (a) and by entering IRES site directly (b)

图 6.14　通过扫描寻找 AUG（a）和通过直接进入 IRES 位点（b）启动翻译

IRESs may allow two proteins to be made from one mRNA. For example, a longer protein can be made using the normal start codon, and a shorter protein can be

IRES 可以使一条 mRNA 用于产生两种蛋白质。例如，使用正常的起始密码子产生一条长的蛋白质链，而从 IRES 处

6.3　Translation in Eukaryotes | 133

made by beginning translation at the IRES. IRESs are also useful because they may allow translation of an mRNA even when the normal translation machinery, which initiates translation by scanning, is suppressed. This property is used by some viruses. While the virus suppresses a cell's ability to initiate translation normally, the virus' own mRNAs contain IRESs and therefore are translated at a normal rate (Figure 6.15).

开始的翻译产生一条较短的蛋白质链。除此以外，IRES 也是相当有用的，因为在通过扫描启动翻译的正常翻译方法被抑制的情况下，它们仍可启动mRNA的翻译。这一性质为一些病毒所利用。病毒在抑制了细胞正常的翻译启动能力之后，它们自身的mRNA因含有 IRES 而能以正常的速率进行翻译（图 6.15）。

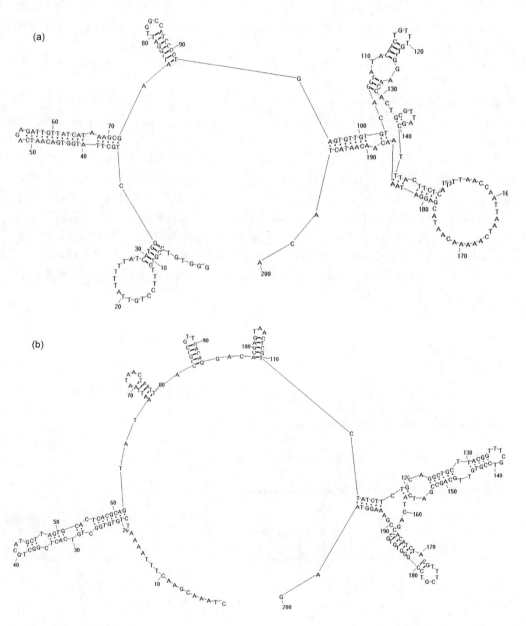

Figure 6.15 IRES structure in (a) poliovirus and (b) SARS-CoV-2 (Source: https://pubmed.ncbi.nlm.nih.gov/32704018/)

图 6.15 （a）脊髓灰质炎病毒与（b）新型冠状病毒中的 IRES 结构（来源：https://pubmed.ncbi.nlm.nih.gov/32704018/）

Normal translation initiation in eukaryotes depends on a complex assortment of factors, called **eukaryotic initiation factors**, or **eIFs** (Figure 6.16). eIF3 prevents the association of the two ribosomal subunits when translation is not in progress, so that isolated 40S is available for translation initiation. eIF2 is involved in binding the initiator tRNA to the 40S subunit. eIF4F is a complex of three factors: eIF4A, eIF4E, and eIF4G. eIF4F binds to the mRNA cap, allowing translation to begin at the 5′ end of the transcript. eIF1 and eIF1A stabilize the 48S initiation complex while scan-

真核生物翻译起始依赖于一种复杂的称为**真核起始因子**或 **eIFs** 的因子混合体（图 6.16）。eIF3 用来在没有发生翻译的时候防止核糖体两个亚基之间的结合，因此可以分离到翻译起始阶段的 40S 亚基。eIF2 在起始 tRNA 结合到 40S 亚基的过程中起作用。eIF4F 是三种因子（eIF4A、eIF4E 和 eIF4G）的复合体。eIF4F 结合到 mRNA 的帽上，使翻译从 mRNA 的 5′末端开始。eIF1 和 eIF1A 在扫描过程中起稳定 48S 起

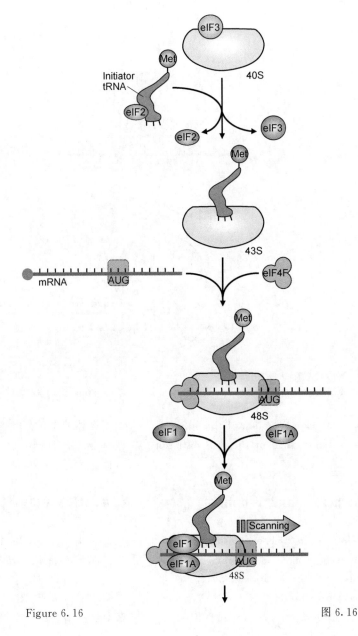

Figure 6.16　　　　　　　　　　　图 6.16

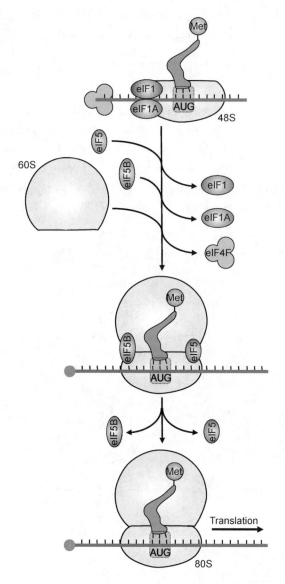

Figure 6.16 Involvement of eukaryotic initiation factors (eIFs) in eukaryotic translation

图 6.16 真核起始因子（eIF）在真核翻译中的作用

ning. eIF5 and eIF5B allow association of the 60S subunit with the 40S subunit to form the 80S initiation complex.

始复合体的作用。eIF5 和 eIF5B 使 60S 亚基与 40S 亚基结合形成 80S 起始复合体。

6.4 tRNA Structure and Wobble

6.4 tRNA 的结构与摇摆

In many ways, tRNA is the basis of translation, acting as a converter between codons and amino acids. To simplify our discussion of translation above, we left out some important considerations about tRNA, which are explored in this section.

在许多情形下，tRNA 是翻译的基础，它扮演着将密码子转换成氨基酸的角色。在上面的讨论中，为简明起见我们忽略了一些关于 tRNA 的重要事项，本节对这些事项做一番探究。

6.4.1 Anti-codons

The structure of tRNA is often drawn in two dimensions so that it looks like a cloverleaf (Figure 6.17). This is not the real structure of tRNA, which is more L-shaped (Figure 6.18). However, the cloverleaf depiction helps to emphasize the two most important parts of the molecule: the amino acid attachment site, and the anticodon.

6.4.1 反密码子

tRNA的结构常以二维形式勾画所以看起来像三叶草（图6.17）。这并不是tRNA的真实结构，tRNA的真实结构更像是L形的（图6.18）。然而，三叶草结构的刻画帮助强调了tRNA分子两个最重要的部分：氨基酸连接位点和反密码子。

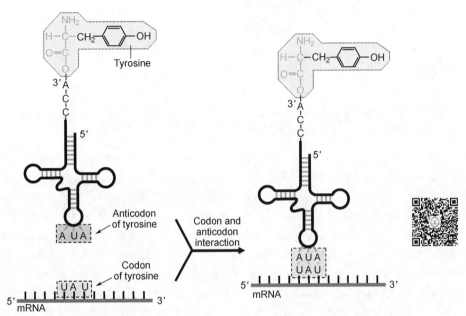

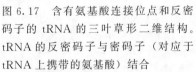

Figure 6.17 Two-dimensional cloverleaf structure of tRNA showing amino acid attachment site and anticodon. Anticodon of a tRNA binds to codon corresponding to the amino acid carried by the tRNA

图6.17 含有氨基酸连接位点和反密码子的tRNA的三叶草形二维结构。tRNA的反密码子与密码子（对应于tRNA上携带的氨基酸）结合

The **anticodon** is a sequence of three bases within the tRNA molecule that is complementary to an mRNA codon. A tRNA's anticodon will only bind to a codon that corresponds to the amino acid that it is carrying (Figure 6.17). Thus, a tRNA carrying the amino acid tyrosine will have an anticodon that only binds to the UAU and UAC codons. This explains how a tRNA matches codons with the precise amino acid that they represent.

One consequence of this translation strategy is that a tRNA with a certain anticodon must always be paired with the proper amino acid. A tRNA with an amino acid is called 'charged' or 'acylated' tRNA. During

反密码子是tRNA分子中与mRNA上的密码子互补的一种三碱基序列。一个tRNA的反密码子只会与它上面所携带的氨基酸对应的密码子结合（图6.17）。这样，一个携带了酪氨酸的tRNA将具有只能与密码子UAU和UAC结合的反密码子。这就解释了为什么tRNA能够将氨基酸准确地匹配到代表这一氨基酸的密码子上。

执行这种翻译策略的一个后果是要求具有一个特定反密码子的tRNA必须总是与正确的氨基酸配成对。携带了氨基酸的tRNA称为"负载"tRNA或"酰基

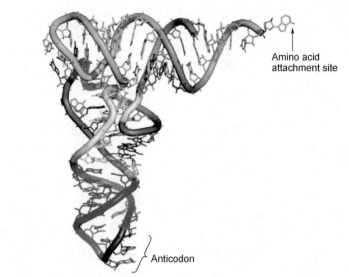

Figure 6.18　Three-dimensional structure of a tRNA (Source：http://www.wikipedia.org)

图 6.18　tRNA 的三维结构 (来源：http://www.wikipedia org)

translation, a tRNA donates its amino acid to the growing polypeptide. After leaving the ribosomes, the tRNA is no longer connected to an amino acid. It is called 'uncharged'. In order to become charged again, **aminoacyl-tRNA synthetase** is required (Figure 6.19). This enzyme recognizes tRNA molecules and matches them with the amino acid that corresponds to their anticodon. It is critically important for the enzyme to work accurately. If a tRNA is matched with the wrong amino acid, then a wrong amino acid will be incorporated at the codon to which the tRNA binds during translation.

化"tRNA。在翻译过程中，tRNA 将它的氨基酸送到生长中的肽链上。离开核糖体后它上面已经没有氨基酸了。这时它被称为"未负载"tRNA。为了再次变成负载 tRNA，需要**氨酰-tRNA 合成酶**的帮助（图 6.19）。这种酶能识别 tRNA 分子并将它们与对应于相应的反密码子的氨基酸进行匹配。这种酶进行准确的工作是极其重要的。如果一个 tRNA 与一个错误的氨基酸匹配，那么在翻译过程中这一 tRNA 所对应的位置就会被加上一个错误的氨基酸。

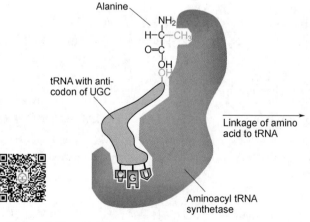 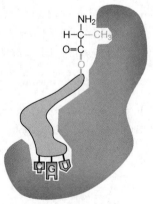

Figure 6.19　Aminoacyl-tRNA synthetase recognizes tRNA anticodon and adds the appropriate amino acid to the tRNA

图 6.19　氨酰 tRNA 合成酶识别 tRNA 反密码子并将合适的氨基酸加到 tRNA 上

6.4.2 Wobble

Recall that there are 61 codons that code for amino acids (64 codons total－3 stop codons). It may be surprising to learn that many cells only have 20 to 30 tRNA molecules. How is it possible to translate so many codons with so few tRNAs? The answer is that one tRNA may recognize more than one codon.

In order to understand how this works, examine the table of codons in Figure 6.3. Notice that the first two bases of a codon are the most important for determining a particular amino acid; the last base can often be altered without changing the amino acid that the codon corresponds to. With valine, for example, the first two bases of the codon, GU, are critical for identifying the amino acid; change any of these and you will get a different amino acid. But change the last base from U to C, A, or G, and the codon will still correspond to valine.

The reason for this property is that the first base in the anticodon of a tRNA usually base pairs loosely with its complement in the codon (Figure 6.20). The position of the anticodon's first base within the ribosome allows it to make unusual base pairs: G can pair with U as well as with C, and U can bind with A as well as G. Thus, a tRNA with the anticodon GGG could recognize the codon CCC as well as CCU.

In addition, the first base of an anticodon is sometimes an unusual base called **inosine**. Inosine has much more flexible binding ability than the normal four bases. It can pair with A, U and C. Thus, a tRNA with the anticodon IGC can pair with GCA, GCU, and GCC codons (Figure 6.20 and Table 6.2). It does not matter that all three codons are recognized by one tRNA, because they all code for the same amino acid. The ability of one tRNA to recognize more than one codon is called **wobble**.

6.4.2 摇摆

回忆一下，有 61 个密码子编码了氨基酸（总共 64 个减去 3 个终止密码子）。当得知许多细胞只含有 20 至 30 种 tRNA 分子时也许我们很惊讶。怎么可能用这么少的 tRNA 来翻译这么多的密码子呢？答案是：一种 tRNA 可以识别不止一个密码子。

为了理解这是怎么回事，请看一下图 6.3 中的密码子表。请注意，一个密码子的前两个碱基在决定一个特殊的氨基酸上是最重要的；第三位碱基的变化通常不会改变密码子所对应的氨基酸。例如，对缬氨酸来说，密码子的前两个碱基 GU 对于确定这个氨基酸来说是最重要的；两个当中的任何一个发生改变都会得到不同的氨基酸。而第三位碱基无论是从 U 改成 C、A 还是 G，这一密码子对应的仍旧是缬氨酸。

这一性质出现的原因是 tRNA 反密码子中第一位碱基通常与密码子中那个互补碱基进行较为松弛的碱基配对（图 6.20）。反密码子第一位碱基在核糖体中的位置允许它产生不寻常的碱基配对：G 能与 U 配对也能与 C 配对，U 能与 A 配对也能与 G 配对。因此，具有反密码子 GGG 的 tRNA 既能识别密码子 CCC 也能识别 CCU。

此外，反密码子的第一位碱基有时是一个不寻常的碱基，称为**次黄苷**。次黄苷比四种正常的碱基具有更灵活的结合能力。它能与 A、U 和 C 配对。这样，一个具有反密码子 IGC 的 tRNA 能与密码子 GCA、GCU 和 GCC 配对（图 6.20 和表 6.2）。所有三个密码子被一种 tRNA 识别是没有关系的，因为它们编码的是同一种氨基酸。一种 tRNA 具有识别不止一个密码子的能力称为**摇摆**。

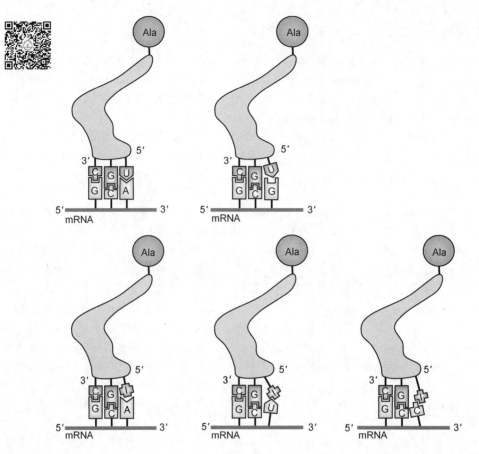

Figure 6.20 Wobble. The first base in an anticodon can often bind to more than one kind of base, allowing one anticodon to recognize more than one codon. When inosine (I) is present in the first position, the anticodon is particularly versatile

图 6.20 摇摆。反密码子的第一位碱基通常可以与不止一个碱基结合，使一个反密码子能够识别不止一个密码子。当次黄苷（I）出现在第一位时，反密码子的功能尤其广泛

Table 6.2　Allowed base pairings from wobble hypothesis 表 6.2　摇摆假说允许的碱基配对

Base at the first position in the anticodon	Bases at the third position in the codon
U	A, G
G	C, U
I	A, U, C

6.5　Experiments

6.5　实验研究

6.5.1　Deciphering the Genetic Code

After DNA base sequence was found to code for proteins, many questions remained as to how the code works. One of the first questions answered was how many amino acids each base is involved in coding for. This was answered by making single base mutations in a DNA sequence, and

6.5.1　破译遗传密码

在 DNA 的碱基序列被发现能为蛋白质编码后，关于这样的密码如何进行编码工作留下了许多问题。最初弄清楚的一个问题是每个碱基参与编码几种氨基酸。这是通过在 DNA 序列中造成单碱

examining the protein produced. It was found that single base mutations never caused changes in more than one amino acid. This indicated that each base was only involved in coding for one amino acid.

But how many bases code for each amino acid? To answer this, pieces of RNA with defined sequence were artificially produced and then translated into protein. For example, an RNA with the repeating sequence UCUCUCUCUCUCU… was synthesized and translated. If you divide this sequence into groups with an even number of bases, each group is the same. Therefore, if each amino acid was coded for by an even number of bases, translation of the sequence would produce a protein with only one kind of amino acid. By contrast, if the bases are divided into groups with an odd number of bases, translation would produce a peptide with two alternating amino acids. For example groups of three would alternate between UCU and CUC. Indeed, proteins with two alternating amino acids were observed. Similar experiments showed that each codon is made of three bases, not another odd number of bases, like five or seven.

6.5.2 Direction of Translation

We have seen that proteins are made in the N-terminal to C-terminal direction. This was shown by isolating a mixture of ribosomes, each at a different point in the translation of an mRNA; some ribosomes had just begun translating the mRNA, others were near the end (Figure 6.21). Before restarting translation, radioactive amino acids were added to the mixture. Any piece of a peptide polymerized after addition of the amino acid was thus labeled by radioactivity.

When the proteins produced from this reaction were analyzed, it was found that they all had radioactivity at the C-terminal end, but very few had radioactivity at the N-terminal end. This is understandable if the protein is made in the N-terminal to C-terminal direction. Only the very few ribosomes that had just begun translation of the mRNA at the time the amino acid was added will have incorporated it near the N-

基突变并检查所产生的蛋白质而弄清楚的。研究发现，单碱基突变从来不会引起多于一种氨基酸的变化。这表明，每个碱基只用于编码一种氨基酸。

但是多少个碱基在一起编码一种氨基酸呢？为了回答这一问题，用人工的方法按设计好的序列合成出一些 RNA 片段，然后将它们翻译成蛋白质。例如，合成出了一条具有重复顺序 UCUCU-CUCUCU… 的 RNA 并用于翻译。如果以偶数对这一序列的碱基进行划分，则得到的各个组都是一样的。因此，如果每个氨基酸是以偶数个碱基编码的，则得到的蛋白质中只会有一种氨基酸。相反，如果碱基以奇数分组，则翻译得到的肽中会交替出现两种氨基酸。例如三个碱基一组会得到 UCU 和 CUC 交替出现的结果。实验中得到的蛋白质确实具有两种交替出现的氨基酸。类似的实验最终显示每个密码子由三个碱基组成，而不是其他的奇数如五或七。

6.5.2 翻译的方向

我们已经看到，蛋白质是以 N-端向 C-端的方向合成的。这是通过采用将核糖体混合物分离出来的方法得到证明的。核糖体混合物中的每个核糖体位于所翻译mRNA的不同位置；有些核糖体才刚刚开始翻译mRNA，有些则已接近完成（图 6.21）。在重新开始翻译之前，将放射性氨基酸加入混合物中。任何在加入了放射性氨基酸之后合成的肽片段都将带有放射性。

对经这样处理后产生的蛋白质进行分析发现大多数蛋白质的 C-末端具有放射性，而很少在 N-末端具有放射性。如果蛋白质的合成方向是从 N-末端到 C-末端，这就很好理解了。只有那些少数刚刚开始翻译mRNA的核糖体才会在靠近 N-末端的位置整合进带有放射性的氨基酸。然而，因为 C-末端是最后生

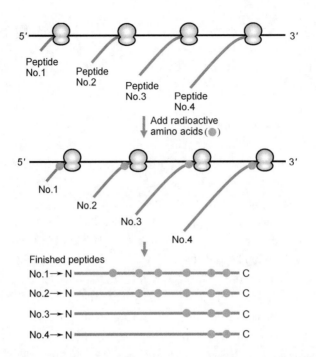

Figure 6.21　Determination of the direction of translation

terminal. However, because the C-terminal is produced last, it will contain radioactivity no matter where the ribosome was in the process of producing protein when the radioactive amino acid was added.

Summary

The base sequence of an mRNA is translated into protein at ribosomes. tRNAs pair groups of three bases, called codons, with the amino acid that they represent. There are more codons than amino acids because each amino acid can be coded for by more than one codon. Also, there are fewer tRNAs than there are codons. This is possible because the anti-codon of many tRNAs is able to recognize more than one codon. Translation begins at a start codon, and ends at one of the three stop codons, which do not code for amino acids. Start codons are generally recognized by a Shine-Dalgarno sequence in prokaryotes, and by scanning in eukaryotes. Peptides are fabricated in the N-terminal to C-terminal direction. Each new amino acid to be added to a growing peptide enters at the A-site in the ribosome. Any part of the peptide already fabricated, which is attached to a tRNA in the P-site, is detached from this tRNA and added to the amino acid in the A-site. Translation initiation, elongation, and termination in prokaryotes and eukaryotes are aided by multiple protein factors—initiation factors, elongation factors, and release factors, respectively.

起始、延伸和终止都需要多种蛋白质因子的帮助，分别包括起始因子、延伸因子和释放因子。

Vocabulary 词汇

alkylation [ˌælkɪˈleɪʃən]	烷基化（作用）	initiator tRNA	起始 tRNA
aminoacyl site (A site) [ˈæmɪnəʊˌæsɪl]	氨酰基位（A 位）	internal ribosome entry sequence (IRES)	内部核糖体进入序列
aminoacyl-tRNA synthetase	氨酰 tRNA 合成酶	N-formyl methionine [ˈfɔːmɪl] [meˈθaɪəniːn]	N-甲酰甲硫氨酸
anticodon [ˌæntɪˈkəʊdən]	反密码子	peptidyl site (P site) [ˈpeptaɪdɪl]	肽基位（P 位）
cloverleaf [ˈkləʊvəliːf]	三叶草		
decipher [dɪˈsaɪfə]	破译	peptidyl transferase	肽基转移酶
elongation factor	延伸因子	release factor	释放因子
eukaryotic initiation factor (eIF)	真核起始因子	Shine-Dalgarno sequence [ʃaɪn] [dælgɑːnəʊ]	SD 序列
exit site (E site)	退出位（E 位）		
genetic code	遗传密码	start codon [ˈkəʊdən]	起始密码子
initiation factor	起始因子	wobble [ˈwɒbl]	摇摆

Review Questions 习题

Ⅰ. True/False Questions（判断题）

1. There are three start codons in the genetic code.
2. tRNA is always connected to an amino acid.
3. Inosine is a nitrogenous base that can be present in tRNA.
4. Translation of eukaryotic mRNAs usually starts at the first AUG codon nearest to the cap structure.
5. tRNA is a T-shaped molecule in 3-dimensions.
6. Translation ends when a terminator tRNA binds to a stop codon.
7. The Shine-Dalgarno sequence is not present in eukaryotic mRNAs.
8. Ribosomes are composed of two large proteins that come together.
9. AUG codon can be a start codon and/or a codon for methionine.
10. There are at least three codons able to code for each amino acid.

Ⅱ. Multiple Choice Questions（选择题）

1. If the sequence AAUAAUGUUUUCGGCUGAGUAGCG is translated by a ribosome, how long will the peptide produced be?
   a. 24 amino acids
   b. 8 amino acids
   c. 6 amino acids
   d. 5 amino acids
   e. 4 amino acids

2. What is the role of rRNA in the ribosome?
   a. provides structure to the ribosome
   b. has enzymatic activity
   c. binds to the mRNA
   d. binds to proteins
   e. all of the above

3. An internal ribosomal entry site (IRES) is used to _____.
   a. mark the AUG nearest the CAP
   b. give structure to the Shine-Dalgarno sequence
   c. provide a translation start site other than the first AUG
   d. allow viruses to enter the cell
   e. none of the above

4. The tRNA with anticodon of 5′-AAA-3′ is changed in a cell so that it now has the anticodon 5′-CAA-3′. What will be the effect on translation?

a. It will incorporate leucine at the proper locations.

   b. It will incorporate phenylalanine in the proper locations.

   c. It will incorporate phenylalanine where it should incorporate leucine.

   d. It will incorporate leucine where it should incorporate phenylalanine.

   e. It will incorporate asparagines improperly.

5. If a Shine-Dalgarno sequence is 5'-AGGAGGU-3', what sequence might we expect to find on the 16S rRNA?

   a. 3'-AGGAGGU-5'   b. 3'-AGGAGGT-5'   c. 3'-UCCUCCA-5'

   d. 3'-TCCTCCA-5'   e. Not possible to predict.

Exploration Questions / 思考题

1. How many instances can you think of in which RNA has a function aside from coding for protein?
2. What are the various ways in which ribosomes choose the start site of translation?
3. What parts of an mRNA are not used to make protein? What function do these parts serve? Answer the question for both prokaryotes and eukaryotes.
4. If you wanted CAU to code for glutamine, what change would you make to which gene in the cell?

Chapter 7　Regulation of Gene Expression in Eukaryotes

第 7 章　真核生物基因表达调控

With very few exceptions, every cell in the human body contains exactly the same DNA. The hundreds of different cell types in our body, from nerves, to blood, to muscle, are able to use the same basic set of genes to achieve completely distinct functions and forms. This is because the ultimate determinant of a cell's properties is not so much which genes it contains, but how proteins are made from the genes—the process called gene expression. Which proteins are produced? When? How often? Where? Without understanding these subtleties, we cannot understand why a eukaryotic cell looks and acts the way it does.

We have already encountered regulation of gene ex-

除极少数例外，人体中的每个细胞都具有完全相同的 DNA。在我们身体里有几百种不同的细胞类型，从神经细胞、血细胞到肌肉细胞，它们都能使用相同的基本基因成分去实现完全不同的功能、表现出完全不同的形式。这是因为细胞的性质最终并不是由它含有什么基因决定的，而是由这些基因产生了哪些蛋白质（即基因表达过程）决定的。需要生产什么蛋白质？什么时间生产？多久生产一次？在哪儿生产？如果对这些细微之处不了解的话，我们就不能理解为什么一个真核细胞看上去是那种样子并具有那样的功能。

我们已经遇到了转录水平的基因表达调

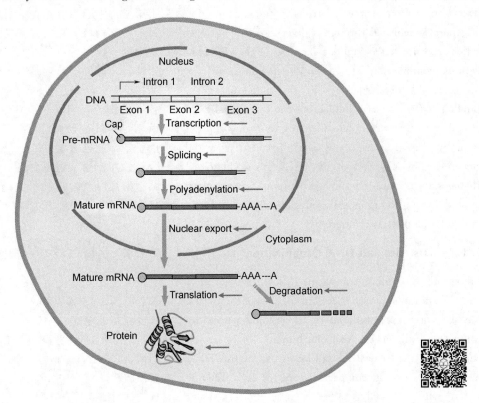

Figure 7.1　There are examples of regulation at nearly every step of protein production in eukaryotes

图 7.1　真核生物蛋白质生产的每个步骤几乎都存在调控的实例

pression at the level of transcription, notably in chapters 3 and 4. Essentially all regulation of gene expression in prokaryotes occurs at the level of transcription. In eukaryotes, the multiple steps between transcription and translation, such as post-transcriptional modifications and mRNA export, allow many more opportunities for regulation. Eukaryotic cells take advantage of these opportunities to exert very fine control over gene expression. Indeed, almost every step of the central dogma has been found to be regulated (Figure 7.1).

7.1 Histones and Transcriptional Regulation

Transcription is the first step of protein production, and also the most important level of regulation—in eukaryotes as well as prokaryotes. If a gene is not to be expressed, it is most economical to cut off its production at the beginning, before energy and nutrients are wasted by making an unnecessary RNA transcript. To review briefly from chapter 4, transcriptional regulation is generally achieved through a diverse set of proteins called specific transcription factors. They bind in various combinations to genes, depending on the precise regulatory sequences at the individual gene. Binding of the transcription factors can cause a range of modulation, from complete silencing to strong expression. Specific transcription factors frequently act by changing the association of DNA with histones. **Histones** are proteins that live in close association with DNA, helping to organize it and playing a critical role in regulation of transcription.

7.1.1 Histones and DNA Organization

Eukaryotes are more complicated than prokaryotes, and contain much more DNA. The *E. coli* chromosome measures a couple millimeters in circumference; by contrast, if all the DNA of one human cell is stretch end-to-end, it is over 2 meters long! All of this DNA must fit into just one compartment of the cell, the nucleus, which is only 10~20 micrometers long. Proteins are used to organize DNA and to make it more compact; instead of being an overstuffed mess, the

控（第3章和第4章）。原核生物的基因表达调控基本都在转录水平进行。而真核生物在转录与翻译之间具有多个步骤（比如转录后修饰与 mRNA 转运），这为调控提供了更多的机会。真核细胞利用这些机会对基因表达调控实施了非常精确的调控。的确，现已发现中心法则的几乎每一步骤都是处于调控之下的（图7.1）。

7.1 组蛋白与转录调控

转录是蛋白质生产的第一个步骤，也是真核生物与原核生物中最重要的调控水平。如果不需要表达一个基因，那么最经济的方式是在一开始就切断它的生产线，防止消耗能量和营养成分去生产一条不需要的 RNA 转录本。简单回顾一下第4章中的内容，转录调控一般是通过一些称为特异转录因子的多种蛋白质实现的。这些蛋白质以不同的组合与基因结合，取决于单个基因有哪些准确的调控序列。转录因子的结合能产生各种各样的影响，有些引起基因的完全沉默，有些则强烈地促进表达。特异转录因子常常通过改变 DNA 与组蛋白的联合而发挥作用。**组蛋白**是与 DNA 紧密结合的蛋白质，它们帮助组织 DNA 并且在转录调控中起着关键的作用。

7.1.1 组蛋白与 DNA 组织

真核生物要比原核生物复杂得多，它们含有更多的 DNA。大肠杆菌的染色体周长也就是几毫米；相反，如果一个人类细胞的所有 DNA 伸展后头尾相接连起来的话，它足足有 2m 长！所有这些 DNA 必须被装进细胞中一个称为细胞核的区域中，而细胞核直径只有 10~20μm。蛋白质被用来组织 DNA 以便使它的结构更紧密；细胞核并不是一个填得满满的混

nucleus is a fairly orderly place.

On the most basic level, DNA is packaged somewhat like thread around a spool. Histone proteins form a core around which a DNA molecule can be wrapped (Figure 7.2). There are four histones that form this core. They are H2A, H2B, H3 and H4. These are collectively called the **core histones**. Histones are basic proteins, meaning that they have many positive charges under normal conditions. These positive charges allow DNA, which is negatively charged at its phosphodiester bonds, to stick to the protein.

在最基础的水平，DNA 的包装有点像是细线缠绕在线轴上。组蛋白形成了一个核心，它们的外面缠绕着 DNA 分子（图 7.2）。有四种组蛋白参与形成这一核心，它们是 H2A、H2B、H3 和 H4。它们一起被称为**核心组蛋白**。组蛋白是碱性蛋白，意味着在正常条件下它们带有许多正电荷。DNA 上的磷酸二酯键带有负电荷，因此这些正电荷有助于 DNA 黏附到组蛋白上。

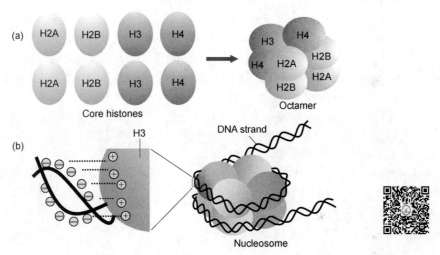

Figure 7.2 (a) Two of each core histone come together to form an octamer; (b) The negative charges on DNA are attracted to the positive charges on the histones. DNA is wrapped around a core of histones to form a nucleosome

图 7.2 (a) 每种核心组蛋白各两个一起形成八聚体；(b) DNA 上的负电荷被吸引到组蛋白的正电荷上。DNA 缠绕在组蛋白核心上形成核小体

For organizing DNA, two of each kind of core histone come together to form an octamer (a collection of 8 subunits). When DNA is wrapped around this octamer the structure is called a **nucleosome**. Approximately 150 base pairs of DNA are associated with one nucleosome. This is almost two wraps of the double-stranded molecule around the protein core. Histones are spaced fairly regularly along the DNA, with approximately 60 bases of 'linker DNA' connecting each of the nucleosomes (Figure 7.3). This simple arrangement is called the **beads-on-a-string model**.

为了组织 DNA，每种核心组蛋白各两个聚集到一起形成八聚体（8 个亚基的组合）。DNA 缠绕到这个八聚体上形成的结构称为**核小体**。一个核小体中大约有 150 个碱基对的 DNA，这些双链分子几乎环绕了组蛋白核心两圈。组蛋白在 DNA 上的间距相当有规律，每个核小体由约 60 个碱基对长的"连接 DNA"串接在一起（图 7.3）。这种简单的排列方式称为**线珠模型**。

DNA in the beads-on-a-string form is still not very

以线珠方式存在的 DNA 还不是很紧密。

7.1 Histones and Transcriptional Regulation | 147

Figure 7.3 The beads-on-a-string level of DNA organization

图 7.3 DNA 组织的线珠结构水平

condensed. In the next higher level of organization, nucleosomes pack together and are coiled into a tube-like arrangement, or solenoid. This is called the **30nm fiber.** Formation of this more organized structure is aided by another histone called H1. H1 binds outside of the nucleosome core, and helps to bring some of the linker DNA into closer association with the core histones (Figure 7.4). The 30nm fibers can be subject to even more folding, especially during cell division. They can be bent into a structure of repeating loops, and this looped structure can be further twisted into a large helix (Figure 7.5).

在下一级更高水平的组织中，核小体进行进一步包装，卷曲成像螺线管那样的管状结构，称为 **30nm 纤维**。形成这种组织性更好的结构需要另外一种称为 H1 的组蛋白的帮助。H1 结合在核小体核心的外面，帮助将连接 DNA 的某些部分拉到靠近核心组蛋白的位置使它们发生更紧密的结合（图 7.4）。特别是在细胞分裂过程中，30nm 纤维可以发生进一步的折叠。它们可以弯曲形成一种重复的环状结构，这种环状结构可以进一步卷曲形成更大的螺旋结构（图 7.5）。

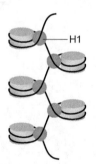

Figure 7.4 H1 histone helps to package nucleosomes more tightly

图 7.4 组蛋白 H1 帮助将核小体包装得更紧密

Eukaryotic DNA in association with protein is called chromatin. There are two kinds of chromatin in the cell: **euchromatin** and **heterochromatin**. Heterochromatin is more tightly condensed than euchromatin (Figure 7.6). Its formation requires extra proteins in addition to the histones. Heterochromatin is generally found in regions of the chromosomes that have structural importance but do not contain many genes. For example, heterochromatin is present at the centromere and

与蛋白质结合在一起的真核 DNA 称为染色质。在细胞中有两种染色质：**常染色质**和**异染色质**。异染色质比常染色质更紧密（图 7.6）。它的形成需要除组蛋白以外其他蛋白质的参与。异染色质通常出现在对维持染色体的结构起重要作用的区域，这些区域含有的基因数很少。例如，异染色质出现在着丝粒和端粒处，分别在染色体的中部和尾部较有

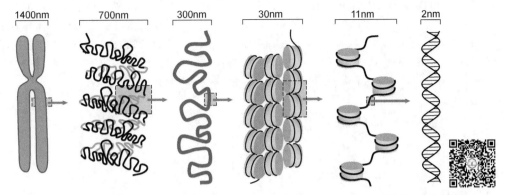

Figure 7.5 Levels of DNA organization

图 7.5 DNA 组织的不同水平

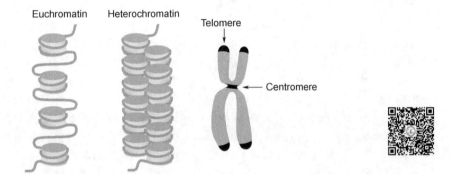

Figure 7.6 Heterochromatin is more tightly packaged than euchromatin. Both kinds of chromatin are associated with more protein than shown in this figure. Location of telomeres and centromere, regions of heterochromatin in a chromosome, is indicated

图 7.6 异染色质比常染色质包装更紧密。两种染色质实际上与比图中显示的更多的蛋白质发生联合。图中显示了异染色质区域、端粒和着丝粒在染色体上的位置

telomere, rigid regions in the middle and ends of the chromosome, respectively. The few genes that are found within heterochromatin are rarely expressed. By contrast, euchromatin contains the vast majority of genes that are transcribed. Most genes in euchromatin are still tightly associated with nucleosomes.

7.1.2 Histones and Transcription

Even in euchromatic regions not all genes are transcribed. As we learned in previous chapters, different genes are expressed at different levels depending on how much of the gene product is required at a given time. Nucleosomes play a key part in such transcriptional regulation.

Generally, nucleosomes have a repressive effect on

刚性的区域。在异染色质中发现的少数基因很少会表达。相反,常染色质含有许多基因,绝大多数基因能得到转录。常染色质中的大多数基因仍然与核小体紧密结合。

7.1.2 组蛋白与转录

即使在常染色质区域也不是所有的基因都会被转录。如我们在前面章节里学到的那样,不同的基因以不同的水平进行表达,表达水平取决于在给定的时间里需要有多少基因产物。在这样的转录调控中核小体起着主要作用。

一般来说,核小体对转录有阻遏效应。

transcription. Genes associated with a normal density of nucleosomes are usually transcribed weakly. This is because histones block access of the general transcription factors and RNA polymerases to the DNA. This is especially true when histones are in contact with promoter regions (Figure 7.7).

位于正常密度核小体区域的基因通常转录较弱。这是因为核小体妨碍了通用转录因子和 RNA 聚合酶接近 DNA。这在组蛋白与启动子区域保持接触的情况下尤其如此（图 7.7）。

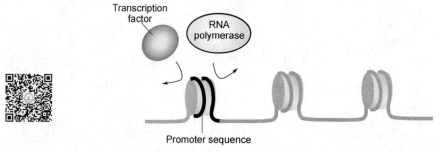

Figure 7.7　DNA associated with nucleosomes is difficult to transcribe

图 7.7　与核小体联合在一起的 DNA 很难转录

Histones can be altered at particular genes to allow transcription to occur at a higher rate. It is important to emphasize that histones are not just simple spools around which DNA wraps. Their structures are quite sophisticated，and can have different physical properties and different binding partners under different conditions.

在一些特殊基因处的组蛋白可以被改变而使转录以更高的速率发生。需要强调的是，组蛋白并不仅仅是提供 DNA 缠绕的简单轴心。它们的结构相当精巧，在不同条件下会具有不同的物理性质和不同的结合对象。

The fundamental structure of each histone is three α helices connected by two short loops（Figure 7.8）.

每种组蛋白的基本结构是由两个短环连接起来的三个 α 螺旋（图 7.8）。N-端

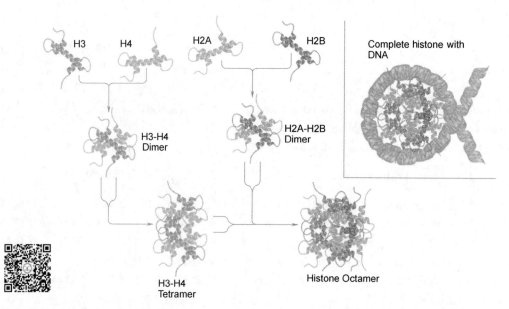

Figure 7.8　Detailed structure of histones showing histone tails (Source：http://www.wikipedia.org)

图 7.8　组蛋白的详细结构（显示组蛋白尾）（来源：http://www.wikipedia.org)

Extending from the N-terminal helix are long unstructured regions that, unlike a helix or a sheet, do not have a regular shape. These are called the **histone tails**. Histone tails branch out from the histones and closely associate with DNA and other nucleosomes. Because of these associations, histones can have important effects on chromatin organization.

7.1.3 Covalent Modification of Histones

Histone tails are fundamental to histone regulation because some of their amino acid residues can be altered. Chemical groups such as acetyl groups, methyl groups, and phosphates can be covalently attached to various places on histone tails (Figure 7.9). This kind of covalent modification has profound effects on the properties of the histone, and consequently on the structure of chromatin and the rate of transcription. Histones provide a particularly rich example of how a cell can regulate the activity of proteins using post-translational modifications.

螺旋的延伸段是长的非结构化区域, 不同于螺旋或折叠结构, 它没有固定形状。它们被称为**组蛋白尾**。组蛋白尾从组蛋白中向外突出并与DNA和其他核小体紧密结合。由于存在这样的结合, 组蛋白对染色质的组织就能施加重要影响。

7.1.3 组蛋白的共价修饰

组蛋白尾对组蛋白发挥调控作用是十分重要的, 因为它上面的一些氨基酸残基可以发生改变。乙酰基、甲基和磷酸等化学基团可以被共价连接到组蛋白尾的不同位置上 (图7.9)。这种共价修饰对组蛋白的性质来说具有深刻影响, 也由此而使染色质的结构和转录速率发生改变。关于细胞如何应用翻译后修饰作用来调控蛋白质的活性, 组蛋白提供了一个特别丰富的实例。

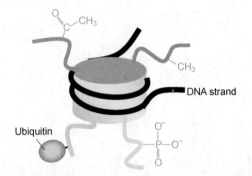

Figure 7.9　Common covalent modifications to histone tails

图7.9　常见的组蛋白尾共价修饰

One kind of covalent modification of histones that can increase transcription is lysine **acetylation**, the addition of an acetyl group to lysine residues. Recall that histones associate tightly with DNA in part because they have many positive charges, which attract the negative charges on DNA. Many of these charges are due to lysine, a positively charged amino acid that is common in the histone tail.

When an acetyl group is added to the lysine R group, the amino acid loses its positive charge and becomes neutral (Figure 7.10). As a result, DNA becomes less attracted to the histone, and therefore less tightly

一类能够增强转录的组蛋白共价修饰作用称为赖氨酸**乙酰化**, 即在赖氨酸残基上加上一个乙酰基团。回想一下, 组蛋白与DNA之所以发生紧密结合, 部分原因就是它们带有许多正电荷, 正是这些正电荷吸引了DNA上的负电荷。而这些正电荷大多数来源于赖氨酸, 赖氨酸是在组蛋白尾中普遍存在的带正电荷的氨基酸。

当乙酰基加到赖氨酸的R基团上时, 赖氨酸失去正电荷从而变成中性 (图7.10)。结果, 组蛋白对DNA的吸引力下降, 因此相互之间的结合也变得不

bound. This ultimately facilitates transcription by giving the transcriptional machinery easier access to the DNA. Acetylated lysines can also allow specific transcription factors to bind at a gene.

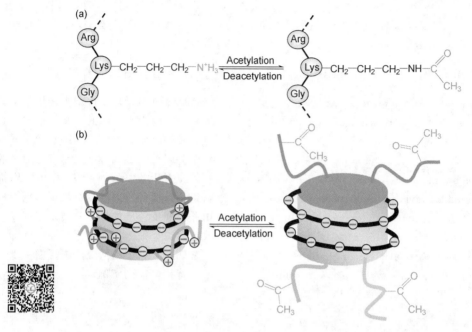

Figure 7.10 Acetylation. (a) Acetylation removes positive charge on lysine; (b) Lysine acetylation loosens association of DNA with histone tails, facilitating transcription

Generally, the effect of histone covalent modification on transcription is not so straightforward. One type of modification can often have different effects depending on which amino acid is modified. For example, addition of a methyl group to H3-K9 (the 9th lysine residue of histone H3) usually represses transcription. By contrast, methylation of H3-K4 usually activates transcription.

Covalent modifications also have different effects depending on whether other modifications are present. For example, the combination of H3-H10 phosphorylation with H4-K8 and H3-K14 acetylation activates transcription much more than any one of these modifications alone. The functional combinations of modifications at a histone are called the **histone code**. Sometimes these modifications have a direct effect on the function of the nucleosome, as with lysine acetylation. However, more often, proteins that recognize histone modifications are required for interpreting the meaning of each combination of modifications and carrying-out its effect.

7.1.4 Proteins that Recognize and Modify Histones

One class of proteins that responds to the histone code is **chromatin remodeling proteins** (Figure 7.11). The best known member of this group is the protein complex called **Swi/Snf**. These proteins use energy from ATP to change the interaction of DNA with the nucleosome, and allow the nucleosome to be moved to a more desirable location. Although Swi/Snf usually activates transcription, it is also capable of repressing transcription, depending on its cellular context.

7.1.4 识别和修饰组蛋白的蛋白质

来识别这些修饰作用以弄清不同组合修饰的含义并将修饰所造成的影响表现出来。

能对组蛋白密码发生响应的一类蛋白质是**染色质重塑蛋白**（图 7.11）。这类蛋白质中最有名的一员是称为 **Swi/Snf** 的蛋白质复合体。这些蛋白质使用 ATP 作为能量来改变 DNA 与核小体的相互作用，从而将核小体移到更合适的位置。虽然 Swi/Snf 通常激活转录，但在一定的细胞环境中它也能阻遏转录。

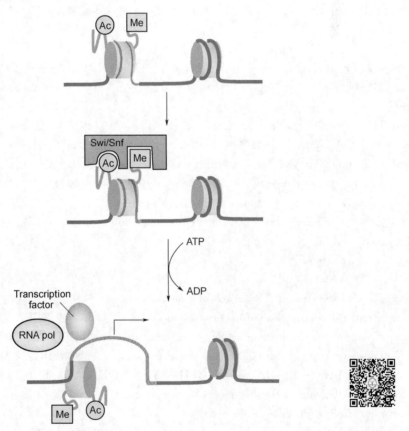

Figure 7.11 Histone remodeling complex Swi/Snf is attracted to genes by specific modifications to histone tails at the gene

图 7.11 组蛋白重塑复合体 Swi/Snf 被在基因的组蛋白尾上发生的特殊修饰吸引了过来

Histone modifications may also help general and specific transcription factors to bind at a promoter. For these reasons, the set of covalent modifications to his-

组蛋白修饰作用也可以帮助通用转录因子和特异转录因子结合到启动子上。正是由于这些原因，在每个基因的位置发

tones at each gene is crucial for proper transcription of the gene. Different genes require histones with very different covalent modifications. There are many kinds of enzymes that can modify histones specifically, depending on which gene they are associated with.

Histone acetyltransferases (HATs) acetylate the protein, while **histone deacetylases (HDACs)** remove acetyl groups. **Histone methyltransferases (HMTs)** are responsible for methylation of histones. Proteins called **kinases** add phosphate groups to histones, while ubiquitin ligases add small proteins called ubiquitin. Each of these groups of proteins has many members. These proteins recognize specific genes by the specific transcription factors that bind to the regulatory region of the gene.

Let us review the process and effect of histone modification. Each gene has a regulatory region to which unique sets of specific transcription factors can bind. These specific transcription factors attract various enzymes, such as HATs or HMTs that can covalently modify histones. The chemical modifications on the histones sometimes directly affect transcription, altering the histones' association with DNA. More often, the modifications attract extra proteins, like the chromatin remodeling complexes that activate or repress transcription.

A classic example for this sequence of events is the regulation of **INF-β gene** (Figure 7.12). This gene has a group of enhancers which attract a set of activator proteins. These activator proteins bind two more proteins, a kinase and an HAT. The HAT acetylates H4-K8, H3-K9 and H3-K14 of histones at the gene. The kinase phosphorylates H3-S10. The chromatin remodeling complex Swi/Snf can bind at the nucleosomes of the gene only after these modifications to the histones have taken place. Swi/Snf then changes the position of DNA relative to the nucleosomes. This makes the TATA box more accessible, so that the transcription factor TFⅡD can bind to the promoter. Finally, the pre-initiation complex can form and transcription can begin.

生的组蛋白共价修饰对这一基因的转录就有着至关重要的影响。不同的基因需要在它们的组蛋白上发生非常不同的共价修饰。有许多种酶能特异性地修饰组蛋白,这取决于这些组蛋白与哪个基因结合在一起。

组蛋白乙酰基转移酶（HAT）使组蛋白乙酰化,而**组蛋白脱乙酰基转移酶（HDAC）**将乙酰基移走。**组蛋白甲基转移酶（HMT）**负责组蛋白的甲基化。称为**激酶**的蛋白质为组蛋白加上磷酸基团,而泛素连接酶将称为泛素的小蛋白加到组蛋白上。每一类这样的蛋白质都有很多成员。它们通过结合到基因调控区域的特异转录因子而识别特殊的基因。

我们来回顾一下组蛋白修饰的过程和影响。每个基因都具有一个调控区域,它可以结合几套独特的特异转录因子。这些特异转录因子吸引不同的酶（如HAT或HMT）而对组蛋白进行共价修饰。组蛋白共价修饰有时会改变组蛋白与DNA的结合状态从而直接影响转录。更多时候则是这些修饰通过吸引其他蛋白（如染色质重塑复合体）而进一步激活或阻遏转录。

这一系列事件的经典实例是**INF-β基因**的调控（图7.12）。这一基因具有一组增强子,它们能吸引一套激活蛋白。这些激活蛋白又与另两种蛋白（激酶和HAT）相结合。HAT乙酰化该基因所处位置组蛋白的H4-K8、H3-K9和H3-K14,激酶磷酸化组蛋白的H3-S10。染色质重塑复合体Swi/Snf只有在组蛋白已得到上述修饰的情况下才会结合到基因的核小体上。之后,Swi/Snf改变DNA与核小体的相对位置,使TATA框更容易接近,这样转录因子TFⅡD就能结合到启动子上。最终,前起始复合体得以形成并启动转录。

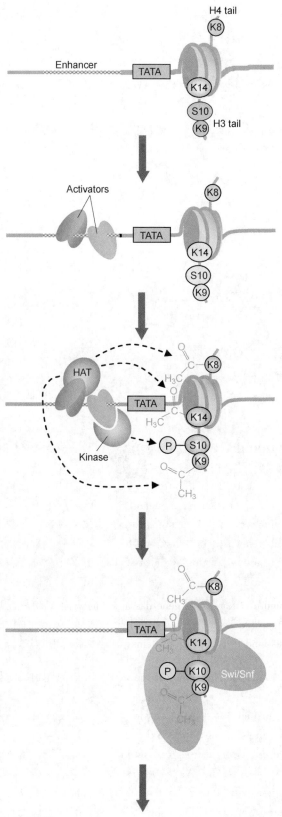

Figure 7.12 图 7.12

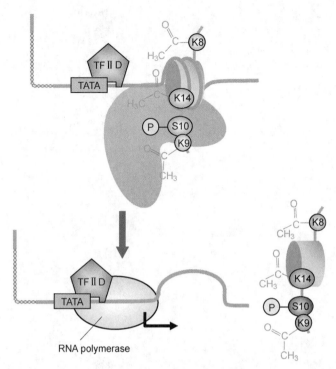

Figure 7.12 Regulation of INF-β gene, classic example of histone modification and chromatin remodeling

图 7.12 INF-β 基因的调控——组蛋白修饰和染色质重塑的经典实例

7.2 Post-Transcriptional Regulation

We have already discussed the use of alternative splicing to make more than one protein from just one coding region. The cell often regulates this step, enhancing the expression of one protein over another. Several examples of alternative splicing were given in chapter 5. Here we give another example that shows particularly well how the cell uses alternative splicing to choose a correct program of gene expression given a specific genetic signal.

Sex determination in *Drosophila* depends on regulation of alternative splicing (Figure 7.13). Flies that express a functional **Sxl protein** become females, others become male. In simple splicing of the Sxl premRNA, an exon containing a premature stop codon is included in the middle of the gene. As a result, only part of the coding region will be translated, and the Sxl protein will be non-functional. These flies become male.

7.2 转录后调控

我们已经讨论过应用可变剪接从一个编码区产生不止一种蛋白质的实例。细胞常常对这一步骤进行调控，借此增强一个又一个蛋白质的表达。第 5 章已经给出了几个可变剪接的实例。在此我们给出另一个例子，它特别能说明细胞是如何根据特殊遗传信号而应用可变剪接去选择基因表达的正确程序的。

果蝇的性别决定依赖于对可变剪接的调控（图 7.13）。表达有功能 **Sxl 蛋白**的果蝇变成雌蝇，其他的变成雄蝇。在对 Sxl 前体 mRNA 的简单剪接中，一个带有提早终止密码子的外显子被留在了基因中部。结果，只有部分编码区被翻译出来，产生没有功能的 Sxl 蛋白。这些果蝇变成雄蝇。

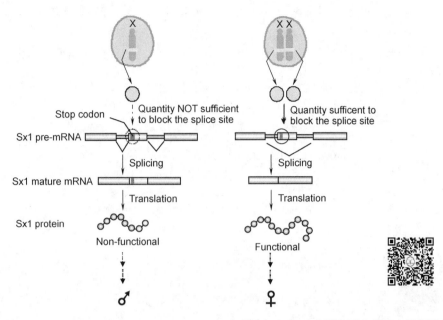

Figure 7.13 Alternative splicing of the *Sxl* gene is used for sex determination in *Drosophila*

图 7.13 *Sxl* 基因的可变剪接在果蝇性别决定中的应用

In females, a gene on the X-chromosome is expressed in sufficient quantities that it can alter the splicing of Sxl pre-mRNA (Females have two X-chromosomes, while males only have one). The protein covers the splice site on the 5′ end of the exon that contains a premature stop codon. As a result, the exon is removed along with its surrounding introns. The Sxl mRNA can then be translated to produce a full length, functional Sxl protein, and the fly becomes female. Thus, alternative splicing can be used to regulate the expression of a gene. Sxl itself goes on to regulate splicing of other proteins required for sex determination, but we will not explain the details of this mechanism here.

Polyadenylation, another level of post-transcriptional modification, can also be regulated. Recall that polyadenylation begins with cleavage of the pre-mRNA near a specific sequence, followed by addition of a poly(A) tail. Many pre-mRNAs contain more than one possible site of cleavage. In fact, more than half of genes in humans have such variable polyadenyaltion sites! **Alternative polyadenylation** can change the length of the coding region or of the 3′ untranslated region (Figure 7.14). In the former case, proteins of different func-

在雌蝇中，X 染色体上的一个基因表达出足够的蛋白以至于能够改变对 Sxl 前体 mRNA 的剪接（雌蝇有两条 X 染色体，而雄蝇只有一条）。这一蛋白质盖住了那个含有提早终止密码子的外显子 5′端。结果，这一外显子与它附近的内含子一起被剪接掉。之后 Sxl mRNA 就能被翻译产生全长的有功能的 Sxl 蛋白，果蝇成为雌蝇。因此，可变剪接可以用于调控基因的表达。Sxl 蛋白自身又会去调控其他在性别决定中起作用的蛋白的剪接，其详细机理我们就不在这儿解释了。

聚腺苷酸化是另一水平的转录后修饰作用，它也能被调控。回忆一下，聚腺苷酸化开始于在一段特殊的序列附近切断前体 mRNA，紧接着加上一条 poly(A) 尾。许多前体 mRNA 含有多个可能的切割位点。事实上，一半以上的人类基因都具有这种可变的聚腺苷酸化位点！**可变聚腺苷酸化**能够改变编码区或 3′非编码区的长度（图 7.14）。前者能产生具有不同功能的蛋白质，后者能够导

tionality can be generated. The latter case can result in altered expression of a protein, as the 3'-UTR often contains sequences important for mRNA stability and translation efficiency.

致蛋白质表达情况的改变，因为 3'-UTR 常常含有对维持 mRNA 稳定性和提高翻译效率起重要作用的序列。

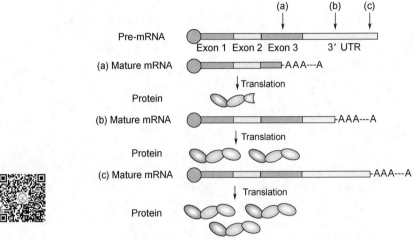

Figure 7.14 (a) Alternative polyadenylation can cause variation in protein size if alternative cleavage site occurs within an exon; (b) and (c) If alternative cleavage site occurs within UTR, variation in mRNA processing and translation efficiency can result (Longer UTRs do not necessarily correlate with increased protein production)

图 7.14 (a) 如果可变切割位点在外显子内，可变聚腺苷酸化能够引起蛋白质大小的变化；(b) 和 (c) 如果可变切割位点在 UTR 中，会导致 mRNA 加工和翻译效率的变化（较长 UTR 不一定与增加蛋白质产量相关）

There are few examples of regulation of gene expression at the level of capping. However, as we will see, some proteins that regulate translation bind at or near the cap structure.

在加帽水平进行基因表达调控的例子较少。但是，如我们将要看到的那样，确实有一些蛋白质通过结合在帽上或结合在帽的附近而调控翻译。

7.3 Nuclear Export

7.3 细胞核输出

After mRNA is modified, it must be transported from the nucleus to the cytoplasm. The mRNA crosses the nuclear envelope through large protein complexes called **nuclear pores**. The pores can selectively choose which molecules exit the nucleus, and, therefore, which get to be expressed.

mRNA 被修饰后必须从细胞核转运到细胞质中。mRNA 从称为**核孔**的蛋白质复合体中穿过核膜。核孔能够选择性地让一些分子离开细胞核，从而让它们得到表达。

mRNA export may be regulated during exposure of cells to extreme conditions (Figure 7.15). When yeast cells are exposed to conditions such as high heat, export of most mRNAs to the cytoplasm is blocked as a way of limiting cell activity. However, mRNAs for

将细胞置于极端条件下可以对 mRNA 的转运进行调控（图 7.15）。处于高温条件下的酵母细胞为了限制细胞的活性会阻断大多数 mRNA 的转运。然而，那些编码能帮助细胞在极端条件生存下

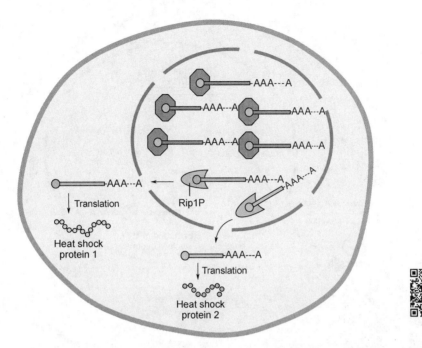

Figure 7.15 Under high heat conditions in yeast, only heat-shock protein mRNAs are allowed export through the nuclear pores. The protein Rip1P escorts heat-shock protein mRNAs to the pore

图 7.15 在高温条件下只有热休克蛋白的 mRNA 允许通过酵母细胞的核孔。Rip1P 蛋白护送热休克蛋白的 mRNA 来到核孔

proteins that help the cell survive extreme conditions, called **heat-shock proteins**, are still allowed to pass through the nuclear pore. This apparently requires a protein (Rip1P) that is activated during shock conditions and binds to heat-shock protein mRNAs to escort them to the nuclear pore.

HIV virus presents a classic example of regulated nuclear transport (Figure 7.16). After incorporating its DNA into a host's genome, the HIV virus produces a variety of mRNAs, some of which include introns with coding sequences. The nuclear pore complex generally blocks the export of mRNA with introns to the cytoplasm, as a way to prevent translation of incorrectly spliced mRNAs. In order to circumvent this blockage, the virus produces a protein called Rev that binds to sequences within the intron and allows the mRNA to pass through the pore. Although this is not an example of regulation in normal cells, it illustrates the principle that nuclear transport can be modulated to determine whether or not a protein is produced.

来的蛋白质（称为**热休克蛋白**）的 mRNA 仍然被允许通过核孔。一般来说这需要一种在休克条件下激活的蛋白质（Rip1P）来结合到热休克蛋白 mRNA 上并护送它们通过核孔。

HIV 病毒给出了一个细胞核转运调控的经典实例（图 7.16）。在将它的 DNA 整合到宿主基因组中去之后，HIV 病毒产生了好几种 mRNA，其中有些在编码序列中还带有内含子。核孔复合体一般不允许带有内含子的 mRNA 转运到细胞质中去，这是为了防止翻译那些没有经过正确剪接的 mRNA。为了躲避这样的阻拦，HIV 病毒产生一种称为 Rev 的蛋白质，Rev 蛋白能结合到内含子序列上从而使 mRNA 得以通过核孔。虽然这不是一个普通细胞中的调控实例，但它说明了一个原理，即：能够通过调整细胞核转运来决定是否生产某种蛋白质。

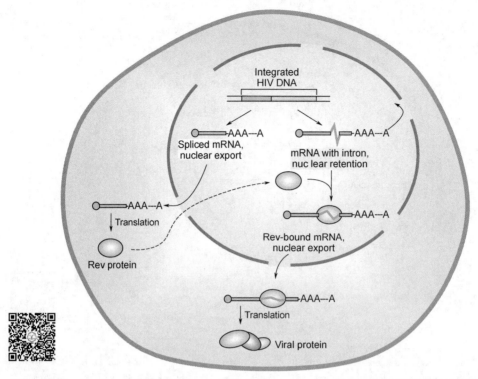

Figure 7.16 Rev protein, produced by the HIV virus, allows unspliced HIV mRNAs to be exported through nuclear pores

图 7.16 HIV 病毒产生的 Rev 蛋白能够使未剪接的 HIV mRNA 转运到核孔外

7.4 RNA Stability

How much protein is made from a gene depends in large part on the stability of its mRNA. Generally, the longer an mRNA lives, the more times it can be translated to make protein. mRNA degradation in the cytoplasm usually begins with removal of the poly(A) tail by enzymes. Following this, the mRNA cap may be removed and degradation proceeds $5' \rightarrow 3'$. Alternatively, degradation can continue in the $3' \rightarrow 5'$ direction, led by a protein complex called the **exosome** (Figure 7.17). Despite the relatively consistent styles of mRNA degradation, mRNAs in the cell experience a wide range of life spans before degradation; some are degraded within seconds, others last for hours or days.

7.4.1 mRNA Stability Regulation by Proteins

Sequence elements are usually central to the regulation of an individual mRNA's stability. Many regulated

7.4 RNA 稳定性

从一个基因生产出多少蛋白质在很大程度上取决于它的 mRNA 稳定性。一般来说，mRNA 存活的时间越长，它可以被用来翻译产生蛋白质的次数就会越多。在细胞质中 mRNA 的降解通常从酶切除 poly(A) 尾开始。接着，mRNA 帽可能被切除并且发生 $5' \rightarrow 3'$ 方向的降解。另一种可能是，降解在 $3' \rightarrow 5'$ 方向上还在继续，这由称为**外来体**的蛋白质复合体负责进行（图 7.17）。尽管 mRNA 降解具有较为一致的样式，但细胞中的 mRNA 还是会具有很不相同的存活期限；有些在几秒钟内就被降解掉，有些则可存活几小时或几天。

7.4.1 蛋白质调控 mRNA 稳定性

序列元件常常对单个 mRNA 分子稳定性的调控起着极为重要的作用。许多被

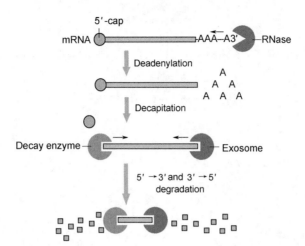

Figure 7.17 Common mechanism for mRNA degradation. Degradation usually occurs either $3'\rightarrow 5'$ or $5'\rightarrow 3'$

图 7.17 mRNA 降解的常见机理。降解通常按 $3'\rightarrow 5'$ 或 $5'\rightarrow 3'$ 方向进行

mRNAs have an **AU-rich element（ARE）** in the $3'$-UTR to which proteins bind to increase or decrease degradation. Often, these regulatory proteins can themselves be regulated. One example is the protein TTP, which binds to mRNA ARE sequences and recruits decay enzymes that increase the rate of mRNA degradation (Figure 7.18). When the cell is exposed to growth signals, a protein causes TTP to be phosphorylated. This modification results in the exclusion of TTP from sites of mRNA degradation, thus stabilizing the mRNAs and allowing the increased protein production necessary for cell growth.

调控的 mRNA 在 $3'$-UTR 具有**富含 AU 元件（ARE）**，可以结合蛋白质来增强或削弱对 mRNA 的降解。在很多情形下，这些调控蛋白自身也是在受调控之下产生的。一个实例是蛋白质 TTP，它结合到 mRNA 的 ARE 序列上并召集降解酶来提高对 mRNA 的降解速率（图 7.18）。当细胞接收到生长信号时，有一种蛋白质会使 TTP 发生磷酸化。这一修饰使 TTP 不能存在于能导致 mRNA 发生降解的位置，因此增强了 mRNA 的稳定性，使它能够为细胞生长产生更多的蛋白质。

Regulation of mRNA stability is central to iron metabolism (Figure 7.19, top row). **Transferrin** is a receptor protein that brings iron into the cell (Figure 7.20). When iron levels in the cell are low, more transferrin protein is needed. Under these conditions, a protein called **aconitase** binds to transferrin mRNA at the $3'$-UTR and stabilizes the transcript. By increasing the lifetime of the transferrin mRNA, more transferrin protein can be made. When the concentration of iron is satisfactory, iron binds to aconitase. This causes the protein to unbind from the transferrin mRNA, allowing it to be degraded. Preventing the production of more transferrin receptors keeps iron levels from rising too high.

mRNA 稳定性的调控对铁代谢也很重要（图 7.19，上半部分）。**运铁蛋白**是将铁带进细胞里的受体蛋白（图 7.20）。当细胞中铁含量低时，细胞需要更多的运铁蛋白。在这些条件下，一种称为**顺乌头酸酶**的蛋白质结合到运铁蛋白 mRNA 的 $3'$-UTR 从而稳定了这一转录产物。通过延长运铁蛋白 mRNA 的寿命便可以产生更多的运铁蛋白。当铁的含量已经满足需要时，铁会结合到顺乌头酸酶上使它从运铁蛋白 mRNA 上脱离，mRNA 便遭到降解。防止生产出过多的运铁蛋白受体能使铁含量保持在一定水平，而不会上升到过高。

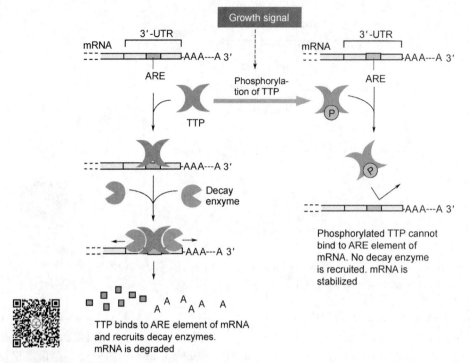

Figure 7.18 TTP protein normally binds to mRNAs with ARE sequences, and then recruits enzymes to degrade the mRNA. In presence of growth signal, TTP is phosphorylated, and no longer has access to mRNAs. As a result, mRNAs with the ARE sequence are stabilized and can be used to produce more protein

图 7.18 TTP 蛋白正常情况下结合在 mRNA 的 ARE 序列上，之后召集酶来降解 mRNA。在存在生长信号的情况下，TTP 被磷酸化，它不再能接近 mRNA。结果，具有 ARE 序列的 mRNA 稳定性提高并用来生产更多的蛋白质

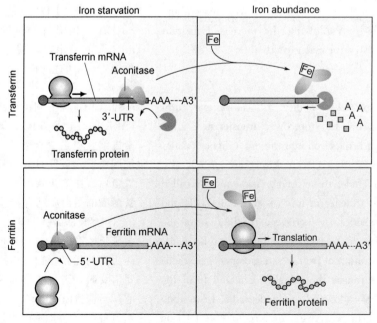

Figure 7.19 Regulation of transferrin and ferritin by aconitase under iron starvation and iron abundance

图 7.19 在铁缺少和铁丰富条件下顺乌头酸酶对运铁蛋白和铁蛋白的调控作用

162 | Chapter 7 Regulation of Gene Expression in Eukaryotes

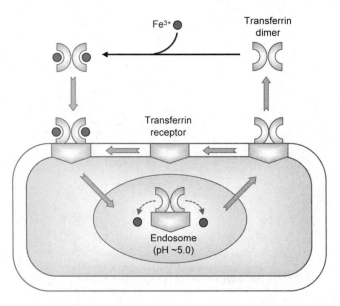

Figure 7.20 Function of transferrin

图 7.20 运铁蛋白的功能

7.4.2 mRNA Stability Regulation by Small RNAs

mRNA stability is not always regulated by proteins. In fact, the most important source of regulation appears to come from very small RNA molecules. The most common among these are **microRNAs (miRNA)** and **small interfering RNAs (siRNA)**. The former are short hairpin loops, the latter are double-stranded RNAs.

The mode of action of siRNA and miRNA is very similar (Figure 7.21). They are both incorporated into a complex of proteins called **RISC (RNAi silencing complex)**. This complex chooses one of the two strands of the small RNA and uses it to bind to target mRNAs. The concept is similar to snRNPs of the spliceosome, which use small RNA strands to bind to pre-mRNA.

If the miRNA or siRNA strand chosen by the RISC complex binds to the target mRNA with high complementarity, the target mRNA is usually destroyed by endonucleases. This process is called **slicing**, and should not be confused with splicing. The use of small RNAs to degrade an mRNA in this way is commonly known as **RNA interference (RNAi)**.

miRNAs and siRNAs may also interfere with proper

7.4.2 小 RNA 调控 mRNA 稳定性

mRNA 稳定性并不总是由蛋白质调控的。事实上，最重要的调控物质看来是那些很小的 RNA 分子。其中最常见的有微小 RNA（miRNA）和小干涉 RNA（siRNA）。前者是短的发夹环，后者是双链 RNA。

siRNA 和 miRNA 起作用的模式非常相似（图 7.21）。它们两者都会被整合到一种称为 RISC（RNAi 沉默复合体）的蛋白复合体中。这一复合体选用小 RNA 两条链中的一条并用它来与目标 mRNA 结合。这一概念类似于剪接体的 snRNP 使用其中的 RNA 短链与前体 mRNA 结合。

如果 RISC 选中的 miRNA 或 siRNA 能与目标 mRNA 结合并且互补性很高，那么目标 mRNA 通常会被内切核酸酶摧毁。这一过程叫作**切片**（slicing），请不要与剪接（splicing）混淆。小 RNA 以这种方式应用于降解 mRNA 的过程称为 RNA 干涉（RNAi）。

miRNA 和 siRNA 还有可能干涉 mRNA

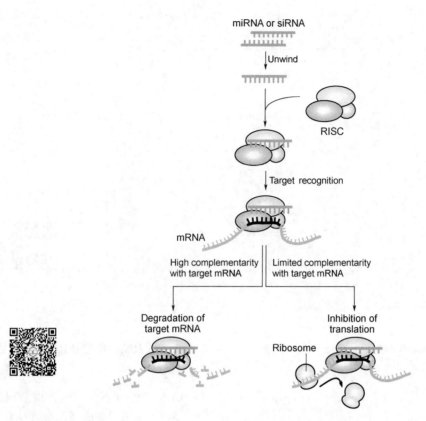

Figure 7.21　siRNA and miRNA can control mRNA stability and translation efficiency

图 7.21　siRNA 和 miRNA 能控制 mRNA 稳定性和翻译效率

translation of an mRNA. This generally occurs when the small RNAs bind to their targets with limited complementarity. The RISC complex probably represses translation by one of two mechanisms: the complex may interfere with proteins required for initiation of translation, or it may recruit proteins that sequester the mRNA.

Although miRNAs and siRNAs function through similar mechanisms, their origins and targets are quite distinct. miRNA is derived from cellular genes, which express large RNA transcripts from which miRNA is cut out. miRNAs then fold onto themselves to produce the small hairpin structures. The hairpins are then further processed by RNase Ⅲ enzymes called **Drosha** and **Dicer**, before being incorporated into the RISC complex (Figure 7.22). Generally, miRNAs are used to control the expression of a variety of genes unrelated to themselves. By contrast, siRNAs are usually not encoded as distinct genes; they derive from processing of protein-coding mRNAs by Dicer, without involvement by Drosha

分子的正常翻译。这通常在小 RNA 与它们的目标 mRNA 结合并且互补性较低的情况下发生。RISC 复合体可能通过下述两种机理中的一种阻遏翻译：干涉翻译起始所需要的蛋白质，或者召集一些蛋白质来隔离 mRNA。

虽然 miRNA 和 siRNA 通过类似的机理发挥作用，但它们的起源与用途却相当不同。miRNA 源自细胞基因，是从细胞基因转录出的 RNA 长链中切割出来的。切割出来之后才对自身进行折叠而产生小的发夹结构。在被整合到 RISC 复合体中之前，这些发夹结构将由Ⅲ型 RNase 酶 **Drosha** 和 **Dicer** 进行进一步加工（图 7.22）。一般来说，miRNA 用于控制一些与它们自身无关的基因的表达。相反，siRNA 通常不是由不同的基因编码而来的；它们来源于 Dicer 酶对编码蛋白质的 mRNA 的加工，这当

(Figure 7.23). They may also be introduced into the cell through viruses. They usually are used to control the expression of genes that are closely related to them in sequence and in origin.

中没有 Drosha 酶的参与（图 7.23）。它们也可能通过病毒带进细胞。它们通常用来控制与它们在序列和起源上密切相关的基因的表达。

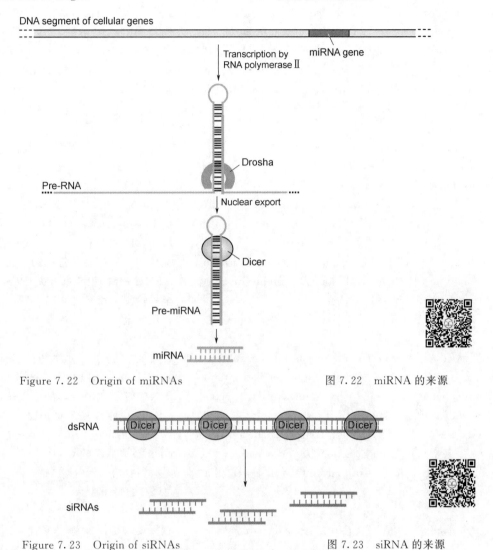

Figure 7.22　Origin of miRNAs

图 7.22　miRNA 的来源

Figure 7.23　Origin of siRNAs

图 7.23　siRNA 的来源

7.5　Translational Control

7.5　翻译调控

Translation is a critical level of control in gene expression, and its regulation goes far beyond small RNAs. Although it may seem wasteful to regulate protein production at the last step, this does carry certain advantages. Control at the level of translation produces quick changes in protein production—immediate halts, and swift resumptions. Indeed we know that translation

翻译是控制基因表达的一个关键水平，它的调控远远超出了小 RNA 涉及的范围。虽然看起来对蛋白质生产的最后一步进行调控有些浪费，但它确实有一定的优点。在翻译水平实施的控制能在蛋白质生产中产生很快的变化——立即停止和迅速恢复。我们确实也知道翻译肯定是

must be regulated because the concentration of various mRNAs in the cell does not match perfectly with concentrations of proteins produced from them.

Translational control may be divided into two broad categories, **global control** and **mRNA-specific control**. The latter is used to control the translation of individual mRNAs. By contrast, global control is used to regulate the translation of all mRNAs in the cell indiscriminately.

7.5.1 Global Control

Global control is important, for instance, when a cell is under low nutrient conditions and needs to decrease the overall production of proteins. There are also examples of viruses causing global repression of host mRNA translation, to give an advantage to its own mRNAs, which initiate translation via a distinct mechanism (see IRES, chapter 6). Global control is usually enacted by regulation of proteins involved in translation initiation.

eIF2 is a eukaryotic initiation factor briefly encountered in chapter 6. It is responsible for bringing initiator tRNA to the 40S initiation complex. eIF2 function depends on the breakdown of the small molecule GTP to GDP. After this hydrolysis, GDP must be replaced with a new GTP. This function is performed by the protein eIF2B, which is present in the cell in very limited quantities.

Under conditions that require a global reduction in translation, eIF2 may become phosphorylated by signaling pathways (Figure 7.24). Phosphorylated eIF2 binds very tightly to eIF2B, preventing the eIF2B from functioning at other sites of translation. Even small quantities of phosphorylated eIF2 can sequester all the eIF2B in the cell, globally slowing translation.

Another initiation factor, eIF4E can also mediate global control of translation (Figure 7.25). eIF4E is a cap-binding protein that is required for binding of an mRNA transcript to the ribosome. In order to function properly, eIF4E must associate with another initiation

受到调控的，因为细胞中不同 mRNA 的浓度与从它们那儿生产出的蛋白质的量并不相称。

翻译调控可以被分成两大范畴：**全局控制**和 **mRNA 特异性控制**。后者用来控制单一 mRNA 的翻译。相反，全局控制是对细胞中所有 mRNA 实施不加区别的调控。

7.5.1 全局控制

举例来说，当细胞处于低营养条件之下、需要降低蛋白质的总体生产时，全局控制就显得很重要。也有病毒引起宿主 mRNA 翻译全局阻遏的例子。由于病毒自身 mRNA 的翻译以不同的机理启动，因此对宿主 mRNA 的翻译实施这样的全局控制非常有利于病毒自身的 mRNA（参阅第 6 章有关 IRES 的内容）。全局控制一般通过调控在翻译起始中起作用的蛋白质而发挥效用。

eIF2 是我们在第 6 章里简要地涉及的真核生物起始因子。它负责将起始 tRNA 带到 40S 起始复合体的位置。eIF2 的功能依赖于将小分子 GTP 分解成 GDP 的反应。在这一水解反应之后，GDP 必须被新的 GTP 替代。这一功能是由 eIF2B 蛋白实施的，在细胞中 eIF2B 的量很少。

在需要对翻译实施全局性减弱的时候，eIF2 可以被某些信号途径磷酸化（图7.24）。磷酸化的 eIF2 能与 eIF2B 非常紧密地结合在一起，防止了 eIF2B 在翻译的其他位置上发挥作用。甚至很少量的磷酸化 eIF2 就能隔离细胞中的所有 eIF2B，从而全局性地减慢翻译速率。

另一个起始因子，eIF4E，也能够促成翻译的全局控制（图 7.25）。eIF4E 是一种帽结合蛋白，需要用它来将 mRNA 转录产物结合到核糖体上去。为了正常发挥作用，eIF4E 必须与另一个起始因子

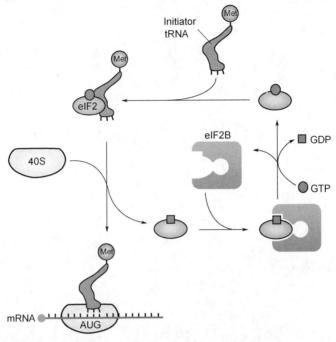

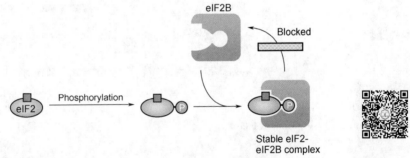

Figure 7.24 (a) Function of eIF2B in translation initiation; (b) Global control through phosphorylation of eIF2

图 7.24 (a) eIF2B 在翻译起始中的作用；(b) 通过对 eIF2 的磷酸化实现的全局控制

factor, eIF4G. However, another class of proteins called 4E-binding proteins (4E-BPs) can bind to eIF4E at the same domain and block binding of eIF4G to eIF4E. 4E-BPs only bind to eIF4E effectively when they are not highly phosphorylated. Thus, when a cell requires higher levels of translation, 4E-BPs are well-phosphorylated. When a cell requires lower levels of translation, 4E-BPs are dephosphorylated.

7.5.2 mRNA-Specific Control

Mechanisms of translation regulation also exist to target

eIF4G 结合。然而，另一类称为 4E 结合蛋白（4E-BP）的蛋白质也能结合到 eIF4E 的同一个结构域上，从而阻断 eIF4G 与 eIF4E 的结合。4E-BP 只有在没有被高度磷酸化的情况下才能与 eIF4E 有效地结合。因此，当细胞需要高水平翻译时，4E-BP 的磷酸化程度就高。当细胞需要低水平翻译时，4E-BP 就会被去磷酸化。

7.5.2 mRNA 特异性控制

翻译调控机理也存在于对特殊的目标

7.5 Translational Control | 167

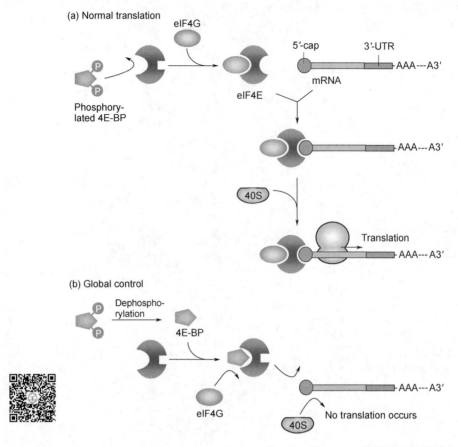

Figure 7.25 Example of global control of translation. (a) Normal translation initiation; (b) Global control, 4E-BP blocks binding site of eIF4G on eIF4E, preventing proper function of these initiation factors

图 7.25 翻译的全局控制实例。(a) 正常的翻译起始；(b) 全局控制，4E-BP 阻断了 eIF4G 在 eIF4E 上的结合位点，阻止了这些起始因子发挥正常功能

specific mRNAs or groups of mRNAs, rather than all transcripts. This is known as mRNA-specific control, and it depends very much on sequences within individual mRNA transcripts. We have already seen examples of this in the previous section, which discussed small RNAs, especially miRNA, and their effect on translation. Other mechanisms for regulating specific mRNA translation exist.

RNAs with a uracil-rich region called a CPE, for example, can be regulated by a mechanism very similar to the one just described for global control (Figure 7.26). The CPE region binds to a protein, CPEB. CPEB subsequently attracts a protein called Maskin. Maskin acts like a 4E-BP, binding to eIF4E and pre-

mRNA 分子或目标 mRNA 群的调控，而不仅仅用于调控所有的转录产物。这就是 mRNA 特异性控制，它在很大程度上依赖于特定 mRNA 转录产物内部的序列。在上一节中我们已经看到了一些例子，其中我们讨论了小 RNA（特别是 miRNA）和它们对翻译的影响。除此以外，还有其他的机理可以对特异 mRNA 的翻译实施调控。

例如，富含尿嘧啶区域的 RNA 称为 CPE，它可以被一种我们刚刚描述过的与全局控制非常相似的机理所调控（图 7.26）。CPE 区域与 CPEB 蛋白结合，CPEB 紧接着吸引一种称为掩蔽蛋白的蛋白质。掩蔽蛋白的作用类似于 4E-BP，它能与

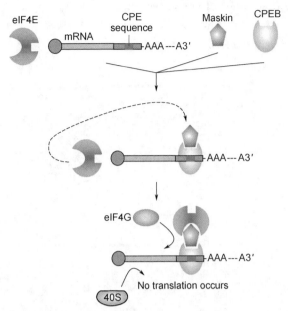

Figure 7.26 Example of mRNA-specific regulation of translation. CPEB attracts Maskin, which blocks binding of eIF4G to eIF4E, preventing proper function of these initiation factors. CPEB only binds to mRNAs with a CPE sequence

图 7.26 mRNA 特异性翻译调控实例。CPEB 吸引掩蔽蛋白，掩蔽蛋白阻止了 eIF4G 与 eIF4E 的结合，防止这些起始因子发挥正常作用。CPEB 只能与具有 CPE 序列的 mRNA 结合

venting its association with eIF4G. As a consequence, many mRNAs with the CPE sequence have greater difficulty forming a complex with a ribosome, and less protein is made from these mRNAs. In other instances, presence of a CPE can activate translation of a transcript. This occurs because when CPEB binds to certain CPE sequences, it attracts polyadenylation factors that extend the poly(A) tail, which promotes translation and stability of the transcript.

Another example of mRNA specific control relates to iron metabolism (Figure 7.19, bottom row). When iron levels in the cell are high, **ferritin** is used to bind and sequester iron from the cytoplasm (Figure 7.27). When iron levels are low, the activity of ferritin is counterproductive. As in the regulation of transferrin, another iron-related protein discussed above, aconitase is used to regulate ferritin levels. However, regulation of ferritin occurs at the level of translation, not mRNA stability. When iron concentration is low, aconitase binds to the 5'-UTR of ferritin mRNA and inhibits translation,

eIF4E 结合从而防止 eIF4E 与 eIF4G 的结合。结果，许多具有 CPE 序列的 mRNA 与核糖体形成复合体的难度大大增加，从这些 mRNA 产生的蛋白质自然也就少了。在其他例子中，CPE 的出现能够激活转录产物的翻译，这是因为当 CPEB 结合到相应的 CPE 上后会吸引聚腺苷酸化因子，从而延长 poly(A) 尾。poly(A) 长尾有利于转录产物的翻译和提高其稳定性。

另一个 mRNA 特异性控制的例子与铁代谢有关（图 7.19，下半部分）。当细胞中铁的含量高时，**铁蛋白**被用来从细胞质中结合并隔离铁（图 7.27）。当铁含量低时，铁蛋白的作用正好相反。与上面讨论过的另一个铁相关蛋白（运铁蛋白）的调控一样，顺乌头酸酶也用于调控铁蛋白的水平。然而，铁蛋白的调控发生在翻译水平，而不是 mRNA 稳定性的水平。当铁浓度低时，顺乌头酸酶结合到铁蛋白 mRNA 的 5'-UTR 并

physically preventing the association of mRNA with ribosomal subunits. By contrast, when iron is abundant, aconitase does not bind to ferritin mRNA and translation proceeds normally.

抑制翻译，从物理空间上防止 mRNA 与核糖体亚基的结合。相反，当铁含量丰富时，顺乌头酸酶不与铁蛋白 mRNA 结合，翻译正常进行。

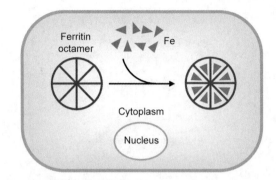

Figure 7.27　Function of ferritin

图 7.27　铁蛋白的功能

To review the importance of aconitase for iron metabolism: when iron is scarce, aconitase binds to ferritin and transferrin mRNAs, blocking ferritin translation but stabilizing transferrin mRNAs. The result is that ferritin levels fall and transferrin levels rise. When iron is plentiful, aconitase unbinds from the mRNAs. Ferritin translation may now proceed normally but transferrin mRNA is quickly degraded. The result is that ferritin levels rise, while transferrin levels fall.

总结一下顺乌头酸酶对铁代谢的重要性：当铁元素稀少时，顺乌头酸酶结合到铁蛋白和运铁蛋白的 mRNA 上，阻断铁蛋白的翻译但增强运铁蛋白 mRNA 的稳定性。结果是铁蛋白水平下降而运铁蛋白水平上升。当铁元素丰富时，顺乌头酸酶从两种 mRNA 上脱离。铁蛋白的翻译现在可以正常进行但运铁蛋白 mRNA 被快速降解。结果是铁蛋白水平上升而运铁蛋白水平下降。

7.6　mRNA Localization

7.6　mRNA 定位

We have now seen that nearly every level of protein production can be controlled to determine whether or not a protein is made, and in what quantity. Regulatory mechanisms also exist to control where proteins are made. Cells are not uniform mixtures of molecules. Some proteins are needed in certain parts of the cell, and some must be absent from certain parts. This is especially true for morphologically complex cells like neurons, which can be very long and consist of distinct regions.

我们已经看到，可以对蛋白质生产的几乎每个水平实施控制来决定是否生产某一种蛋白质以及生产多少。调控机理还能用来决定在哪儿生产蛋白质。细胞并不是均匀的分子混合物。某些蛋白质需要在细胞的某些部位出现，而有些则一定不能出现在那里。在像神经元这样的形态极为复杂的细胞中这种情况更为明显，神经元细胞可以很长并由互不相同的区域组成。

Many proteins contain signals that cause them to localize to specific areas after translation. However, many other proteins do not contain such signals. Their localization depends to a great extent on where in the cell their mRNA is translated.

许多蛋白质都带有信号，使它们在翻译后被定位到特定的区域中去。然而，还有许多蛋白质并不含有这样的信号。它们的定位在很大程度上依赖于其 mRNA 在细胞中进行翻译的位置。

mRNA localization is achieved through various mechanisms (Figure 7.28). An mRNA may contain sequences that cause it to be degraded in some regions of the cell but stabilized in others. Alternatively, some regions of the cell may contain proteins that trap specific mRNAs so they are preferentially translated in that location. Most commonly, mRNAs are actively transported by motor proteins to sites where the proteins they code for are needed.

mRNA 定位通过多种不同的机理实现（图 7.28）。一条 mRNA 可能含有会让它在细胞中的一些区域中被降解掉而在其他区域保持稳定的序列。另一种情况是，细胞中的某些区域可能具有能捕捉特异 mRNA 的蛋白质，因此这些 mRNA 优先在那个地点翻译。更常见的情况是，mRNA 由动力蛋白主动运送到需要它们所编码蛋白质的特定位置。

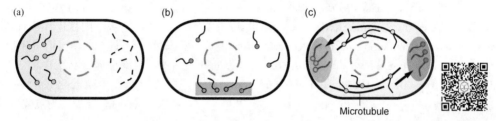

Figure 7.28 Three mechanisms for localizing an mRNA at specific site. (a) Degradation of the mRNA everywhere except at the site; (b) Trapping of the mRNA at the site; (c) Active transport of the mRNA to the site

图 7.28 mRNA 定位到特定位置的三种机理。(a) 除了特定位置外其他地点的 mRNA 都发生降解；(b) 在特定位置捕捉 mRNA；(c) mRNA 向特定位置的主动转运

An example of the latter mechanism occurs in fibroblasts (Figure 7.29). These cells are able to move due to structural changes in the cytoskeleton. The moving edge of a fibroblast depends on high concentrations of

后一种机理的实例发生在成纤维细胞中（图 7.29）。在细胞骨架中发生的结构变化使这些细胞能够移动。成纤维细胞的移动头依赖于高浓度的 β-肌动蛋白，

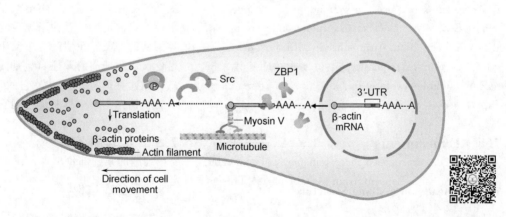

Figure 7.29 Example of regulation of mRNA localization. β-actin is specifically transported to the motile edge of fibroblasts. Its translation is prevented by ZBP1 before arrival at the edge. Src protein removes ZBP1 from β-actin mRNA when it reaches the moving edge. This acts to concentrate β-actin protein where it is needed most

图 7.29 mRNA 定位调控实例。β-肌动蛋白被特异性地转运到成纤维细胞的移动头中。在到达移动头前 ZBP1 防止发生翻译。Src 蛋白在 β-肌动蛋白 mRNA 到达移动头后将 ZBP1 移走。通过这样的动作将 β-肌动蛋白集中到最需要的地方

β-actin, the principle component of the cytoskeletal fibers that move the cell. To meet this demand, motile fibroblasts specifically localize β-actin mRNA to the moving edge of the cell.

β-actin mRNA contains a short sequence in the 3′-UTR. In fibroblasts, a protein called ZBP1 is expressed that can bind to this sequence. ZBP1 prevents translation of β-actin mRNA as it is transported, by a motor protein, from the nucleus to the moving edge of the fibroblast. Once the β-actin mRNA has reached the moving edge, it encounters a high concentration of a protein called Src. Src phosphorylates ZBP1, causing it to detach from the 3′-UTR, and allowing translation to occur in the desired location.

7.7 Protein Regulation

If we really want to understand how genetic information is transformed into a functional group of proteins, we also have to discuss the fate of proteins after translation. Many mechanisms exist in the cell to modify the function of proteins. We have seen examples of this throughout the book; in this chapter alone we have seen that phosphorylation of eIF2 affects eIF2B function, and that eIF4E function is repressed by binding of another protein, 4E-BP. We have also seen numerous examples of allosteric regulation. Although we will not devote a special section to regulation of protein function, the reader should be aware of this kind of regulation and its importance as we encounter it throughout the book.

7.8 Experiments

7.8.1 Beads-on-a-string Structure

The beads-on-a-string organization of DNA around histones can be seen directly using **electron microscopy**. Normal microscopes use amplification of light to visualize small structures. But light can only be used to clearly observe structures that are larger than its wavelength, and macromolecules like DNA are much smaller than the wavelength of light. To overcome this limitation, electron microscopes use electron beams instead of light rays.

β-肌动蛋白是细胞运动骨架纤维的主要成分。为了满足这一需要，移动性成纤维细胞将 β-肌动蛋白 mRNA 特异性地定位到细胞的移动头上。

β-肌动蛋白在 3′-UTR 中含有一段短序列。在成纤维细胞中，称为 ZBP1 的蛋白质被表达并结合到这一序列上。ZBP1 在 β-肌动蛋白 mRNA 的运输途中防止它被翻译。运输是由动力蛋白完成的，运输的方向是从细胞核运往成纤维细胞的移动头。一旦 β-肌动蛋白 mRNA 到达移动头，它就会遇到高浓度的 Src 蛋白。Src 使 ZBP1 发生磷酸化并从 3′-UTR 脱离，这样就能在所需要的地点发生翻译。

7.7 蛋白质调控

如果我们想要真正理解遗传信息是如何被转化成有功能的蛋白质集群的话，我们还必须讨论蛋白质在翻译后的命运。细胞中存在着多种修饰蛋白质功能的机理。我们在本书中看到了不少这样的例子；单是在本章我们就已经看到了 eIF2 的磷酸化影响 eIF2B 的功能，还有 eIF4E 的功能被另一个蛋白 4E-BP 的结合所阻遏。我们也已经看到了无数的别构调控的例子。虽然我们没有单独用一节来讨论蛋白质功能的调控，但是读者应该意识到这种调控作用以及它的重要性，因为我们在本书中不断地遇到它。

7.8 实验研究

7.8.1 线珠结构

应用**电子显微术**能够直接看到 DNA 绕在组蛋白外面的线珠组织形态。普通显微镜使用光的放大作用来看清微小的结构。但光只能用来清楚地观察大于光的波长的物体，而像 DNA 这样的大分子要比光的波长小得多。为了克服这一局限，电子显微镜采用电子束而不是光线。电子具有很短的波长，因此可以用

Electrons have a very small wavelength, and therefore can be used to visualize much smaller structures. The structure of most proteins still cannot be detected by electron microscopes, but large complexes, like DNA wrapped around a core of histones, can be seen.

The beads-on-a-string structure was also detected by exposing chromatin to **deoxyribonuclease (DNase)** enzymes that randomly cut DNA (Figure 7.30). The chromatin was then put through gel electrophoresis, a technique that separates DNA fragments according to size. Many pieces of DNA approximately 150 base pairs in size were observed. This could be explained if the DNA were wrapped into nucleosomes as in the beads-on-a-string model. According to the model, each nucleosome contains approximately 150 bp of

来对更微小的结构成像。大多数蛋白质还不能用电子显微镜进行检测，但是像DNA缠绕在组蛋白外面形成的大型复合体可以用电子显微镜进行观察。

线珠结构也能通过将染色质暴露在**脱氧核糖核酸酶（DNase）**进行检测，DNase能随机地切断DNA链（图7.30）。将经过DNase处理的染色质进行凝胶电泳，凝胶电泳是一种能根据分子大小将DNA片段分开的技术。从电泳结果中可以看到许多大小约150碱基对长的片段。如果DNA是以线珠模型缠绕成核小体结构的话，这一现象就解释得通了。根据这种模型，每个核小体含有紧密地缠绕

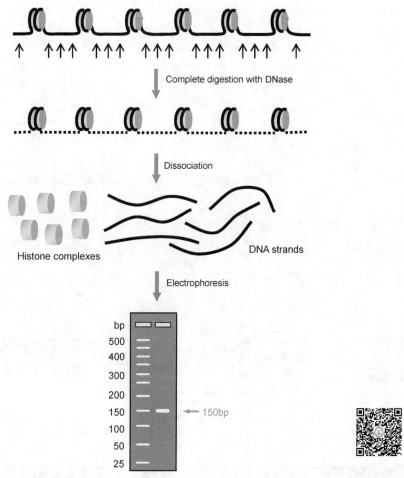

Figure 7.30 Demonstration of bead-on-a-string structure with DNase treatment

图7.30 用DNase处理证实线珠结构

DNA wrapped tightly around a core of proteins. These stretches of DNA would be protected from the DNase by their association with histones. Cutting would occur at the stretches of DNA between nucleosomes, which are not associated with histones and are therefore vulnerable. When the DNA is put through a gel, the only DNA observed would be the 150 bp long pieces that were protected by nucleosomes.

7.8.2 Repression of Gene Expression in Heterochromatin

The lack of transcription in regions of heterochromatin was easily demonstrated in yeast. When yeast cells do not transcribe the gene *ADE2* they become red. Normal yeast colonies are white and always transcribe the gene. When the *ADE2* is moved from its normal location in euchromatin to the tip of a chromosome, which is packaged as heterochromatin, yeast colonies turn red (Figure 7.31). This means that the *ADE2* gene is not transcribed in heterochromatic regions. There are many other examples of genes that are not transcribed when moved to regions of heterochromatin.

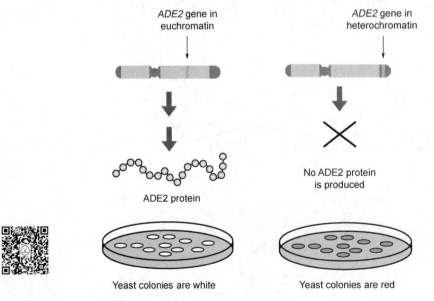

Figure 7.31 Expression of yeast *ADE2* gene is repressed in heterochromatin

Summary

The large number of eukaryotic genes and the existence

of multicellularity, among other factors, require eukaryotes to have a particularly fine ability to regulate gene expression. Much regulation of gene expression occurs at transcription. One of the important classes of DNA-binding proteins in the eukaryotic cells is histones. These proteins were originally recognized for their ability to organize DNA, but histones also play a fundamental role in regulating transcription of genes. Histone tails can be covalently modified, altering the association of DNA with the histones, and allowing specific genes to be recognized by transcription factors. Unlike prokaryotes, eukaryotes also regulate gene expression at many other steps in protein production. Alternative splicing and alternative polyadenylation can be controlled by the cell to produce one version of a gene over another. Regulation of nuclear export and mRNA stability can be used to determine whether the mRNA ever gets to be translated, and for how long. mRNAs can be localized to specific regions of the cell to ensure that proteins produced are appropriately localized for their functions. Finally, translational control can globally or specifically alter the production of protein.

性，再加上一些其他原因，真核生物需要具备特别精确的基因表达调控能力。许多基因表达调控在转录水平发生。组蛋白是真核生物中重要的 DNA 结合蛋白之一。最初发现这些蛋白质具有组织 DNA 的能力，后来证实它们在调控基因的转录方面也有着十分重要的作用。组蛋白尾可以被共价修饰，改变它们与 DNA 的联合状况，从而使特定的基因能够被转录因子识别。与原核生物不同，真核生物还在蛋白质生产过程的许多其他步骤中调控基因的表达。细胞可以利用可变剪接和可变聚腺苷酸化来选择使用基因的某个版本或另一个版本。细胞核转运和 mRNA 稳定性调控能够用来决定 mRNA 是否应该翻译以及翻译多长时间。mRNA 可以被定位到细胞的特定区域以确保产生出的蛋白质为行使它的功能已到达了适宜的位置。最后，翻译控制还能全局性地或特异性地改变蛋白质的生产水平。

Vocabulary 词汇

acetylation [əˌsetiˈleiʃən]	乙酰化（作用）
acetyl group [ˈæsitil]	乙酰基
aconitase [əˈkɔniteis]	顺乌头酸酶
alternative polyadenylation [ˌpɔliˈədenileiʃən]	可变聚腺苷酸化
AU-rich element	富含 AU 元件
centromere [ˈsentrəˌmiə]	着丝粒
chromatin remodeling protein [ˈkrəumətin]	染色质重塑蛋白
core histone	核心组蛋白
cytoplasmic polyadenylation element (CPE)	细胞质聚腺苷酸化元件
deacetylation [diəˌsitiˈleiʃən]	脱乙酰化（作用）
deoxyribonuclease (DNase) [diːˈɔksiˌraibəuˈnjuːklieis]	脱氧核糖核酸酶
Dicer [ˈdaisə]	切丁酶
euchromatin [juːˈkrəumətin]	常染色质
exosome [ˈeksəsəum]	外来体
ferritin [ˈferitən]	铁蛋白
fibroblast [ˈfaibrəublɑːst]	成纤维细胞
global control	全局控制
heat-shock protein	热休克蛋白
heterochromatin [ˌhetərəuˈkrəumətin]	异染色质
histone acetyltransferase (HAT)	组蛋白乙酰基转移酶
histone [ˈhistəun]	组蛋白
histone code	组蛋白密码
histone deacetylase (HDAC)	组蛋白脱乙酰基酶
histone methyltransferase (HMT)	组蛋白甲基转移酶
histone tail	组蛋白尾
interferon-β (INF-β) [ˌintəˈfiərɔn]	β 干扰素
kinase [ˈkineis]	激酶
ligase [liˈgeis]	连接酶
Maskin [ˈmɑːskin]	掩蔽蛋白
methylation [ˌmeθiˈleiʃən]	甲基化（作用）
methyl group [ˈmeθil]	甲基
microRNA (miRNA)	微小 RNA
mRNA-specific control	mRNA 特异性控制
nuclear export	核输出
nuclear pore	核孔

nucleosome ['njuːklɪəsəum]	核小体	sex determination	性别决定
octamer ['ɔktəmə]	八聚体	small interfering RNA (siRNA)	小干涉 RNA
phosphorylation [ˌfɔsɪfɔri'leiʃən]	磷酸化（作用）	solenoid ['səulinɔid]	螺线管
		telomere ['tiːləmiə]	端粒
post-translational modification	翻译后修饰	transferrin [træns'ferin]	运铁蛋白
RNA interference (RNAi)	RNA 干涉	ubiquitin [ˌjubi'kwitin]	泛素（蛋白）
RNAi silencing complex (RISC)	RNAi 沉默复合体		

Review Questions   习题

Ⅰ. **True/False Questions**（判断题）

1. DNA associated with a histone is more difficult to digest with nucleases.
2. A nucleosome is composed of 8 different kinds of histone.
3. There are 150 nucleosomes on each chromosome.
4. Methyl groups can be added to the tails of histones.
5. Histones have an overall positive charge.
6. miRNA can cause mRNA to be degraded，but has no effect on translation.
7. eIF2B is present in relatively small concentrations in the cell.
8. Highly phosphorylated 4E-BP blocks translation.
9. When iron levels in the cell are low，cells need more transferrin.
10. Polyadenylation is commonly regulated in prokaryotes.

Ⅱ. **Multiple Choice Questions**（选择题）

1. Methylation of a histone tail _____ .
   a. represses transcription
   b. activates transcription
   c. has no effect on transcription
   d. alters the structure of DNA
   e. cannot say without more information

2. Which of the following is not true of heterochromatin?
   a. It is found at telomeres.
   b. It is found at centromeres.
   c. It contains histones.
   d. It contains few transcribed genes.
   e. None of the above.

3. What is true of miRNA and siRNA?
   a. They usually target the same mRNAs.
   b. They both work in conjunction with the RISC complex.
   c. They both are processed by Drosha.
   d. They are coded for the same genes.
   e. They are present in prokaryotes.

4. Iron levels in a cell become very low. What is likely to happen?
   a. Translation of transferrin mRNA will decrease.
   b. Ferritin mRNA will be degraded.
   c. Transferrin mRNA will be degraded.
   d. Aconitase will detach from transferrin mRNA.
   e. None of the above.

5. If you add a CPE sequence to an mRNA，what is likely to happen?
   a. It will be transcribed less frequently.
   b. It will not be spliced correctly.
   c. It will not be able to exit the nucleus.
   d. It will be degraded more quickly.
   e. It will be translated less efficiently.

Exploration Questions 思考题

1. What are the two major functions of histones? How does the structure of histones make them so well suited for these two functions?
2. Histone structure is very conserved between different eukaryotes. Is this surprising? Why or why not?
3. What are some of the differences between siRNA and miRNA?
4. By looking at the DNA sequence of a protein, what aspects of the regulation of its expression might we be able to predict?

Chapter 8　DNA Replication

Until this point we have largely considered the flow of genetic information from genes to mRNA to protein. Genetic information is also passed in another direction: from generation to generation. A crucial property of life is that even the most complex organisms are able to reproduce. When one of your cells divides by mitosis, for example, it makes two new cells that are essentially perfect replicas of the parent cell. This property is astounding, and totally unique within nature. Yet life's seemingly miraculous ability to replicate derives directly from the properties of one simple inanimate molecule, DNA. DNA's remarkable structure provides a very convenient basis for its own replication. In this chapter we examine how DNA replicates, allowing the replication of life itself.

8.1　Semi-Conservative Replication

Recall from chapter 2 that DNA in the cell is double-stranded. Each strand is a perfect complement of the other strand. Where there is C in strand one, for example, there must be G in strand two. The important consequence of this property is that if you separate the two strands, each one holds all the information for how to make its complementary strand. We say that each strand is a template for regenerating a complete double helix (Figure 8.1).

DNA replication in cells proceeds roughly along these lines. The two strands of the helix are separated, and complementary nucleotides are bound to each individual strand. The nucleotides are also covalently joined to each other so that they form a strand. Thus each original strand becomes paired to a completely new strand. This style of replication is called **semi-conservative replication**; each new double stranded molecule contains one new DNA strand, but conserves one old strand that was used as a template.

第8章　DNA复制

直到现在为止，我们考虑的主要是遗传信息从基因到mRNA再到蛋白质的流动。实际上遗传信息也在另一个方向上进行着传递：从一个世代到另一个世代。生命至关重要的特性是：即使是最复杂的生命也能够被繁殖出来。例如，当你的细胞进行有丝分裂时，它能产生两个本质上是母细胞完美复制品的新细胞。这一特性让人叹为观止，在自然界中也完全是独一无二的。而生命看起来极为神奇的复制能力却来自一种简单而单调的分子所具有的特性，它就是DNA。DNA不寻常的结构为它自身的复制提供了一个非常便利的基础。本章我们审视DNA是如何进行复制的，正是DNA复制使生命的繁衍成为可能。

8.1　半保留复制

回忆第2章中讲到的，细胞中DNA是双链的。每条链都是另一条链的完美互补链。例如，如果一条链的某个位置是C，那么另一条链的对应位置一定是G。这一性质的重要作用是，如果两条链被分开，那么每条链都保留着如何产生它的互补链的所有信息。因此我们说每条链都是产生完整双链的模板（图8.1）。

细胞DNA的复制基本上是按照图8.1中的线路进行的。螺旋的两条链被分开，之后互补的核苷酸结合到每条单链上。核苷酸彼此之间也以共价键相连接，所以它们形成的是一条链。这样，每条原始链与一条全新的链配成了一对。这种复制类型称为**半保留复制**；每个新的双链分子中都含有一条DNA新链，但还保留了一条用作模板的老链。

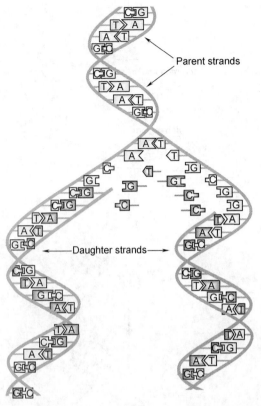

Figure 8.1 Semi-conservative replication. Each double-stranded DNA after replication conserves one parent strand but adds one new daughter strand

图 8.1 半保留复制。复制后每条双链 DNA 保留了一条母链但加上了一条新的子链

From the above description, it may seem that DNA replication is quite a simple process. In reality, DNA replication is complicated because none of the steps just mentioned occur either quickly or spontaneously without the help of proteins. Some of the proteins required contain intrinsic restrictions that greatly complicate the mechanism of replication. Let us now examine DNA replication as it occurs in *E. coli*.

从上面的叙述看，DNA 复制似乎是一个相当简单的过程。实际上，DNA 复制是相当复杂的，因为刚才提到的那些步骤如果没有蛋白质帮助的话则没有一个步骤是能够快速或自发地进行的。其中的一些蛋白质还具有内在的限制，这又大大增加了复制机理的复杂性。让我们来探究发生在大肠杆菌中的 DNA 复制过程。

8.2 Initiation of Replication

8.2 复制的起始

The start site of replication is often called the **origin of replication**. *E. coli* replication begins at a site on the chromosome called the ***OriC*** (Figure 8.2). 'Ori' refers to the word 'origin' and 'C' refers to chromosome. The *OriC* possesses a distinct sequence structure, including a sequence of 9 nucleotides (9-mer) with consensus sequence TTATCCACA repeated four times.

复制的起始位点常被称为**复制起点**。大肠杆菌 DNA 复制开始于染色体上称为 ***OriC*** 的位点（图 8.2）。"Ori"指单词"origin"，"C"指单词 chromosome。*OriC* 拥有独特的序列结构，包括一段由 9 个核苷酸组成（9-mer）并重复四次的序列，9-mer 的共有序列是 TTATCCACA。

There is also a 13-mer repeated three times that is rich in A and T bases.

OriC 还具有一段由富含碱基 A 和 T 的 13-mer 重复三次的序列。

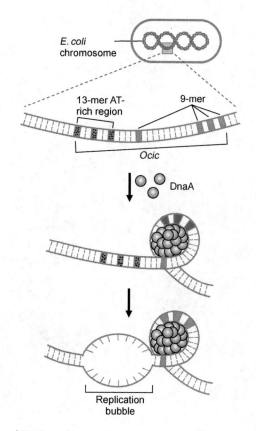

Figure 8.2   Initiation of DNA replication in *E. coli* at the *OriC*

图 8.2   大肠杆菌 DNA 复制在 *OriC* 位置的起始

At the beginning of replication, a protein called **DnaA** binds to the 9-mers, causing a loop in the DNA. As a result, the double stranded DNA at the 13-mers separates into individual strands. This separation is facilitated by the fact that the 13-mers have many A-T base pairs. A-T base pairs are more easily broken because they are held together by only two hydrogen bonds, as opposed to three for G-C base pairs. Strand separation provides exposure for each strand to be used as template, and space for the replication proteins to function. This space is called the **replication bubble**.

Replication proceeds in both directions from the origin of replication. This is called **bidirectional replication** (Figure 8.3). The enzyme DNA polymerase runs along each of the single-stranded templates, matching complementary nucleotides and joining them together to

在复制开始的时候，一种称为 **DnaA** 的蛋白质结合到 9-mer 结构上，使 DNA 形成一个环。结果，双链 DNA 在 13-mer 的区域分开成为单链。因为事实上 13-mer 区域含有许多 A-T 碱基对，所以这种分离比较容易发生。A-T 碱基对更容易被打开，因为使它们保持在一起的氢键只有两个，而不是 G-C 碱基对中的三个。链的分离使每条链暴露出来作为复制的模板，也为即将参与复制的蛋白质发挥作用提供了空间。这种空间叫作**复制泡**。

复制从复制起点开始朝着两个方向进行，这叫作**双向复制**（图 8.3）。DNA 聚合酶沿着每条单链模板移动，将互补的核苷酸加上并将它们连接在一起产生新的双链 DNA。在复制进行的时候，

generate new double-stranded DNA. As replication proceeds, the replication bubble must grow to gradually expose more and more of the DNA template.

复制泡必须逐步扩大以暴露出越来越多的 DNA 模板。

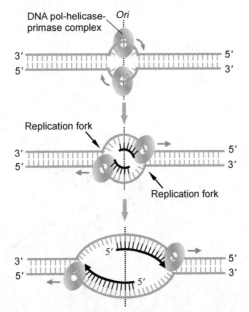

Figure 8.3　Bidirectional replication. Two replication forks move in opposite directions from the origin of replication

图 8.3　双向复制。两个复制叉从复制起点开始沿相反方向移动

There are two centers of polymerization, one at each corner of the growing bubble. Each center is called a **replication fork** because at these sites the double-stranded DNA diverges into two, separate single strands. To simplify our discussion of replication, we will focus on the mechanism of replication at the fork, rather than in the bubble as a whole.

在生长中的复制泡的两个角落分别有两个聚合反应中心。每个中心称为一个**复制叉**，因为在这些位置双链 DNA 分叉成了两条分开的单链。为了简化对复制的讨论，我们将把目光集中在复制叉位置发生的复制机理，而不是对整个复制泡的复制机理进行描述。

8.3　Semi-Discontinuous Replication

8.3　半不连续复制

There are two DNA polymerases at each replication fork, one to synthesize each strand. In reality, the two proteins are physically connected as part of a large polymerizing complex, but for now imagine that they can move independently of each other. As the DNA polymerases moves along each template strand, pairing complementary nucleotides and synthesizing DNA, two major factors govern the mechanism of replication.

在每个复制叉处有两个 DNA 聚合酶，各自合成自己的新链。事实上，这两个蛋白质在空间上是结合在一起的，它们共同形成一个大的聚合反应复合体，但现在请想象它们能够互相独立地移动。在 DNA 聚合酶沿着各自的模板链移动、匹配互补核苷酸和合成 DNA 的时候，有两个因素支配着复制的机理。

First, DNA polymerase is only able to create new

首先，DNA 聚合酶只能沿 $5' \rightarrow 3'$ 方向合

strands in the $5'\rightarrow3'$ direction (Figure 8.4). Recall that double stranded DNA is always anti-parallel. If one strand runs $5'\rightarrow3'$ with respect to a fixed position, the complementary strand must run $3'\rightarrow5'$. A DNA polymerase creating a $5'\rightarrow3'$ strand must be running along a template in the $3'\rightarrow5'$ direction if the final product is to be anti-parallel.

成新链（图 8.4）。回忆一下，双链 DNA 总是反向平行的。如果一条链相对于一个固定的位置来说是 $5'\rightarrow3'$ 的走向，那么它的互补链必定是 $3'\rightarrow5'$ 的走向。因此，如果终产物也是反向平行的话，那么一个以 $5'\rightarrow3'$ 方向合成新链的 DNA 聚合酶必须沿着模板按 $3'\rightarrow5'$ 的方向移动。

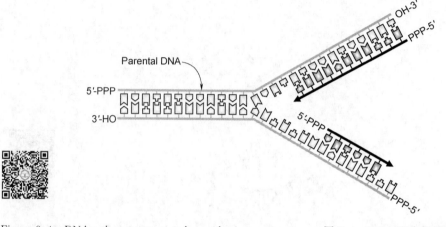

Figure 8.4　DNA polymerase can only synthesize new DNA in the $5'\rightarrow3'$ direction. For the double-stranded DNA to be anti-parallel, it must move down the template DNA in $3'\rightarrow5'$ direction

图 8.4　DNA 聚合酶只能以 $5'\rightarrow3'$ 方向合成新的 DNA。为了得到反向平行的双链 DNA，它必须在模板 DNA 上沿 $3'\rightarrow5'$ 方向移动

A second crucial factor governing the replication mechanism is that the replication fork is constantly widening (Figure 8.5). As replication proceeds, more of the template strands must be exposed. In order to do so, the two strands of the

第二个支配复制机理的重要因素是：复制叉在不断地扩大（图 8.5）。在复制进行之中，必须暴露出更多的模板链。为了达成此目的，原先 DNA 分子的两

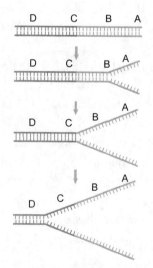

Figure 8.5　Each replication fork must gradually grow wider to expose more template DNA

图 8.5　每个复制叉必须逐渐扩大以暴露出更多的模板 DNA

original DNA molecule are increasingly pulled apart, and the vertex of the fork moves.

The combination of these two factors necessitates a style of replication called **semi-discontinuous replication** (Figure 8.6). On the template strand that has a 3′ end behind the moving fork, DNA polymerase can synthesize DNA continuously. It creates a new strand 5′→3′ that begins at the origin of replication and continues in the same direction as the moving replication fork.

条链被拉得越来越开,复制叉的顶点也随之移动。

这两个因素的综合作用需要复制采取一种称为**半不连续复制**的方式进行(图8.6)。对那条在移动叉后面是3′端的模板链来说,DNA 聚合酶能够连续地合成 DNA。它以 5′→3′方向合成新链,新链合成从复制起点开始并沿着与复制叉移动的相同方向持续进行。

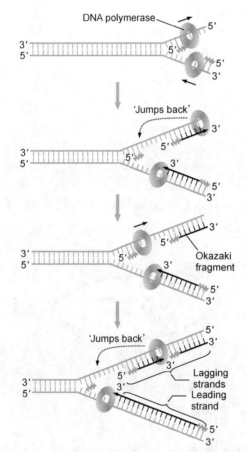

Figure 8.6 Discontinuous replication. On one of the strands, DNA polymerase moves in the opposite direction as the replication fork. To synthesize newly exposed template DNA, the polymerase must periodically jump back to the vertex of the fork

图 8.6 不连续复制。在其中一条链上,DNA 聚合酶沿着与复制叉相反的方向移动。为了合成新暴露出来的模板 DNA,聚合酶必须定期地跳回到复制叉的顶点

However, on the other template strand, which has a 5′ end behind the moving fork, DNA polymerase can only move in the opposite direction as the fork. As the fork continues to open wider, newly exposed portions of the template strand will be unreplicated. To compensate,

然而,另一条模板链在移动叉的后面是5′端,DNA 聚合酶只能沿着与复制叉相反的方向移动。随着复制叉继续扩大,新暴露出来的模板链没有得到复制。为了弥补这一点,DNA 聚合酶每隔一段时

DNA polymerase periodically jumps backwards, returning to the vertex of the fork. It then continues in its usual direction, filling in the unreplicated portion of the strand.

The strand that is polymerized continuously is called the **leading strand**. It is comprised of one long, unbroken DNA strand. The other strand is called the **lagging strand**. The lagging strand is not synthesized as one long strand, but rather a collection of small fragments caused by DNA polymerase jumping back to the fork, synthesizing a stretch, and then jumping back again. The small fragments are named **Okazaki fragments**, after the scientist who discovered them.

The term semi-discontinuous replication should not be confused with the term semi-conservative replication. Also, looking back now at the entire replication bubble, it should be noted that what is a lagging strand at one fork is a leading strand at the other, and vice-versa (Figure 8.7), because the two forks move in opposite direction.

间就往回跳跃一次以便回到复制叉的顶点。之后再沿着它通常的方向继续，以填补链上还没有复制的部分。

持续合成的那条链称为**先导链**，它是一条没有断点的 DNA 长链。另一条链叫作**后随链**。后随链不是一条一次性合成出的长链，而是由许多短片段组成的，这些短片段是 DNA 聚合酶多次跳跃回到复制叉后合成的，每合成完一个片段后 DNA 聚合酶又会跳回到复制叉。根据发现这些短片段的科学家的姓名把它们命名为**冈崎片段**。

术语半不连续复制不要与半保留复制相混淆。同时，现在我们回过头看一下整个复制泡，应该注意到一个复制叉上的后随链是另一个复制叉上的先导链，反之亦然（图 8.7），因为两个复制叉分别朝相反的方向移动。

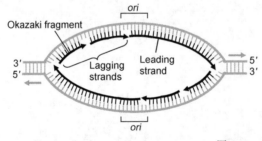

Figure 8.7 Discontinuous replication in the context of the whole replication bubble

图 8.7 在整个复制泡背景中的不连续复制

8.4 Elongation of Replication and its Proteins

For clarity, many details of semi-discontinuous replication were not included in the above section. Here we introduce the enzymes involved in synthesis, as well as refinements to the mechanism introduced above.

8.4.1 Helicase and SSBs

A wide range of proteins are active around the replication fork. The separation of DNA strands as the fork

8.4 复制延伸及其相关蛋白

为清楚起见，上一节中并没有述及半保留复制的许多细节。在此我们介绍 DNA 合成中涉及的酶以及上述机理的一些详细内容。

8.4.1 解旋酶与 SSB

在复制叉附近有许多蛋白质都相当活跃。当复制叉移动的时候 DNA 链的分

grows is accomplished by a **helicase** protein (Figure 8.8). This enzyme is one of the first proteins to bind at the *OriC* after the opening of the replication bubble. It is donut shaped, and rides along one strand of DNA ahead of the polymerases. As helicase proceeds, it breaks hydrogen bonds between the two DNA strands. This function requires energy, supplied in the form of ATP.

离由**解旋酶**这种蛋白质完成（图8.8）。解旋酶是在复制泡打开后首先结合到 *OriC* 位置的几种蛋白质中的一种。它是圆圈形的，在 DNA 聚合酶的前面套在 DNA 的一条链上。在向前推进的时候，解旋酶打断了两条 DNA 链之间的氢键。这一功能需要以 ATP 的形式提供的能量。

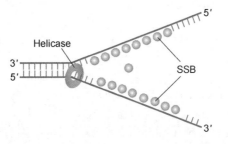

Figure 8.8 Helicase separates the two DNA strands and SSBs protect the single strands and prevent them from rejoining

图 8.8 解旋酶将两条 DNA 链分开，SSB 保护分开了的单链防止它们重新结合在一起

After strands have been separated by the helicase, it is crucial that they do not re-anneal. To ensure this, small proteins called **single-strand DNA binding proteins (SSBs)** cover the separated strands, eliminating their attraction to each other (Figure 8.8). SSBs also protect the strands from breakage and degradation. This is an important function because single strands of DNA are less stable than double-stranded DNA. Though seemingly simple, SSBs are so important that *E. coli* cannot survive if these proteins are inactivated.

在解旋酶分开了 DNA 链之后，防止它们又重新退火非常重要。为了确保这一点，称为**单链 DNA 结合蛋白（SSB）**的小蛋白覆盖到已分开的链上，消除了它们之间的互相吸引（图8.8）。SSB 同时也保护 DNA 单链防止它们发生断裂和降解。这是一项很重要的功能，因为单链 DNA 没有双链 DNA 那么稳定。虽然看起来很简单，但 SSB 是如此重要以至于如果这些蛋白质失去活性的话大肠杆菌就不能生存。

8.4.2 DNA Polymerases

8.4.2 DNA 聚合酶

DNA polymerase follows closely behind the helicase and SSBs, pairing complementary nucleotides and synthesizing new strands of DNA. Two DNA polymerases are essential for replication in *E. coli*, DNA polymerase Ⅰ and Ⅲ. DNA polymerase Ⅰ is discussed below. **DNA polymerase Ⅲ** is used for almost all of replication. DNA Polymerase Ⅲ contains up to 10 subunits, all of which are named by Greek letters (Figure 8.9). The full collection of subunits, which is required for normal replication in the cell, is called the **DNA polymerase Ⅲ holoenzyme**. The minimal collection of subunits needed for polymerization is called the **DNA polymerase Ⅲ core**.

DNA 聚合酶紧跟在解旋酶和 SSB 之后，将互补的核苷酸加上去合成 DNA 新链。有两种 DNA 聚合酶对大肠杆菌复制来说非常重要，它们是 DNA 聚合酶Ⅰ和Ⅲ，DNA 聚合酶Ⅰ将在后面讨论。**DNA 聚合酶Ⅲ**几乎在所有复制过程中都起作用。它含有多达 10 个亚基，所有亚基都以希腊字母命名（图8.9）。所有亚基在一起组成的酶称为 **DNA 聚合酶Ⅲ全酶**，是细胞正常复制所需要的。聚合反应所需要的最少亚基组合称为 **DNA 聚合酶Ⅲ核心酶**。

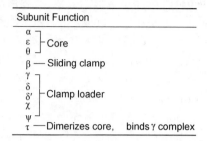

Figure 8.9　Subunits of DNA polymerase Ⅲ

图 8.9　DNA 聚合酶Ⅲ的亚基

The three subunits of the core polymerase are α, ε, and θ. At each replication fork there is only one DNA polymerase Ⅲ complex, but it is composed of two core polymerases, each one polymerizing one strand. Other subunits, described below, are also duplicated (Figure 8.10). The α subunit, part of the core, is responsible for polymerization of nucleotides. The general polymerization reaction was described in chapter 2 (The 3′-OH group of the last nucleotide in the strand reacts with the 5′-triphosphate group of an incoming nucleotide, forming a phosphodiester bond). ε and θ subunits are required for an error correction mechanism of DNA polymerase, described below.

核心聚合酶的三个亚基是 α、ε 和 θ。在每个复制叉处，只有一个 DNA 聚合酶Ⅲ复合体，但它是由两个核心聚合酶组成的，每个核心聚合酶各自合成一条链。其他亚基（后面将对它们进行描述）也是各有两个（图 8.10）。核心聚合酶的 α 亚基负责核苷酸聚合反应。普通的聚合反应已在第 2 章中有过描述（即：DNA 链上最后一个核苷酸的 3′-OH 基团与后面新来核苷酸的 5′-磷酸基团反应形成磷酸二酯键）。DNA 聚合酶需要它的 ε 亚基和 θ 亚基在校正机理中发挥作用，稍后将对它进行描述。

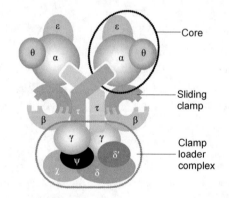

Figure 8.10　Schematic drawing of subunit composition of DNA polymerase Ⅲ

图 8.10　DNA 聚合酶Ⅲ亚基组成示意图

The core polymerase alone is able to polymerize DNA, but it has very little **processivity**. This means that the polymerase frequently falls off the DNA, and is unable to synthesize long, continuous strands. The core polymerase can only polymerize approximately 10 nucleotides before detaching from the template. The other subunits of the holoenzyme deal with this limitation.

光是核心聚合酶就能够聚合 DNA，但它的**持续合成能力**很差。这意味着它会频繁地从 DNA 链上脱落，并且不能合成连续的长链 DNA。核心聚合酶在从模板上脱落前只能聚合大约 10 个核苷酸。聚合酶全酶的其他亚基能弥补这方面的缺陷。

The β subunit, also called the **sliding clamp**, is a donut shaped protein, like helicase (Figure 8.11). Its two subunits clamp around the DNA template and also bind to the core, preventing it from falling off the template. Other subunits of the holoenzyme form a complex called the γ complex, also called the **clamp loader**. This collection of proteins is necessary to attach the sliding clamp onto the DNA (Figure 8.12). The sliding clamp, with help from the clamp loader, allows remarkable processivity. The holoenzyme can polymerize 50 000 nucleotides or more without stopping.

β亚基也被称为**滑行夹**，与解旋酶一样也是圆圈形的蛋白质（图 8.11）。它的两个亚基环绕 DNA 模板链并将其套住，它还与聚合酶核心保持结合，以防止聚合酶核心从模板链上掉落。全酶的其他亚基形成的一个复合体称为 γ 复合体，它也被称为**滑行夹加载器**。这一蛋白质的集合体在将滑行夹套到 DNA 链上去的时候是必需的（图 8.12）。在加载器帮助下起作用的滑行夹使持续合成能力大大增强。聚合酶全酶能够不停地连续聚合 50 000 个或更多个核苷酸。

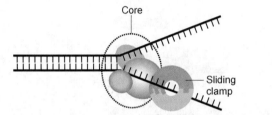

Figure 8.11 The sliding clamp gives DNA polymerase Ⅲ greater processivity

图 8.11 滑行夹大大增强了 DNA 聚合酶 Ⅲ 的持续合成能力

ε and θ, the two components of the core polymerase besides α, are required for $3'\rightarrow 5'$ **exonuclease activity**. This means that they are able to remove nucleotides from the growing strand in the opposite direction of synthesis. This is important because if an incorrect nucleotide has been incorporated into a growing DNA strand, DNA polymerase can uses its $3'\rightarrow 5'$ exonuclease activity to remove the mistake.

除了核心酶的 α 亚基外，ε 亚基和 θ 亚基是发挥 $3'\rightarrow 5'$ **外切核酸酶活性**所必需的。这意味着它们能够按与 DNA 合成相反的方向从生长链上去除核苷酸。这一点很重要，因为如果一个错误的核苷酸被整合到了生长中的 DNA 链上，DNA 聚合酶就能使用它的 $3'\rightarrow 5'$ 外切核酸酶活性去除这一错误的核苷酸。

If DNA polymerase did not have this **proofreading** ability, one incorrect base would be incorporated into the DNA for each 100 000 bases synthesized, approximately. This error rate is much too high. If an *E. coli* genome contains approximately 3 million base pairs, each round of replication would leave 30 errors in the DNA. As we will see in chapter 9, so much as one error in a gene can be lethal to a cell.

如果 DNA 聚合酶没有这种校正能力，则大约每合成 100 000 个碱基就会有一个错误的碱基被整合进 DNA 链中。这样的错误率太高了。如果一个大肠杆菌的基因组含有 3×10^6 个碱基对，那么每一轮复制将在 DNA 中留下 30 个错误。如我们将在第 9 章中看到的那样，在一个基因中就算只有一个错误对细胞来说都可能是致命的。

The proofreading function enacted by ε and θ subunits improves the accuracy of DNA polymerase Ⅲ to approximately one mistake in 10 000 000. Subsequently, another proofreading mechanism in the cell, discussed in chapter 9, can reduce this rate to 1 in a billion.

在 ε 亚基和 θ 亚基的作用下，DNA 聚合酶 Ⅲ 的准确率提高到大约 10 000 000 个碱基中出现一个错误。还有，细胞中另一种校正机理（将在第 9 章中讨论）又能将这一比率降低到 10 亿个碱基中出现一个错误。

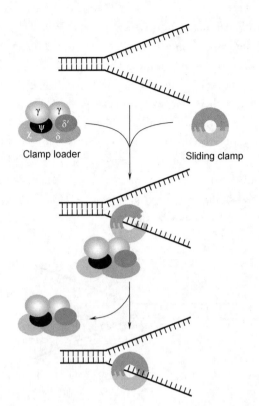

Figure 8.12 The clamp loader attaches the sliding clamp to the DNA template

图 8.12 滑行夹加载器将滑行夹套到 DNA 模板上

This error rate is ideal because it rarely threatens normal gene function in the organism, but also provides occasional mutations necessary for evolution.

这样的错误率是理想的，因为它很少会危及生物中正常的基因功能，同时又能提供偶然突变，生物的进化需要这种偶然突变。

8.4.3 Explanation for $3'\rightarrow 5'$ Synthesis

The existence of a proofreading function helps to explain why DNA polymerase is only able to synthesize DNA in the $5'\rightarrow 3'$ direction (Figure 8.13). Each time a new nucleotide is added to a growing strand, a new $3'$-OH end is produced. Likewise, each time a nucleotide is removed, a $3'$-OH end is regenerated. The hydroxyl group at this $3'$ end is necessary for reacting with the triphosphate group of an incoming nucleotide, allowing chain growth to continue.

Imagine a DNA polymerase that synthesizes in the $3'\rightarrow 5'$ direction (Figure 8.14). Each time a new nucleotide is added to a growing strand, a new $5'$-triphosphate end would be formed. Chain growth could actually continue, because this group could react with the $3'$-OH group of the next nucleotide. But what if the wrong

8.4.3 关于 $3'\rightarrow 5'$ 合成

校正功能的存在帮助解释了为什么 DNA 聚合酶只能以 $5'\rightarrow 3'$ 方向合成 DNA（图 8.13）。每次当一个新的核苷酸被加到生长链上时都产生一个新的 $3'$-OH。同样，每次去掉一个核苷酸时会重新产生一个 $3'$-OH。$3'$ 端的这个羟基需要用来与下一个核苷酸的三磷酸基团反应，以便使链的生长能够继续。

设想一下 DNA 聚合酶以 $3'\rightarrow 5'$ 合成 DNA 的情形（图 8.14）。在这种情形下，每次一个新核苷酸加到生长链上去时会形成一个新的 $5'$-三磷酸末端。链生长应该可以继续，因为 $5'$-三磷酸基团能够与下一个核苷酸的 $3'$-OH 反应。但是如果一个错

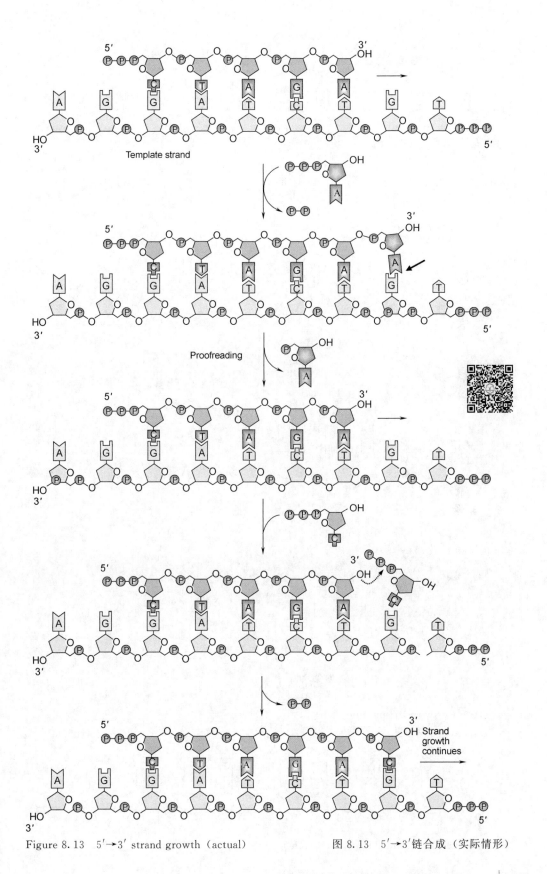

Figure 8.13　5′→3′ strand growth (actual)　　图 8.13　5′→3′链合成（实际情形）

8.4　Elongation of Replication and its Proteins

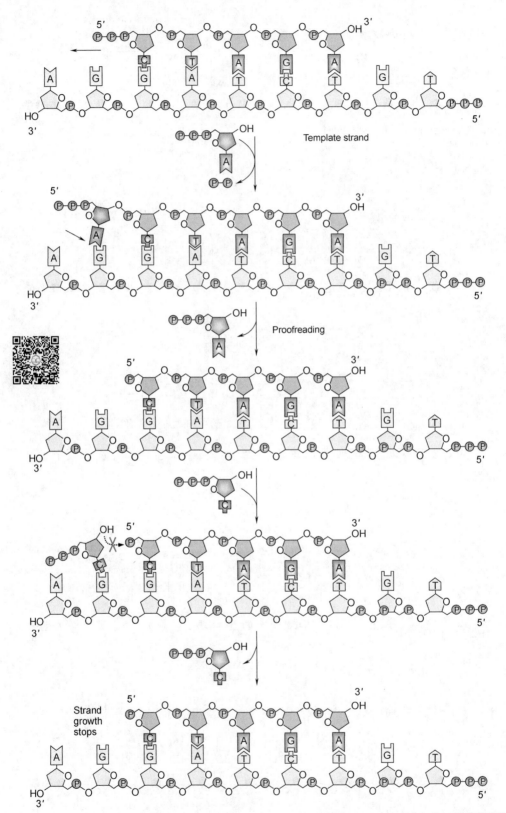

Figure 8.14　3′→5 strand growth (hypothetical)　　图 8.14　3′→5′链合成（假设的情形）

nucleotide were incorporated, and had to be removed by a 5′→3′ exonuclease activity? In this case, a 5′ end of the DNA with only one phosphate would be produced. The triphosphate ending would not be regenerated. This is problematic because a monophosphate could not react with the 3′-OH end of the next nucleotide to be added. With no other recourse, replication would have to terminate. Thus, 5′→3′ polymerization is the only direction of synthesis that allows synthesis to resume easily after the removal of an incorrect base.

8.4.4 Primers

Another apparent consequence of the proofreading function is that DNA polymerase cannot initiate replication on a single-stranded template. If the polymerase is designed to check the accuracy of the last nucleotide added, it presumably cannot put down a first nucleotide because there is no previous nucleotide for it to check. This may sound strange. How can DNA polymerase replicate anything if it cannot initiate replication?

To overcome this problem, the cell relies on RNA polymerase, which has no problem initiating synthesis. A specialized RNA polymerase, called **primase**, puts down short pieces of RNA at sites of replication (Figure 8.15). These pieces are about 10 nucleotides long and are called **primers**. Primers are properly paired to the template, forming short RNA-DNA hybrids. The primers contain a 3′ end, quite similar to a normal DNA strand. DNA polymerase is able to bind to the primer and uses this 3′ end to begin synthesizing DNA.

In leading strand synthesis, a primer is rarely needed beyond the origin of replication. However, in lagging strand synthesis, each Okazaki fragment must begin with a new primer (Figure 8.16). Thus primase must add a primer for every 1~2kb of lagging strand synthesis(1kb=1000 base pairs).

RNA primers must be replaced with DNA before replication is completed. **DNA polymerase** I is specially equipped for this task (Figure 8.17). In addition to 3′→

8.4.4 引物

校正功能另一个明显的后果是DNA聚合酶不能够在一条单链模板上启动复制。如果DNA聚合酶被设计成能够检查刚刚加上去的那个核苷酸是否正确的话，那它就应该不具备将第一个核苷酸放上去的能力，因为那时还没有前面的核苷酸让它来检查。这听起来也许有些奇怪，如果DNA聚合酶不能够启动复制的话，那它又怎么能复制出任何DNA呢？

为了解决这一问题，细胞依靠的是RNA聚合酶，启动合成对它来说不是问题。一种特殊的RNA聚合酶（叫作**引发酶**）能在需要复制的位置合成出RNA短片段（图8.15）。这些片段长约10个核苷酸，叫作**引物**。引物与模板之间正确配对，形成短的RNA-DNA杂交链。引物含有3′末端，与正常的DNA链非常相似。DNA聚合酶能够结合到引物上并使用这一3′末端开始合成DNA。

在先导链的合成中，除去在复制起点以外很少再需要引物。然而，在后随链的合成中，每条冈崎片段必须以一个新的引物开始（图8.16）。因此在后随链合成中引发酶必须每隔1~2kb就加上一个引物(1kb=1000碱基对)。

在复制完成之前RNA引物必须用DNA替换掉。**DNA聚合酶**I具有特殊的装备来完成这一任务（图8.17）。除了3′→

8.4 Elongation of Replication and its Proteins

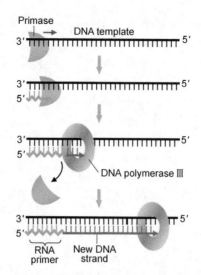

Figure 8.15 RNA primers are required for DNA polymerase to begin synthesis. Primase adds primers to the DNA

图 8.15 DNA 聚合酶启动合成需要 RNA 引物。引发酶能将引物加到 DNA 模板上

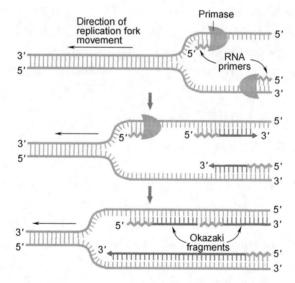

Figure 8.16 Each new Okazaki fragment requires a new primer. Leading strand synthesis only requires a primer at the origin

图 8.16 每条冈崎片段都需要一个引物。先导链合成只需要在起点处有一个引物

5′ exonuclease activity, DNA polymerase Ⅰ also has a **5′→3′ exonuclease activity**. This is a function which the main DNA polymerase used in replication, polymerase Ⅲ, does not possess. After replication, DNA polymerase Ⅰ runs back over the replicated DNA in the normal 5′→3′ direction. When it reaches a primer, it uses its exonuclease activity to degrade it beginning at the 5′ end. Simultaneously, it replaces the primer by synthesizing a short stretch of DNA in its place.

外切核酸酶活性外，DNA 聚合酶Ⅰ也具有 **5′→3′外切核酸酶活性**。这一功能是 DNA 聚合酶Ⅲ（复制中催化聚合反应的主要酶）所不具备的。在复制完成后，DNA 聚合酶Ⅰ以正常的 5′→3′方向回头快速检查一遍复制出的 DNA。当它到达一个引物的位置时，会用它的外切核酸酶活性从 5′端将引物降解。同时合成一小段新的 DNA 将引物替换掉。

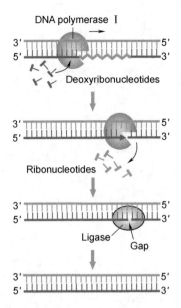

Figure 8.17　DNA polymerase Ⅰ uses its $5'\to 3'$ exonuclease activity to remove primers. It synthesizes DNA to replace the primers. Ligase joins the pieces of DNA

图 8.17　DNA 聚合酶Ⅰ使用它的 $5'\to 3'$ 外切核酸酶活性去除引物，并且合成 DNA 来替换引物。连接酶将 DNA 片段连接在一起

8.5　DNA Topology

As replication proceeds, the double-stranded DNA is increasingly unwound. Because of the helical structure of DNA and the circular shape of the *E. coli* chromosome, this creates a high degree of tension ahead of the replication fork (Figure 8.18). The tension relieves itself by tangling the DNA into a shape called a **supercoil**. You might have observed a similar effect in the tangling of a telephone chord if you rotate the receiver too many times.

If the supercoil is not relieved, it will eventually block replication from proceeding. A group of enzymes called topoisomerases can undo supercoils. **Type Ⅰ topoisomerases** make a cut in one of the DNA strands, allowing tension to be relieved by rotating the chromosome around the single bond in the adjacent uncut strand (Figure 8.19). **Type Ⅱ topoisomerases** create temporary double-stranded cuts in the chromosome, and then pass another region of the chromosome through the gap. The most common topoisomerase Ⅱ in *E. coli* is called **DNA gyrase**.

8.5　DNA 拓扑学

在复制进行的时候，越来越多的双链 DNA 被打开。由于存在 DNA 的螺旋结构和大肠杆菌染色体的环形结构，打开越来越多的双螺旋会在复制叉前带来很大的张力（图 8.18）。这种张力通过把 DNA 扭曲成**超螺旋**的形状而得到释放。如果你把电话听筒转很多圈的话，或许你也会观察到类似的效果。

如果这样的超螺旋不被释放掉，它最终会阻止复制的进行。一组称为拓扑异构酶的酶能够消除超螺旋。**Ⅰ型拓扑异构酶**在 DNA 的一条链上产生一个切口，使染色体围绕邻近未切断链上的那个单键旋转从而释放张力（图 8.19）。**Ⅱ型拓扑异构酶**在染色体上产生暂时的双链切口，让染色体的另一区域穿过这一缺口。大肠杆菌中最常见的Ⅱ型拓扑异构酶是 **DNA 旋转酶**。

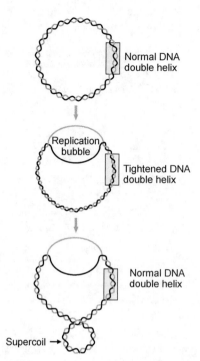

Figure 8.18 Replication causes supercoiling ahead of the replication fork

图 8.18 复制引起复制叉前方产生超螺旋

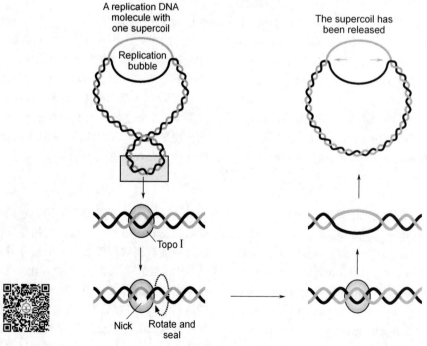

Figure 8.19 Topoisomerase I undoes supercoiling by introducing temporary single-stranded cuts in the DNA. The whole chromosome can rotate around the single-stranded DNA to relieve tension

图 8.19 拓扑异构酶 I 通过在 DNA 上产生暂时的单链切割而消除超螺旋。整个染色体可以绕单链 DNA 旋转以释放张力

Type Ⅱ topoisomerases are also important for the termination of replication (Figure 8.20). In *E. coli*, the replication forks proceed until they reach a set of termination sequences in the DNA. At this point the two forks pause quite near each other. Only a small region of the original chromosome's DNA remains intact and unsynthesized. Next, this region is denatured, so that each single strand can be used as a template to fully complete replication of the chromosome. After synthesis, the two daughter chromosomes are interconnected. To unlink the molecules, topoisomerase Ⅱ makes a double stranded break in one of the chromosomes and releases the other chromosome by passing it through the gap. After the chromosomes are separated, the gap is resealed.

Ⅱ型拓扑异构酶在复制终止过程中也起着很重要的作用（图8.20）。在大肠杆菌中，复制叉一直向前直到它们在DNA上遇到了一组终止序列。此时两个复制叉在相当接近的地方暂停。原先的染色体DNA中只有很少的区域保持完整且没有被合成。下一步，这一区域会发生变性，这样每条单链就能够用来作为模板以便完成全部染色体的复制。合成完成之后，两条子代染色体是互相联结在一起的。为了解开这样的连环分子，Ⅱ型拓扑异构酶在其中的一条染色体上产生一个双链缺口，让另一条染色体从缺口中穿过而将它释放。在染色体被分开后，缺口又被重新封闭。

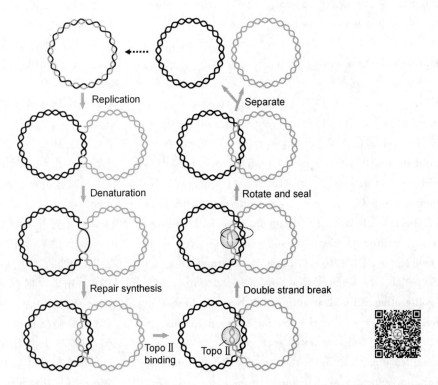

Figure 8.20 Termination of replication in *E. coli*. Topoisomerase Ⅱ unlinks the interconnected chromosomes by introducing a double-stranded cut in one of the chromosomes, and passing the other chromosome through the cut

图8.20 大肠杆菌复制终止。拓扑异构酶Ⅱ通过在其中的一条染色体上引入一个双链切口并让另一条染色体穿过这一切口而拆开连环染色体

8.6 DNA Replication in Eukaryotes

Replication in eukaryotes, though largely similar to

8.6 真核生物DNA复制

真核生物DNA复制虽然与原核生物大

prokaryotic replication, comes with several additional challenges. Notably, the genomes of eukaryotes are much larger that prokaryotic genomes. In addition, eukaryotic replication forks move much more slowly than prokaryotic forks; this is possibly due to the histones associated with eukaryotic DNA.

E. coli DNA polymerase adds up to 1000 nucleotides per second, whereas eukaryotic DNA polymerases add a similar number per minute. At this rate, one might expect it would take many days to replicate a human cell's DNA. Actually, eukaryotic cells can replicate their DNA quickly because eukaryotic replication initiates at hundreds or thousands of different origins, as opposed to only one origin in prokaryotes. In addition, eukaryotic cells contain tens of thousands of DNA polymerase molecules, whereas *E. coli* only have several dozen.

There are numerous DNA polymerase enzymes found in eukaryotes, divided into at least four families-A, B, X, Y (Table 8.1). Of these, enzymes of family B coordinate nearly all of genome replication. Family B is composed of polymerase α, δ, and ε. α is the only polymerase with priming activity, and therefore it initiates synthesis of all leading and lagging strands. However, it has limited processivity and only replicates short stretches of DNA after priming. By contrast, δ and ε have high processivity, aided by their association with **proliferating cell nuclear antigen (PCNA)**, a clamp that tethers the enzyme to the template strand, similar to the sliding clamp of DNA polymerase Ⅲ in prokaryotes. δ and ε differ in the strands that they preferentially replicate, with ε more often associated with the leading strand and δ more often associated with the lagging strand. Family A of polymerases (polymerase γ) can also complete genome replication, but is used for replication of mitochondrial genomes. The other families, X and Y, are primarily involved in DNA damage responses.

体相同，但也遇到了几个额外的挑战。显而易见的是，真核生物的基因组比原核生物的大得多。此外，真核生物的复制叉移动的速率比原核生物的慢得多；这可能是因为真核生物的 DNA 是与组蛋白结合在一起的。

大肠杆菌 DNA 聚合酶每秒钟能加上近 1000 个核苷酸，而真核生物 DNA 聚合酶每分钟才能加上这么多核苷酸。你可能会想，以这种速率去复制一个人类细胞的 DNA 需要花费很多天才能完成。实际上，真核细胞能够很快地复制完它们的 DNA，因为真核生物 DNA 会在几百个或几千个不同的起点开始复制，而不像原核生物那样只有一个复制起点。此外，真核细胞含有几万个 DNA 聚合酶分子，而大肠杆菌中只有几十个。

在真核生物中存在许多 DNA 聚合酶，至少可以将它们分为 A、B、X、Y 四个家族（表 8.1）。其中 B 家族 DNA 聚合酶几乎协调了所有基因组的复制过程。B 家族由聚合酶 α、δ、ε 组成。只有聚合酶 α 具备引发活性，因此它起始了所有先导链与后随链的合成。然而，聚合酶 α 的持续合成能力较弱，在引发 DNA 合成后只能复制一小段 DNA。相反，聚合酶 δ 和聚合酶 ε 具有很强的持续合成能力，因为两者均与**增殖细胞核抗原（PCNA）**结合在一起。与原核生物 DNA 聚合酶Ⅲ的滑行夹相似，PCNA 也是一种夹状结构，能将聚合酶与其模板链维系在一起。聚合酶 δ 和聚合酶 ε 倾向于复制不同的 DNA 链；聚合酶 ε 更多地参与先导链合成，而聚合酶 δ 更多地参与后随链合成。A 家族 DNA 聚合酶（聚合酶 γ）也能完成基因组复制，不过它是用来复制线粒体基因组的。X 家族与 Y 家族的 DNA 聚合酶主要在 DNA 损伤响应中发挥作用。

Table 8.1　Families of eukaryotic DNA polymerases　　　表 8.1　真核生物 DNA 聚合酶家族

Family	Representative members	Number of subunits	Function(s)
A	γ	1 subunit	The only polymerase involved in replication of mitochondrial DNA
B	α, δ, ε	3~4 subunits	α—priming during genome replication, followed by limited DNA synthesis
			δ—genome replication, primarily of lagging strand
			ε—genome replication, primarily of leading strand, regulation of cell cycle checkpoints
X	β, μ, λ, TdT	1 subunit	β—base excision repair
			μ, λ—non-homologous end joining
			TdT—adds nucleotides to breaks following VDJ recombination
Y	η, ι, κ	1 subunit	Enable translesion synthesis during DNA replication

8.6.1　Initiation of DNA Replication in Eukaryotes

Origins of replication in eukaryotes have been relatively hard to identify. They have been most successfully studied in the yeast *Saccharomyces cerevisiae*. Origins in this organism generally contain several 10~15bp conserved sequences spread within a 100~200bp region. One sequence in particular, the **autonomously replicating sequence (ARS)** is important for initiation. It was discovered because pieces of DNA containing this sequence are able to replicate autonomously, even if they are not part of a chromosome.

In higher eukaryotes, it has been much more difficult to pinpoint sequences that serve as origins of replication. In some cases the site of the origin may vary over thousands of bases, in other cases there seem to be no consistent sites of replication origin. Factors other than base sequence, such as nucleosome density, may play important roles in determining origin placement.

In contrast with the diversity of replication origins among eukaryotes, the protein machinery that acts to initiate replication appears to be well conserved. The mechanism has adapted so that replication is coordinated within the life cycle of the cell. Briefly, the **cell cycle** is often divided into four phases: G1, a growth phase; S, a DNA synthesis phase; G2, another growth phase; and M, mitosis. It is critically important that all of the DNA be replicated only once, and only during

8.6.1　真核生物 DNA 复制起始

真核生物的复制起点相对来说较难确定。在酵母 *Saccharomyces cerevisiae* 中，复制起点的研究最为成功。在这种酵母中，复制起点一般含有几个 10~15bp 的保守序列，这些保守序列分布在一个 100~200bp 的区域内。一个特别的序列是**自主复制序列（ARS）**，对启动复制很重要。它被发现的原因是：含有这些序列的 DNA 片段能自动地进行复制，即使它们不是染色体的一部分。

在更高等的真核生物中，要找出作为复制起点序列的确切位置要困难得多。在有些情况下，复制起点的序列可能会相差几千个碱基，在另一些情况下则看起来根本没有统一的复制起点位置。这时，在决定复制起点位置中起重要作用的是一些其他因素如核小体密度等，而不是碱基序列。

与复制起点的多样性相反，真核生物用于启动复制的蛋白质装置看起来倒是相当保守的。该机制已经适应，从而使复制在细胞生活周期内是协调的。简单说来，**细胞周期**常被划分成四个时期：G1，生长期；S，DNA 合成期；G2，另一个生长期；M，有丝分裂期。有一点至关重要，就是所有 DNA 只应在 S 期被复制一次，这样才能避免基因组中出现混乱场

S-phase, to avoid chaos in the genome. The collection of proteins that initiates replication in S-phase is called the **pre-replicative complex（pre-RC）**（Figure 8.21）.

面。在 S 期启动复制的蛋白质集合体称为**前复制复合体（pre-RC）**（图 8.21）。

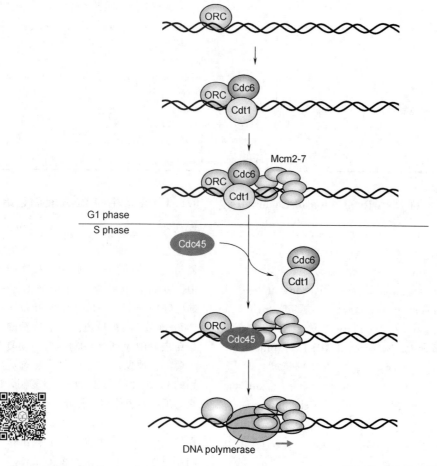

Figure 8.21　Replication initiation in eukaryotes

图 8.21　真核生物复制起始

The first part of the pre-RC to bind to DNA is the **origin recognition complex（ORC）**. This collection of proteins attracts two more proteins, the Cdc6 and Cdt1, which in turn attract the helicase proteins Mcm2-7 to complete the complex. Importantly, this pre-RC complex only forms at the beginning of the G1 phase, when DNA synthesis does not occur.

The complex remains attached to the DNA but inactive until S-phase, at which point a protein called Cdc45 binds. Cdc45 allows the pre-RC to initiate replication by activating Mcm2-7 to function as a helicase, and recruiting DNA polymerases to the pre-RC. How does the Cdc45 know to only activate the pre-RC during S-phase? A protein called Cdk2 is crucial for regula-

pre-RC 结合到 DNA 上的第一个部位是**起点识别复合体（ORC）**。这些蛋白质吸引另外两种蛋白质 Cdc6 和 Cdt1，这两种蛋白质又吸引了解旋蛋白 Mcm2-7 后形成完整的复合体。重要的是，这一 pre-RC 复合体只在 G1 期的开始阶段形成，此时 DNA 合成并没有开始。

一直到 S 期，这一复合体都保持与 DNA 结合但是没有活性，等到了 S 期，称为 Cdc45 的蛋白质结合上来。Cdc45 通过激活 Mcm2-7 使其具有解旋酶功能，并且召集 DNA 聚合酶前来而使 pre-RC 启动复制。Cdc45 是怎么知道只在 S 期才去激活 pre-RC 的呢？一种称为 Cdk2 的

ting the process. During the G1 phase, Cdk2 is inactive; however it becomes active during S phase. Active Cdk2 is necessary for Cdc45 to be able to bind to the pre-RC. Proteins like Cdk2, which vary in activity according to phases of the cell cycle, are very important for coordinating events in the cell.

8.6.2 Telomeres

Another challenge faced by eukaryotes during synthesis is what to do with primers at the end of the chromosome. The end of a eukaryotic chromosome is called the **telomere**. Leading strand synthesis proceeds to the tip of the telomere without any problem. For lagging strand synthesis, the most complete replication requires a primer at the very tip of the telomere, with synthesis proceeding backwards from the tip. These 10nts or so of RNA primer can be removed, but it is impossible to replace the gap with DNA using normal DNA polymerase.

Eukaryotic cells employ a special enzyme, **telomerase**, to deal with this problem (Figure 8.22). Telomerase begins at the staggered end of the chromosome, where removal of the primer has left the parental strand slightly longer than the new strand. Telomerase then extends the parental strand by several dozen nucleotides. It is able to do so without a DNA template because within the enzyme is a short piece of RNA that serves as a kind of template.

The extra extension of the parental strand allows enough space to add another primer and complete synthesis of the original chromosome. Although this primer will eventually be removed, leaving a staggered end again, the final result is that the chromosome is at least as long as in the previous round of replication.

Telomerase is used in many, but not all, animal cells. Most human cells do not express telomerase, and as a result chromosomes shrink by 50～100 nucleotides during every round of replication. This DNA loss limits the number of times our cells can divide, and may contribute to the aging process.

蛋白质在调控这一过程中起决定性的作用。在 G1 期，Cdk2 是无活性的；而在 S 期它变成有活性的了。有活性的 Cdk2 对 Cdc45 能够结合到 pre-RC 上去是必需的。像 Cdk2 这样能根据细胞周期不同阶段改变活性的蛋白质在协调细胞活动方面是非常重要的。

8.6.2 端粒

真核生物 DNA 合成面临的另一个挑战是如何处理位于染色体末端的引物。真核生物染色体的末端称为**端粒**。先导链的合成可以一直进行到端粒的末尾而不会有问题。但对于后随链来说，最完整的复制要求在端粒的最末尾处出现一个引物，以引导 DNA 的合成从末尾往回进行。这些 10nt 或更长一点的 RNA 引物能够被去除，但留下的空缺却不可能由普通的 DNA 聚合酶来填补。

真核细胞采用**端粒酶**这种特殊的酶来解决这一问题（图 8.22）。端粒酶从染色体的交错末端开始，交错末端上由于去除引物而使母链比新链稍微长了一点。之后端粒酶将母链延长几十个核苷酸。它可以在没有 DNA 模板的情况下进行这样的反应，因为在酶的内部有一小段 RNA 可以作为模板。

母链的额外延长使在它上面加上另一个引物并全部完成原始染色体的合成有了足够的空间。虽然这一引物最后也要被去除，又会留下一个交错末端，但最终的结果是：复制出的染色体至少与前一轮复制时一样长。

端粒酶在许多动物细胞但不是所有细胞中发挥作用。绝大多数人类细胞并不表达端粒酶，结果在每一轮复制中染色体都会缩短 50～100 个核苷酸。这种 DNA 损失限制了我们的细胞可以分裂的次数，也因此可能与衰老过程有关。

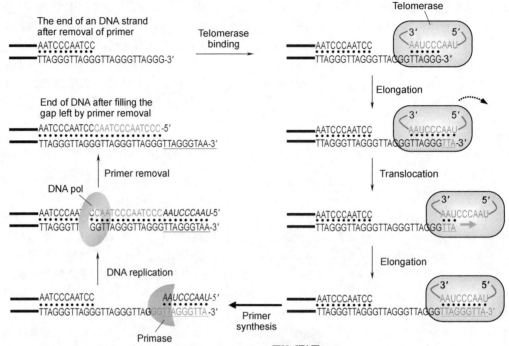

Figure 8.22 Telomerase elongates the parent strand at the telomere. This allows replication of the entire length of the parent strand

图 8.22 端粒酶在端粒处延长母链。这使得母链的全部都能被复制出来

8.7 Experiments

DNA replication was originally thought to occur by one of three possible mechanisms: conservative replication, semi-conservative replication, and dispersive replication. Semi-conservative replication is discussed above. **Conservative replication** is a model in which an entirely new copy of the DNA is made, leaving the original DNA perfectly intact. The idea is similar to making a copy using a Xerox machine. In the **dispersive replication** model, the original DNA is fragmented, and new DNA is synthesized to make two double-stranded molecules from the set of fragments.

A famous experiment, called the **Meselson-Stahl experiment**, determined which model was correct (Figure 8.23). In the experiment, cells were grown on an isotope of nitrogen, ^{15}N. An isotope is an atom with a different number of neutrons but same number of protons. Normal nitrogen, ^{14}N, has 14 neutrons and 14 protons. ^{15}N has 15 neutrons and 14 protons. In the case of ^{15}N, the extra neutron does not affect the reactivity

8.7 实验研究

DNA 复制最初被认为以三种可能的机理进行：保留复制、半保留复制和散乱复制。半保留复制已在上面讨论过了。**保留复制**模型认为复制产生的是一个全新的 DNA 拷贝，母本 DNA 完完全全以原样保留。这种想法跟使用复印机制作一个拷贝相似。在**散乱复制**模型中，母本 DNA 成为零碎的片段，新的 DNA 双链分子又以这些片段为基础合成。

一个著名的实验，称为**麦塞尔逊-斯托尔实验**，检验了哪一个模型是正确的（图 8.23）。在实验中，细胞生长在含有氮的同位素^{15}N的培养基中。同位素是一种具有不同中子数但是相同质子数的原子。普通氮原子^{14}N具有 14 个中子和 14 个质子。同位素^{15}N则具有 15 个中子和 14 个质子。在^{15}N中，多出的中子并不影响原

of the atom. However, the atom is denser than normal, and makes any molecules that incorporate it, like DNA, denser than normal.

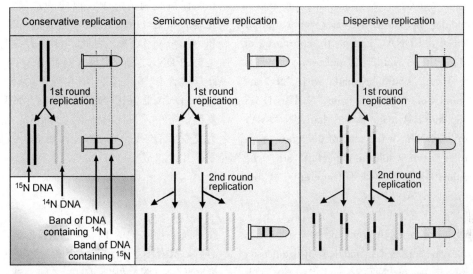

Figure 8.23 The Meselson-Stahl experiment proved that the semi-conservative model of replication is used by cells

After growing cells for a long time using ^{15}N, all DNA in the cells included ^{15}N. The density of the DNA could be verified by a method called **density ultracentrifugation**. This method spins molecules very fast in a special solution so that they become arranged at distinct levels in a tube. The position of each molecule in the tube after spinning indicates its density. Accordingly, ^{15}N DNA appears lower in the tube than ^{14}N DNA.

Next, the cells that had been grown for many generations on ^{15}N growth medium were suddenly switched to a medium with ^{14}N. The cells were allowed to grow for only one generation—one round of DNA replication—on the ^{14}N. Any new DNA made in the cell during this round of replication would have ^{14}N, not ^{15}N. Then the density of the DNA from the new generation of cells was examined using ultracentrifugation.

In the case of conservative replication, the expected result would be that half of the DNA after replication is completely new, and half is completely unchanged. In other words, half of the DNA would be made entirely from ^{15}N,

子的反应活性。然而,这个同位素的密度比普通元素更大,会使任何含有它的分子(如 DNA)具有比普通分子更大的密度。

图 8.23 麦塞尔逊-斯托尔实验证明了半保留复制是细胞采用的复制模型

在用 ^{15}N 将细胞培养很长一段时间后,细胞中的所有 DNA 都会含有^{15}N。DNA 密度可以通过一种称为**密度超速离心**的方法来验证。这一方法将分子放在一种特殊的溶液中以极高的速度旋转,使它们停留在离心管中不同的位置。每种分子在离心管中的位置指示了它的密度。相应地,含有^{15}N 的 DNA 比含有^{14}N 的 DNA 出现在更低的位置。

下一步,已经在含^{15}N 的培养基中生长了很多代的细胞被突然转换到含^{14}N 的培养基中。设法让细胞只在^{14}N 上生长一代,也就是只发生一轮的 DNA 复制。细胞中任何新合成的 DNA 将具有^{14}N 而不是^{15}N。之后用密度超速离心检查新一代细胞中 DNA 的密度。

在保留复制情形中,预计的结果是:复制后有一半 DNA 是全新的,另一半 DNA 完全保持原样。换句话说,有一半 DNA 全部由^{15}N 组成,而另一半 DNA

and the other half would be made entirely from ^{14}N. After ultracentrifugation, two bands would be visible, one for DNA made from each isotope.

In the case of semi-conservative replication, the expected result would be that all DNA after replication is partly composed of old DNA, and partly composed of new DNA. Thus, there would be only one kind of DNA. The new strand would be made with ^{14}N, and the parental strand would be made from ^{15}N. This DNA molecule would have a density lower than ^{15}N DNA, but higher than ^{14}N DNA. In the case of dispersive replication, a similar result would be expected, since the DNA after replication consists of fragments of old DNA and fragments of new DNA.

After ultracentrifugation of DNA from the new generation of cells, only one band appeared. In addition, it had density intermediate between that of ^{15}N DNA and ^{14}N DNA. This result strongly favored semi-conservative replication and dispersive replication models over conservative replication. In order to distinguish between the first two models, semi-conservative and dispersive, the cells were grown for one more generation on the ^{14}N medium. In the case of semi-conservative replication, the cells from the first generation of ^{14}N growth have DNA with one ^{14}N strand and one ^{15}N strand. Thus, for the next round of replication the 'old strand' can be ^{14}N or ^{15}N, and all new strands are ^{14}N. Therefore, after replication half of the cells will receive DNA that is made only of ^{14}N. The other half will receive DNA that is ^{15}N on one strand and ^{14}N on the new strand.

In the case of dispersive replication, the cells from the first generation of ^{14}N growth have DNA with a mix of ^{14}N and ^{15}N on both strands. If this DNA is replicated again by breaking up the parental DNA into fragments and filling in the rest with new DNA, all DNA after synthesis is expected, again, to have the same density.

After ultracentrifugation of DNA from cells, two bands appeared, one at the level of DNA made with only ^{14}N, and one in between the ^{14}N DNA and ^{15}N DNA density. Thus, the semi-conservative model of replication was proved to be the one used by cells.

全部由^{14}N组成。在超速离心后,将看到两条带,它们来自各自的同位素。

在半保留复制情形中,预计的结果是:复制后所有DNA含有一部分老DNA和一部分新DNA。因此,将只有一种类型的DNA。新的DNA链将含有^{14}N,而母链含有^{15}N。这一DNA分子的密度比^{15}N DNA低但比^{14}N DNA高。在散乱复制情形中,预计的结果与此相似,因为复制后DNA由老DNA和新DNA的零碎片段组成。

在对新一代细胞的DNA进行了超速离心后,只看到了一条带。此外,它的密度介于^{15}N DNA和^{14}N DNA之间。这一结果有力地支持了半保留复制模型和散乱复制模型,而不支持保留复制模型。为了弄清到底是前面两种模型(半保留复制和散乱复制)中的哪一种,细胞被放到含^{14}N的培养基上再培养一代。在半保留复制情形中,从^{14}N生长中得到的第一代细胞DNA含有一条^{14}N链和一条^{15}N链。这样,下一轮复制中,"老链"含有^{14}N或^{15}N,而所有的新链含有^{14}N。这样,复制后有一半细胞的DNA全部由^{14}N组成,另一半细胞DNA中的一条链含^{15}N而另一条新链含^{14}N。

在散乱复制情形中,从^{14}N生长中得到的第一代细胞DNA在两条链上都含有^{14}N和^{15}N的零碎片段。如果这样的DNA再次被打断并用新DNA填补余下的位置,那么预计合成后所有的DNA又会具有相同的密度。

在对这些细胞中的DNA进行超速离心后,出现了两条带,一条位于只含^{14}N的DNA水平,另一条介于^{14}N DNA和^{15}N DNA密度之间。因此,半保留复制模型被证明确实是细胞所采用的复制方式。

Summary

The basis of DNA replication lies in the structure of the double helix. Each strand contains the necessary information to regenerate a full double-stranded molecule. Indeed, after DNA replication in cells, each DNA molecule contains one strand of parental DNA and one new strand, a mechanism called semi-conservative replication. Although the double helix provides a basis for replication, many proteins are required for replication to actually occur. The inherent restrictions of some of these proteins complicate the process of replication, requiring for instance the use of discontinuous replication and primers. Replication is similar in eukaryotes and prokaryotes, although eukaryotes have to deal with certain additional challenges such as larger genomes and the presence of histones and telomeres.

小结

DNA复制的基础建立在它的双螺旋结构上。每条链都含有产生一条完整双链分子的必需信息。事实上，细胞DNA在复制后每个DNA分子含有一条母链DNA和一条新链，这一机理称为半保留复制。虽然双螺旋提供了复制的基础，但复制要真正进行还需要许多蛋白质的参与。其中一些蛋白质的内在限制使复制过程更为复杂，造成了比如需要采用不连续复制和引物这样的情况。在真核生物和原核生物中复制的过程是相似的，虽然真核生物必须处理一些额外的挑战，例如更大的基因组以及组蛋白和端粒的存在。

Vocabulary 词汇

autonomously replicating sequence (ARS)	自主复制序列	pre-replicative complex (pre-RC)	前复制复合体
bidirectional replication [ˌbaidi'rekʃənəl]	双向复制	primase [prai'meis]	引发酶
cell cycle	细胞周期	processivity [prəu'sesiviti]	持续合成能力
clamp loader	滑行夹加载器	replication bubble [ˌrepli'keiʃən]	复制泡
DNA gyrase ['dʒaiəreis]	DNA旋转酶	replication fork	复制叉
DNA topology [tə'pɔlədʒi]	DNA拓扑学	semi-conservative replication	半保留复制
helicase ['hi:likeis]	解旋酶	semi-discontinuous replication	半不连续复制
lagging strand	后随链	single-strand binding protein (SSB)	单链结合蛋白
leading strand	先导链	sliding clamp	滑行夹
Okazaki fragment ['ɔ:kɑ:'zɑ:ki:]	冈崎片段	supercoil ['sju:pəˌkɔil]	超螺旋
plasmid ['plæzmid]	质粒	telomerase ['ti:ləməreis]	端粒酶
		topoisomerase [ˌtɔpəi'sɔməreis]	拓扑异构酶

Review Questions 习题

Ⅰ. True/False Questions（判断题）

1. There is one origin of replication in prokaryotes and eukaryotes.
2. The sliding clamp is a kind of helicase.
3. DNA polymerase Ⅲ has the ability to correct most of its mistakes.
4. DNA polymerase Ⅲ has a $5'\rightarrow 3'$ exonuclease activity.
5. Topoisomerase Ⅰ introduces single stranded cuts in the DNA.
6. DNA polymerase Ⅲ is used to replace RNA primers with DNA.
7. The pre-replicative complex is inactive until S-phase of the cell cycle.
8. Telomerase contains a short piece of RNA.
9. Okazaki fragments are never joined together.
10. Initiation of DNA replication in *E. coli* occurs at a DNA sequence called DnaA.

Ⅱ. **Multiple Choice Questions**（选择题）

1. Without the sliding clamp, DNA polymerase Ⅲ cannot _____ .
   a. bind to DNA
   b. synthesize DNA
   c. recognize the origin of replication
   d. synthesize long pieces of DNA
   e. all of the above

2. DNA polymerase probably synthesizes in the $5' \rightarrow 3'$ direction because _____ .
   a. that is the opposite direction of the replication fork
   b. that is the only way it can bind to DNA
   c. that allows synthesis to resume easily if one base needs to be removed
   d. this is the only direction permitted by the double helix
   e. none of the above

3. If a eukaryotic organism does not have telomerase, what will happen to its cells after they divide many times?
   a. The telomeres will get shorter and shorter.
   b. The telomeres will get longer and longer.
   c. The telomeres will be replaced with RNA.
   d. The telomeres will stay the same length.
   e. None of the above.

4. On which strand would you expect DNA polymerase Ⅰ to be more active?
   a. the leading strand
   b. the lagging strand
   c. equally active on both strands
   d. neither, only DNA polymerase Ⅲ is involved in synthesis
   e. whichever strand DNA polymerase Ⅲ is not synthesizing

5. Why does the *OriC* have A-T rich sequences?
   a. These are easier for DnaA protein to bind to.
   b. Strands with this composition are more easily separated.
   c. Strands with this composition stay together more stably.
   d. This is the preferred binding site for the pre-replicative complex.
   e. None of the above.

Exploration Questions 思考题

1. What are some of the constraints that make DNA replication so complicated?
2. What are some important similarities and differences between DNA replication and transcription?
3. What are problems that eukaryotic cells face during DNA replication that prokaryotes do not have to deal with?
4. Are lagging strands really synthesized more slowly than leading strands? Why or why not?

Chapter 9 Mutations and Mutation Repair

第 9 章 突变与突变修复

The ability of DNA to hold sequence information accurately is vital for the formation of functional proteins, and therefore, the very survival of cell. DNA is well-suited for the task; it is an unusually stable molecule, due to its compact, double helix structure and its relatively strong covalent bonds. Nonetheless, our genetic material is not invincible. Certain normal cellular processes, as well as chemicals, radiation, and mobile DNA elements pose a constant threat to the integrity of DNA. In this chapter we discuss these dangers and the mechanisms adopted by the cell to counteract their effects.

DNA 对序列信息的准确把握能力对于功能蛋白的形成非常重要，因此，对于细胞的生存至关重要。DNA 很适合完成这一任务；由于 DNA 具有紧密的双螺旋结构和较强的共价键，因此是一种异常稳定的分子。尽管如此，我们的遗传物质并不是牢不可破的。一些正常的细胞过程以及化学物质、物理辐射和可移动 DNA 元件等时时刻刻都在威胁着 DNA 的完整性。本章我们来讨论这些威胁以及细胞采取了什么机理去抵消这些威胁所带来的影响。

9.1 DNA Damage and Mutations

9.1 DNA 损伤与突变

A heritable change in DNA sequence is called a **mutation**. Most frequently, the word mutation is applied to heritable alterations that affect the function of a gene. It is important to distinguish mutations from DNA damage in general, which may or may not be heritable.

DNA 序列中可遗传的改变称为**突变**。大多数时候突变一词用来指影响到基因功能的可遗传改变。将突变与普通的 DNA 损伤区别开来很重要，后者有些是可遗传的，有些是不能遗传的。

The word mutation often carries a negative connotation. This is because the majority of mutations have a neutral or negative effect on the fitness of an organism. However, mutations can occasionally enhance an organism's ability to survive and reproduce. Although rare, advantageous mutations are responsible for the very existence of life on earth. If genomes were perfectly stable, life would never have evolved beyond its earliest stages.

突变一词常带有一点负面的含义，这是因为大多数突变对生物的适应性来说是中性的或不利的。然而，突变偶尔也会增强生物的生存和繁殖能力。有利的突变虽然很少发生，但正是它们才使地球上存在着生命。如果基因组是绝对稳定的话，那么生命就永远也不会从它最早的阶段进化而来。

Our chapter will mostly focus on mutations, and on DNA damage that may lead to mutations, because these changes have the greatest impact on populations of organisms and populations of cells within an organism.

本章我们把注意力放在突变和能引起突变的 DNA 损伤上，因为这些变化对生物种群和生物细胞群体的影响最大。

9.2 Point Mutations

Mutations affecting only one base pair are called **point mutations.** Usually, point mutations involve substitution of one base with another base. If a pyrimidine is replaced with another pyrimidine, or a purine by another purine it is called a **transition.** If a purine is replaced by a pyrimidine or vice-versa, it is called a **transversion.**

9.2.1 Mismatched Base

Mistakes during DNA replication are a common source of transitions and transversions. DNA polymerase occasionally pairs a base on the template strand with a non-complementary base. Usually, the mistake is recognized and removed using the protein's $3' \rightarrow 5'$ exonuclease activity. However, the error is occasionally left unfixed, creating an abnormality called a **mismatched base** or a **DNA mismatch.**

If the mismatched base is not repaired, it will eventually develop into a mutation (Figure 9.1). In the next round of replication the incorrect base will be used in a template strand, and its complementary base will be added in the new strand. Although there will be nothing structurally wrong with this base pair, it will not correspond to the

9.2 点突变

只影响到一个碱基对的突变叫作**点突变**。通常，点突变涉及一个碱基被另一个碱基替换。如果一个嘧啶被另一个嘧啶或一个嘌呤被另一个嘌呤取代，这样的点突变称为**转换**。如果一个嘌呤被一个嘧啶取代或反之，这样的点突变称为**颠换**。

9.2.1 错配的碱基

DNA复制过程中出现错误是转换和颠换的常见来源。DNA聚合酶偶尔会配上一个与模板链上的碱基不互补的碱基。通常，这种错误能被该酶的$3' \rightarrow 5'$外切核酸酶活性识别并加以去除。然而，这样的错误偶尔会留在DNA上从而造成一种异常情况，这被称为**错配碱基**或**DNA错配**。

如果这样的错配碱基没有被修复，最终它将发展成一种突变（图9.1）。在下一轮复制中，这一不正确的碱基出现在一条模板链上，与它互补的碱基会被加到新链中去。虽然就这一碱基对而言没有任何结构上的错误，但它已经与基因

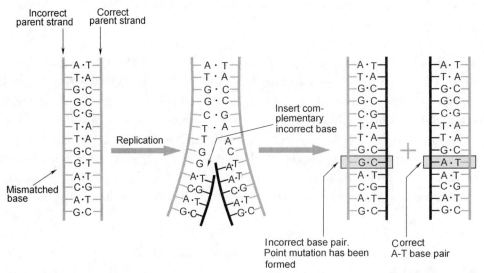

Figure 9.1 A mismatched base can lead to a point mutation after one round of DNA synthesis. The point mutation will occur in only one of the two daughter molecules

图9.1 一个错配的碱基在一轮DNA合成后能导致点突变。点突变只出现在其中一条子链中

original, normal base pair at that position in the gene.

9.2.2 Spontaneous Mutations

The other common source of point mutations is chemical alteration of a base. Some of these alterations occur naturally in the chemical environment of the cell, and are called **spontaneous mutations**.

(1) Deamination

Deamination, the removal of an amino group from a nucleotide, occurs by simple reaction with water (Figure 9.2). When deamination occurs on cytosine, a uracil base is produced. In each human cell, this change naturally occurs approximately 100 times per day. As with a mismatched base, if left uncorrected each uracil will cause a mutation during the next round of replication. Because uracil is very similar to thymine, it is paired to adenine during DNA replication. In the original gene, however, cytosine is paired to guanine. Thus, a point mutation is introduced into the gene.

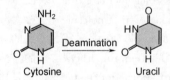

Figure 9.2 Deamination of cytosine produces uracil

(2) Depurination

Another common spontaneous mutation is **depurination**, the detachment of a purine base from the DNA backbone (Figure 9.3). The N-glycosyl linkage between purine bases and deoxyribose is particularly vulnerable to reaction with water. Every day, approximately 5 000~10 000 nucleotides in a human cell's genome lose a purine. Once again, this can cause a mutation and other difficulties during the next round of replication. If DNA polymerase reaches a nucleotide without a base, it may well add the wrong base to the growing strand, as the template is not providing it with any information. Another possible consequence is that replication will simply stop, requiring the cell to employ a bypass mechanism that introduces mutations.

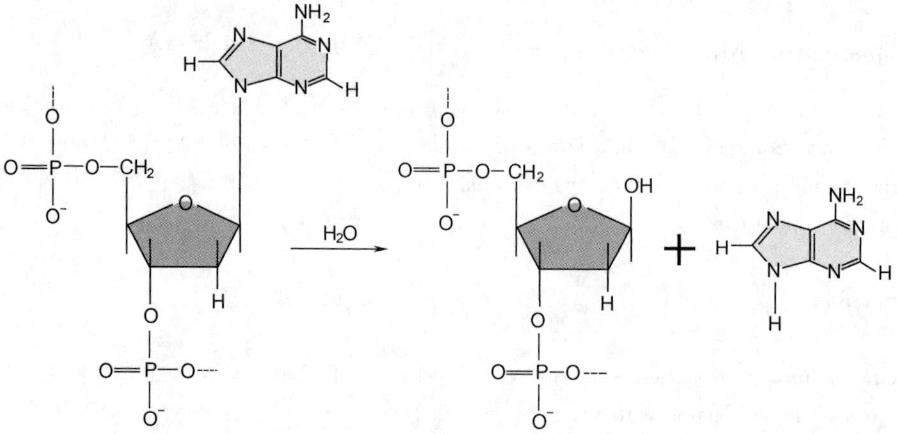

Figure 9.3 Depurination

图 9.3 脱嘌呤

(3) Tautomerization

Not all spontaneous mutations involve chemical reactions. Tautomerization is another source of spontaneous mutation. **Tautomers** are molecules that can interconvert between two or more different forms. A simple example of tautomers is the enol and ketone (Figure 9.4). This conversion occurs because no atomic rearrangement of the molecule is required, and both versions of the molecule are somewhat energetically stable. Normally, however, one form of the molecule is more stable than the other, and the molecule exists predominantly in the more stable form.

(3) 互变异构化

并不是所有的自发突变都涉及化学反应。互变异构化是自发突变的另一个来源。**互变异构体**是能够在两种或更多种不同结构形式之间转换的分子。互变异构体的简单例子是烯醇式和酮式（图9.4）。这种转换之所以发生是因为不需要对分子中的原子进行重排并且两种形式的分子从能量角度来看都是比较稳定的。然而，一般情况下一种形式的分子总是比另一种形式更稳定，分子多数时候会以更稳定的形式存在。

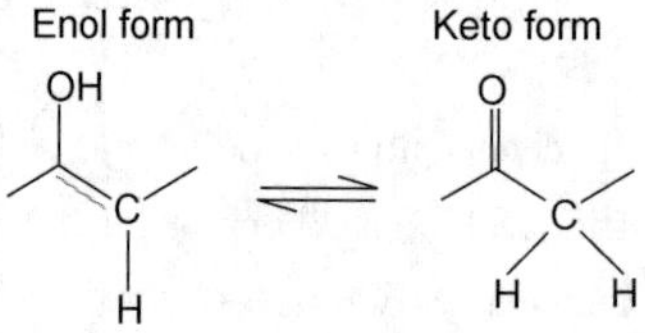

Figure 9.4 Simple example of tautomerization

图 9.4 互变异构化的简单例子

The nitrogenous bases with which we are familiar also happen to be tautomers (Figure 9.5). The more stable form of each base is the structure that we have learned. However, the bases can also interconvert to have other structures. A and C tautomerize to their imino forms, usually written A* and C*. T and G tautomerize to their enol forms, T* and G*. These alternate bases have very different hydrogen bonding prop-

我们所熟悉的含氮碱基恰巧也是互变异构体（图9.5）。每种碱基更稳定的形式是我们前面学习过的结构。然而，它们也可以互相转换成其他结构。A 和 C 可以互变异构化为它们的亚氨基式，通常写成 A* 和 C*。T 和 G 可以互变异构化为它们的烯醇式 T* 和 G*。这些候补碱基具有非常不一样的氢键成键特性：A* 与 C 结合，C* 与

208 | Chapter 9 Mutations and Mutation Repair

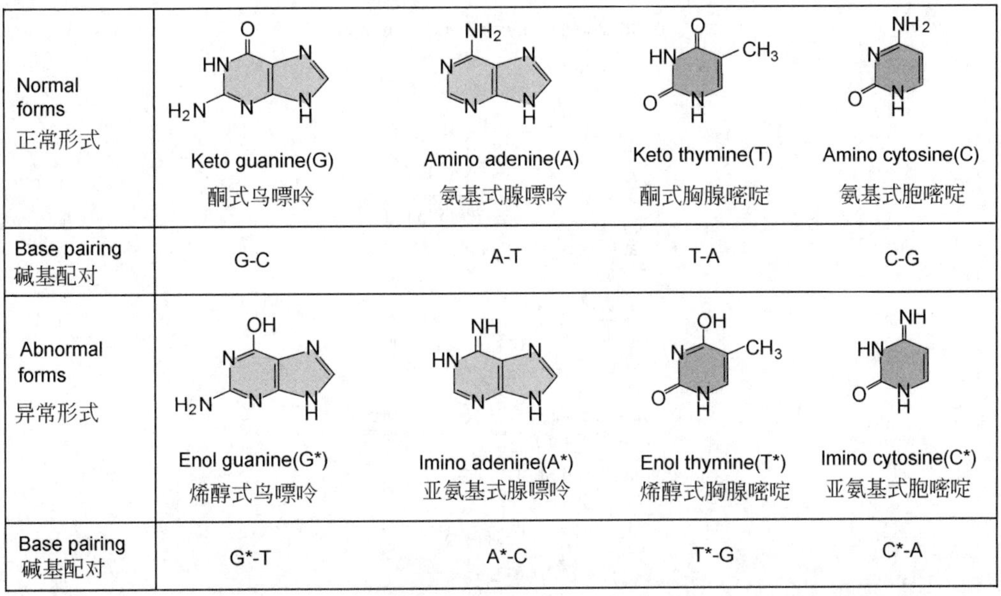

Normal forms 正常形式	Keto guanine(G) 酮式鸟嘌呤	Amino adenine(A) 氨基式腺嘌呤	Keto thymine(T) 酮式胸腺嘧啶	Amino cytosine(C) 氨基式胞嘧啶
Base pairing 碱基配对	G-C	A-T	T-A	C-G
Abnormal forms 异常形式	Enol guanine(G*) 烯醇式鸟嘌呤	Imino adenine(A*) 亚氨基式腺嘌呤	Enol thymine(T*) 烯醇式胸腺嘧啶	Imino cytosine(C*) 亚氨基式胞嘧啶
Base pairing 碱基配对	G*-T	A*-C	T*-G	C*-A

Figure 9.5　Tautomeric forms of the DNA bases

图 9.5　DNA 碱基的互变异构形式

erties: A* binds to C, C* binds to A, T* binds to G, and G* binds to T. Although the normal form of these bases is favored by approximately 10 000 : 1, if the tautomerization occurs during replication, either in the template or in the base being added, it can lead to the incorporation of an incorrect base in the newly synthesized strand. After the base reverts to its normal form, a DNA mismatch will result.

9.2.3　Induced Mutations

In many cases, point mutations are caused by factors that are not integral parts of the cell's environment. These are called **induced mutations**.

(1) Base Analogues

Chemicals that resemble DNA bases, called **base analogues**, are a powerful source of induced mutations (Figure 9.6). One example is **5-bromouracil (BU)**, which strongly resembles thymine, because the bromine atom is approximately the size of thymine's methyl group. During replication, BU can easily be incorporated into DNA instead of thymine.

BU pairs with adenine, just like thymine; however, mutation due to tautomerization is much more

A 结合，T* 与 G 结合，G* 与 T 结合。虽然这些碱基的正常形式以大约 10 000 : 1 的比例得到偏爱，但是如果这种互变异构化发生在复制过程中，那么它不管是发生在模板链上还是发生在将要加上去的碱基中，都会导致新合成链中整合进一个不正确碱基。当这一碱基恢复到它的正常形式后，将出现一个 DNA 错配。

9.2.3　诱发突变

在许多情形下，点突变由并不是构成细胞环境整体所必需的因子引起，这样的突变被称为**诱发突变**。

(1) 碱基类似物

与 DNA 碱基相似的化学物质叫作**碱基类似物**，它们是非常有效的诱发突变根源（图9.6）。一个例子是**5-溴尿嘧啶 (BU)**，它与胸腺嘧啶很相像，因为溴原子大小与胸腺嘧啶上的甲基相近。复制过程中，BU 代替胸腺嘧啶而很容易地被整合到 DNA 中。

跟胸腺嘧啶一样，BU 会与腺嘌呤配对；然而，BU 比胸腺嘧啶更容易引起

9.2　Point Mutations

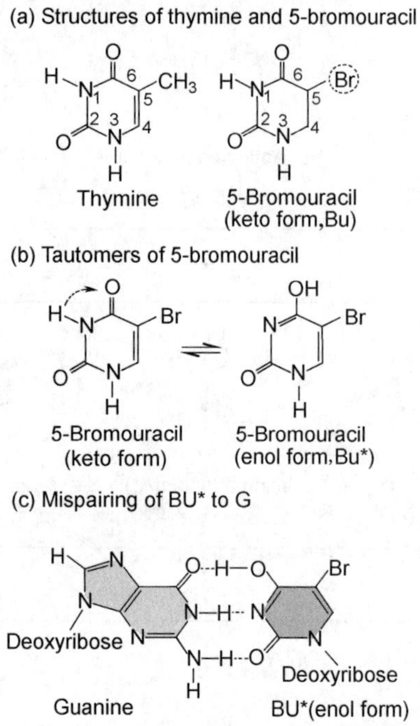

(a) Structures of thymine and 5-bromouracil

Thymine

5-Bromouracil (keto form, Bu)

(b) Tautomers of 5-bromouracil

5-Bromouracil (keto form)

5-Bromouracil (enol form, Bu*)

(c) Mispairing of BU* to G

Guanine

BU*(enol form)

Figure 9.6　5-bromouracil structure and tautomerization. Tautomeric form is common and pairs with G instead of A

likely with BU than with thymine. BU interconverts quite frequently to BU*, a form that pairs to guanine instead of adenine. As with normal tautomeric bases, if this occurs during replication a mutation may occur.

Another example of a mutagenic base analogue is **2-aminopurine**. It resembles adenine, and can be incorporated into DNA as such; however, it is able to pair with cytosine as well as guanine. This leads to mutations in the next round of replication by the mechanism we are now well familiar with.

(2) Alkylating Agents

Another class of molecules with the ability to cause point mutations is **alkylating agent**s (Figure 9.7). These compounds can add alkyl groups to other molecules, including DNA bases. In one extreme, the group might be added in such a way that it is insignificant to the function of the base, or its addition may be easily reversed by the cell. On the other extreme, the

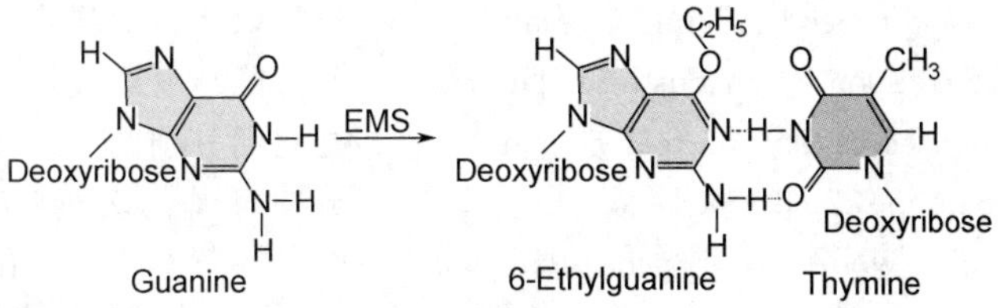

Figure 9.7 Alkylating agent, EMS. Addition of ethyl group to carbonyl of Guanine causes the base to mispair to thymine

图 9.7 烷化剂 EMS。在鸟嘌呤的羰基上加上乙基引起它与胸腺嘧啶发生错配

alteration might be so drastic that it leads to cell death. In between these two extremes is the possibility that alkylation produces a base that is still functional, but has modified function, such as altered base pairing properties. This is the condition most likely to lead to point mutations.

的改变是如此严重以至于细胞因此而死亡。处于两个极端之间的可能情况是，烷化产生的碱基仍旧具有功能，但碱基的功能已被修饰，比如改变了碱基配对性质。这种情况最容易导致点突变。

One example of a mutagenic alkylating agent is **ethylmethane sulfonate (EMS)**. EMS can add an ethyl group ($-CH_2CH_3$), to various sites on a nitrogenous base. The result of this modification depends on where the ethyl group is added. A highly mutagenic modification occurs when the group is added to the ketone of guanine. In this case, the base pairing properties of guanine are changed such that it binds to thymine instead of cytosine. As usual, if unfixed, a full mutation will be present in a DNA molecule after the next round of replication.

具有诱变作用的烷化剂的一个实例是**乙基甲磺酸（EMS）**。EMS 能够将乙基（$-CH_2CH_3$）加到含氮碱基的多个位置上。其后果取决于乙基被加到哪个位置。当乙基被加到鸟嘌呤的酮上时能发生很强的诱变修饰作用。此时，鸟嘌呤的碱基配对特性被改变，它会与胸腺嘧啶配对而不与胞嘧啶配对。同样，如果这样的改变没有得到修复的话就会在下一轮复制产生的 DNA 中出现一个完全的突变。

(3) Nitrous Acid

(3) 亚硝酸

A fairly simple chemical mutagen is **nitrous acid** (Figure 9.8). The oxidative potential of this molecule causes

一种相当简单的化学诱变剂是**亚硝酸**（图 9.8）。这一分子的氧化能力能使胺

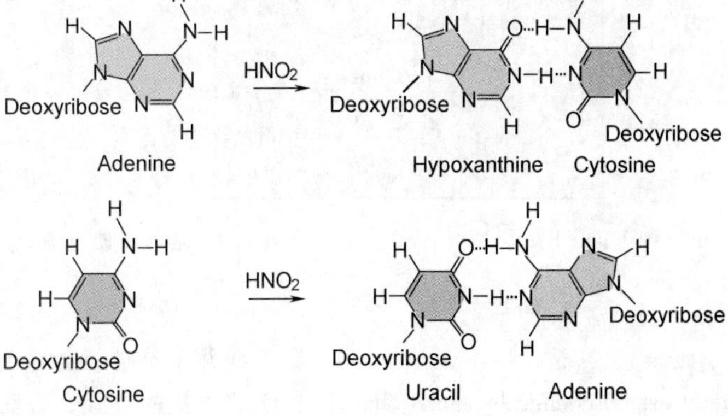

Figure 9.8 Damages caused by nitrous acid, and mispairing of resulting bases

图 9.8 亚硝酸引起的损伤以及损伤碱基的错配

amines to be replaced by ketones. Adenine, guanine, and cytosine can all be acted upon by nitrous acid. The reaction converts them into bases not normally found in DNA: hypoxanthine, xanthine, and uracil, respectively. Hypoxanthine pairs to C, while uracil pairs with A. These binding properties are different from the original base, and may therefore lead to mutation (The conversion of guanine to xanthine does not seriously affect its binding properties). Nitrous acid, EMS, and BU are just a few examples of the many chemicals that may cause point mutations.

(4) Ultraviolet Radiation

Ultraviolet radiation (UV) is a non-chemical means of inducing a point mutation (Figure 9.9). If a photon of UV light is absorbed by two adjacent pyrimidines, a covalent bond can form between the two bases. This is called a **pyrimidine dimer**. If this lesion is left unfixed it can stop DNA replication. Usually, replication can continue, but it cannot tell from the dimerized bases which nucleotides to add in the growing strand. Thus, incorrect nucleotides are frequently incorporated at this point and mutations occur.

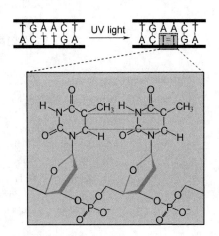

Figure 9.9 UV light causes covalent bonds between pyrimidines, creating a pyrimidine dimer

UV light is not the only way that bases may become covalently linked to each other. Certain chemicals cause the formation of links, between adjacent nucleotides or between nucleotides on opposite strands. In the latter case,

Figure 9.10 Chemicals can cause bases on opposite strands to become covalently bonded, a form of damage called cross-linking

图 9.10 化学物质能引起碱基与相对链上的碱基形成共价键，产生一种称为交联的损伤

the damage is referred to as **cross-linking** (Figure 9.10). A common cross-linking agent is mustard gas, used during several wars of the 20th century.

损伤称为**交联**（图 9.10）。芥子气是一种常见的交联剂，它在 20 世纪的几次战争中曾被使用。

9.3 Insertions and Deletions

9.3 插入和缺失

Instead of alterations of existing bases, **insertions** or **deletions** of bases in the DNA may also occur. As with the DNA damage leading to point mutations, this kind of damage may result from an error during replication or because of chemicals.

除了改变现有碱基外，在 DNA 中也可能会发生碱基的**插入**或**缺失**。与其他能导致点突变的 DNA 损伤一样，这种损伤也可以起因于复制错误或化学物质的作用。

9.3.1 Strand Slippage

9.3.1 链滑动

The main error during replication that causes insertions and deletions is **strand slippage** (Figure 9.11). Normally, a growing DNA strand is tightly base-paired to the template strand. However, when the DNA contains repeated nucleotide sequences, the two strands may slip relative to each other. If they repair incorrectly, a small loop will be produced in one of the strands that cannot bind to the other strand. This loop is called an **insertion/deletion loop** (IDL).

能引起插入或缺失的主要复制错误是**链滑动**（图 9.11）。正常情况下，一条生长中的 DNA 链与模板链发生紧密的碱基配对。然而，当 DNA 中含有核苷酸重复序列时，两条链可能会相对于各自的位置而发生滑动。如果它们不正确地重新配对就会在其中的一条链上产生一个小的、不能与另一条链结合的环。这种环叫作**插入/缺失环**（IDL）。

If the IDL occurs in the template strand, some bases in the template will not be available for synthesis. The new DNA that is made using this template will therefore lack bases, and a deletion is produced. If the IDL occurs in the growing DNA strand, the site of replication will move backwards several bases. When replication continues, these bases will be replicated for a second time, causing extra bases to be added into the growing strand. This causes an insertion.

如果 IDL 出现在模板链上，模板链上的部分碱基将不能用于合成。用这一模板合成出的新 DNA 将因此而少一些碱基，从而产生缺失突变。如果 IDL 出现在 DNA 生长链上，复制的位置将往回退几个碱基。当复制继续进行的时候，这些碱基会又一次被复制，导致在生长链中出现额外的碱基，从而产生插入突变。

Strand slippage during replication is an important source of mutations in mammals. Mammalian genomes contain

复制中发生链滑动是哺乳动物突变的一个重要来源。哺乳动物基因组中含有许

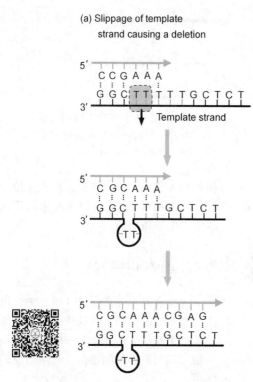

Figure 9.11 Strand slippage occurs in regions with repeat sequences during DNA replication, creating loops that cause insertion or deletions

图 9.11 DNA 复制过程中在含有重复序列的区域发生链滑动，产生能引起插入或缺失的环

many regions with repeated nucleotide sequences of $[A]n$ or $[CA]n$, called **microsatellites**. If slippage in these areas is not corrected by the cell, a phenotype known as microsatellite instability (MSI) is produced, in which the cell rapidly acquires mutations. Many cancerous cells are known to have MSI.

多具有$[A]n$或$[CA]n$这样的重复核苷酸序列的称为**微卫星**的区域。如果在这些区域的滑动没有被细胞修复，将产生一种称为微卫星不稳定性（MSI）的表型，这样的细胞以很快的速度产生突变。已知许多癌细胞都含有 MSI。

9.3.2 Transposons

Insertions and deletions may also be caused by DNA elements called **transposons**. We will discuss these elements in detail in the following chapter. For now it is sufficient to know that transposons are DNA elements of hundreds or thousands of nucleotides that occasionally move from one site of the genome to another. When a transposon relocates to another area of the genome it causes an insertion.

Transposons are usually not selective about where they reintegrate, and may well incorporate themselves into functionally important areas, such as the coding region of a gene. In this case the insertional mutation

9.3.2 转座子

插入和缺失也可能由称为**转座子**的 DNA 元件引起。我们将在下一章中详细讨论这些元件。现在我们只要知道转座子是由几百个或几千个核苷酸组成的、有时会从基因组的一个位置移动到另一个位置的 DNA 元件。当转座子迁移到基因组的另一区域时，它就会引起插入突变。

转座子通常对它们将要重新整合进去的位置并不加以选择，因此很有可能就将它们自己整合进了具有重要功能的区域，比如基因的编码区域。在这种情形

may have serious phenotypic results. Certain transposons can also be responsible for deletions. When a transposon leaves its original location, double stranded cuts must be made on either side of the element. Depending on which kind of mechanism the cell uses to repair the break, a deletion may occur at the site. Specifically, non-homologous end joining may lead to deletion, whereas homologous end joining does not. We will discuss these mechanisms later in the chapter and in chapter 10.

9.3.3 Intercalating Agents

Chemicals may also be responsible for insertions and deletions. **Intercalating agents** are molecules that resemble base pairs, and can insert themselves into the double helix (Figure 9.12). Intercalation causes neighboring base pairs to be pushed apart. During replication, the extra space in the template strand can cause DNA polymerase to add an extra nucleotide in the newly synthesized strand. This extra nucleotide constitutes an insertion. Intercalating agents can also cause insertions and deletions by promoting strand slippage during replication.

下，插入突变可能会导致严重的表型后果。一些转座子还会引起缺失突变。当一个转座子离开它原先的位置时，必须在它的两边都产生双链切割。在这一位置有可能产生缺失突变，而这取决于细胞采取什么样的机理对此断裂进行修复。具体来说，非同源末端的连接可能导致缺失突变，而同源末端的连接不会导致缺失突变。我们将在本章的后面和第 10 章中讨论这些机理。

9.3.3 嵌入剂

化学物质也能造成插入和缺失突变。**嵌入剂**是与碱基对相似、能插入 DNA 双螺旋中的分子（图 9.12）。嵌入作用使邻近的碱基对被挤开。在复制过程中，模板链中出现的额外空间会使 DNA 聚合酶在新合成链中多加一个核苷酸。这一额外核苷酸构成了一个插入突变。嵌入剂还能通过促进复制过程中的链滑动而引起插入和缺失突变。

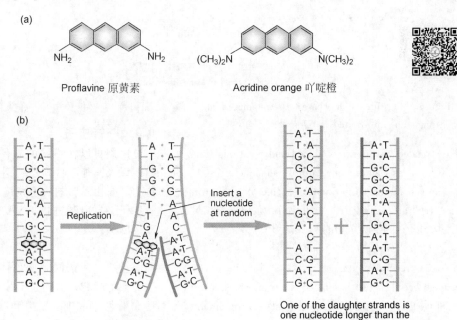

Figure 9.12 (a) Structures of two common intercalating agents; (b) Intercalating agents cause insertions and deletions during DNA replication

图 9.12 (a) 两种常见嵌入剂的结构; (b) 嵌入剂引起 DNA 复制过程中的插入和缺失

9.4 Large-Scale DNA Changes

Aside from the DNA damage we have already covered, larger and more dramatic changes to chromosomes are possible. One example is a **translocation**, in which two regions on distinct, non-homologous chromosomes swap positions (Figure 9.13). Another example is **inversion**, in which the orientation of a chromosomal region is reversed (Figure 9.14). Such rearrangements often result from double-stranded breaks to the DNA.

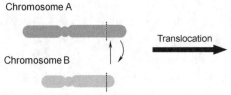

Figure 9.13 In translocations, large pieces of DNA are swapped between chromosomes

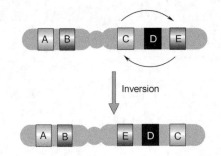

Figure 9.14 Inversion, where large regions of a chromosome are rotated 180 degrees

Double-stranded breaks are produced under various conditions. X-rays, as well as some natural metabolic processes, cause the release of dangerously reactive oxygen atoms called 'free radicals'. Free radicals attacking a DNA molecule can result in double-stranded breaks.

Homologous recombination, the process by which DNA is swapped between two homologous chromosomes, may also malfunction to produce chromosomal rearrangements, as well as deletions, and insertions. Also, mistakes during cell division may cause large scale damage. In extreme cases a cell may completely lose a chromosome, or gain an extra copy of a chromosome.

9.4 大规模 DNA 变化

除了我们已经涉及的 DNA 损伤外，更大和更剧烈的变化也可能在染色体上发生。一个例子是**易位**，即位于非同源染色体上的两个互不相同的区域发生位置交换（图 9.13）。另一个例子是**倒位**，即一个染色体区域的方向颠倒了过来（图 9.14）。这样的重排常常来自 DNA 的双链断裂。

图 9.13 在易位中，大片段 DNA 在染色体之间发生交换

图 9.14 在倒位中，一条染色体上较大的区域发生了 180 度旋转

双链断裂可以在不同的条件下发生。X 射线以及其他一些自然代谢过程能产生称为"自由基"的具有危险反应活性的氧原子。自由基进攻 DNA 分子能导致双链断裂。

同源重组（两条同源染色体上 DNA 发生交换的过程）有时也会发生故障从而造成染色体重排、缺失和插入。还有，细胞分裂中发生错误也会引起大规模损伤。在极端的情况下，细胞可能丢失一整条染色体，或额外得到一条染色体。

9.5 Consequences of DNA Mutations

We have now discussed a wide variety of mutations and their causes, but have not discussed their consequences for the cell. These consequences range from unnoticeable to lethal, depending on the type of mutation and its context.

9.5.1 Consequences of Point Mutations

Point mutations in a coding region are often classified as silent, missense, or nonsense mutations, depending on their effects (Figure 9.15).

9.5 DNA 突变的后果

现在我们已经讨论了很多种突变类型以及它们的起因，但还没有讨论它们对细胞造成的影响。根据突变类型和发生的背景不同，突变造成的影响严重程度差别很大，轻微的可以察觉不到，严重的则足以致命。

9.5.1 点突变的后果

依据它们所产生的影响，在编码区域发生的点突变常被分为沉默突变、错义突变和无义突变（图9.15）。

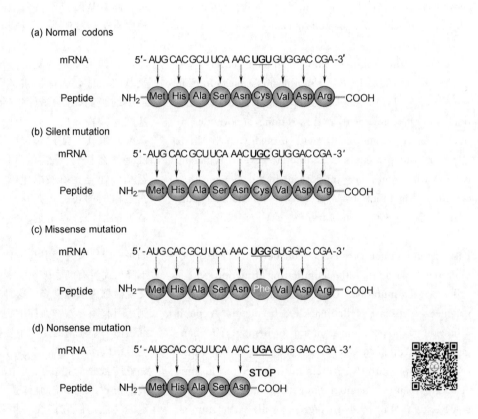

Figure 9.15 Summary of three classes of point mutations

图 9.15 三种类型的点突变小结

Silent mutations are base substitutions which do not alter the amino acid sequence of the protein produced from the mutated gene. This is possible because the genetic code is redundant, and multiple codons may correspond to the same amino acid. In particular, the last base of a codon can often be changed without consequence. A mutation changing a codon to a different codon that

沉默突变是不改变突变基因所编码的蛋白质中氨基酸序列的碱基替换。这是有可能的，因为遗传密码是冗余的，多个密码子对应的是同一个氨基酸。具体来说，密码子的最后一个碱基的改变一般不会带来影响。把一个密码子改变成另一个编码同一个氨基酸的密码子的突变

codes for the same amino acid will have no effect on the protein produced by the gene. There are exceptions to this rule; if the base change alters a splice site, for instance, it could seriously damage the protein produced even if it did not fundamentally alter a codon's information.

More often base substitutions do affect the protein produced from the mutated gene. If the mutation occurs such that one codon is changed to another codon that codes for a different amino acid, it is called a **missense mutation**. This kind of mutation changes only one amino acid in the protein produced from the gene. The severity of the mutation on the protein's function can vary widely, depending on where the amino acid is in the protein and which amino acid replaced the original.

Usually, if the original amino acid is replaced by a similar amino acid, the effect of the missense mutation is minimized. For instance, replacing lysine, a positively charge amino acid, with arginine, another positively charge amino acid, will probably have little effect on the protein's function. However, replacing lysine with a non-polar amino acid like isoleucine may certainly alter function, especially if the change occurs near the active site of an enzyme.

If a missense mutation causes a reduction in the activity of a protein, without fully eliminating function, it is called a **leaky mutation**. An example of such a mutation produces a disease called sickle-cell anemia. A point mutation in the gene for a subunit of hemoglobin changes a GAG codon to a GTG codon. As a result, valine is incorporated into the protein in the place of glutamic acid. The resulting hemoglobin protein still functions to some extent, but tends to aggregate into large clumps. The hemoglobin clumps cause red blood cells carrying the protein to adopt an odd sickle shape. These cells circulate with great difficulty throughout the body and produce a range of severe symptoms.

Occasionally, point mutations change a codon within a coding region to a stop codon. This produces a **nonsense mutation**. The mRNA produced from the mutated gene is only translated as far as the new stop codon, and a trun-

将对这一基因产生的蛋白质没有影响。这一规则也有例外的情况，例如，如果碱基变化改变了一个剪接位点，则即使它没有从根本上改变密码子的信息也会严重损坏所产生的蛋白质。

更常见的情形是，碱基替换确实会影响到突变基因所产生的蛋白质。如果突变发生以后一个密码子改变成了编码不同氨基酸的另一个密码子，这样的突变称为**错义突变**。这种类型的突变只改变突变基因所产生蛋白质中的一个氨基酸。错义突变对蛋白质功能影响的严重程度差别很大，取决于此氨基酸位于蛋白质的什么位置以及原有氨基酸被哪一个氨基酸取代。

通常，如果原来的氨基酸被一个类似的氨基酸取代，则这种错义突变的影响是最小的。例如，赖氨酸（带正电荷的氨基酸）被精氨酸（另一种带正电荷的氨基酸）取代，或许对蛋白质的功能几乎没有什么影响。然而，用一种非极性的氨基酸如异亮氨酸取代赖氨酸无疑将改变蛋白质的功能，尤其是当突变发生在靠近酶活性位点附近的时候。

如果一个错义突变引起一种蛋白质的活性下降，而没有彻底破坏它的功能，这样的突变被称为**渗漏突变**。这种突变的一个实例是引起镰刀型贫血症的突变。在编码血红蛋白一个亚基的基因中发生了一个点突变，密码子从GAG变成了GTG。结果，缬氨酸被整合进蛋白质中原来属于谷氨酸的位置。得到的血红蛋白仍然具有一定的功能，但它们容易聚集在一起形成大的凝块。血红蛋白凝块使得携带了它们的红细胞呈现一种古怪的镰刀形状。这些细胞在整个身体的循环系统中流动非常困难，产生严重程度不同的症状。

有时点突变会将一个在编码区中的密码子变成终止密码子，这时就会产生**无义突变**。从突变基因产生的mRNA只能被翻译到新的终止密码子所在的位置，

cated protein is produced. The size of the protein depends on where the mutation occurs within the coding region. Usually, truncation is so severe that the protein is completely nonfunctional. In many cases, mRNA with a nonsense mutation is degraded before it even has the chance to be translated.

9.5.2 Consequences of Insertions and Deletions

Mutations involving insertions and deletions are more likely to produce nonfunctional proteins than point mutations. This is because the insertion or deletion of even a single base often seriously alters the way an mRNA transcript is translated.

We know that mRNA bases are translated in groups of three bases called codons. However, there are several ways in which a sequence of bases can be grouped into codons. Each possible grouping is called a **reading frame**. The start codon, AUG, determines the reading frame. The ribosome knows that the following three bases after AUG are read as one codon, and the three bases after that are read as one codon, and so forth.

If a nucleotide is inserted into a coding region, all nucleotides downstream of it are shifted over by one. This completely alters their grouping into codons. The effect can be illustrated by the following example.
Consider the sentence:
　　"The cat ate its rat."
If we add one letter near the beginning of the sentence, but still read the letters in groups of three, we get a sentence that makes no sense:
　　"Txh eca tat eit sra t."

Because the reading frame has been moved, such a mutation is called a **frameshift** (Figure 9.16). A frameshift may also occur if a base is deleted from a coding region. In fact, an insertion or deletion of any number of nucleotides other than a multiple of three (which would maintain the reading frame) will have this effect.

Frameshifts have drastic consequences during transla-

产生的是一个截短了的蛋白质。得到的蛋白质大小依赖于突变发生在编码区所在的位置。通常，截短是如此严重以至于蛋白质将彻底失去功能。在很多情形下，具有无义突变的mRNA甚至在它有机会被翻译之前就被降解掉了。

9.5.2 插入和缺失的后果

涉及插入和缺失的突变比点突变更容易产生无功能的蛋白质。这是因为即使是一个碱基的插入或缺失也常常会严重地改变mRNA转录本被翻译的方式。

我们知道，mRNA的碱基以三个为一组构成密码子被翻译。然而，将碱基序列组织成密码子的方式可以有几种。每种可能的组织方式称为**读码框**。起始密码子AUG决定了读码框开始的位置。核糖体知道AUG后面的三个碱基应该被读为一个密码子，它的后面三个碱基被读为另一个密码子，以此类推。

如果一个核苷酸插入到了编码区中，那么所有下游的核苷酸将被向后移动一个核苷酸的位置。这会彻底改变它们组成密码子的顺序。其影响可以用下面的例子来说明。看下面的句子：
　　"The cat ate its rat."（意思是：猫吃老鼠。）
如果我们在句子的开头加一个字母，并且仍然以三个字母为一组来读这些字母，我们得到的是一个没有意义的句子：
　　"Txh eca tat eit sra t."

由于读码框已经被移动过了，这样的突变称为**移码突变**（图9.16）。如果一个碱基从编码区中删去也会发生移码突变。事实上，以任何不是三的倍数的数目插入或缺失核苷酸都会产生这样的后果（以三的倍数插入或缺失核苷酸将保持读码框不变）。

移码突变在翻译中会产生剧烈的影响。

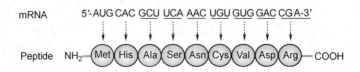

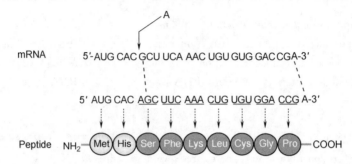

Figure 9.16 Insertions and deletions of bases can cause frameshifts mutations

图 9.16 碱基的插入和缺失能引起移码突变

tion. Codons upstream of the site of mutation are unaffected. However, most codons downstream of the mutation will be changed to completely different codons, that code for completely different amino acids or stop translation. If a protein is even produced, it will almost always be nonfunctional.

9.5.3 Consequences of Translocations

Large scale genomic changes, such as translocations, may have a range of effects on protein production. If the breakpoints in a translocation occur in unimportant DNA sequences, the rearrangement may have little effect on the genes within the relevant regions. On the other hand, if breakpoints occur within genes, the protein produced is generally non-functional.

In rare cases, a translocation breaks two genes within their coding regions, causing one piece of a gene to be fused to a piece of a different gene (Figure 9.17). This mutant gene can produce a **fusion protein**, a polypeptide consisting of sequences from two different proteins. In some cases, these fusion proteins are active, but have new functions. A kind of cancer called chronic myelogenous leukemia is caused by a translocation that creates a fusion protein.

位于突变位点上游的密码子没有受到影响。然而，突变位点下游的大多数密码子将变成彻底不同的密码子，它们编码完全不同的氨基酸或变成终止密码子。如果蛋白质还是被生产出来了，那么它基本上是没有功能的。

9.5.3 易位的后果

大规模的基因组变化，比如易位，会对蛋白质的合成产生不同程度的影响。如果易位的断裂点位于不重要的 DNA 序列上，则易位造成的重排对相关区域的基因几乎不会有影响。而如果断裂点位于基因内部，则产生的蛋白质一般来说是没有功能的。

在罕见的情况下，易位打断了两个基因的编码区，引起一个基因的片段连接到了另一个基因的片段上（图 9.17）。这一突变基因能够产生**融合蛋白**，即由两种不同蛋白质序列组成的多肽。有时，这些融合蛋白具有活性，但具有不同的功能。一种称为慢性骨髓性白血病的癌症就是由易位产生了一种融合蛋白而引起的。

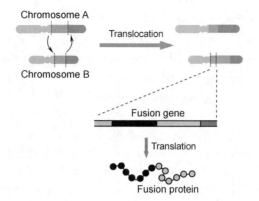

Figure 9.17 Translocations can create fusion proteins when breakage occurs in the coding region of two genes on each chromosome

图 9.17 当链的断裂发生在两条染色体上两个基因的编码区内时，易位能产生融合蛋白

In some instances, a translocation does not disrupt a coding region, but rather places the coding region downstream of a new promoter. The amount of protein made from such a mutation will depend on the properties of the promoter. Another kind of cancer called Burkitt's lymphoma is caused when translocation moves a gene responsible for cell proliferation, *c-myc*, downstream of a very active promoter (Figure 9.18).

在有些例子中，易位没有扰乱编码区，但却把编码区放到了一个新的启动子下游。这种突变得到的蛋白质产量将取决于启动子的性质。另一种称为伯基特淋巴瘤的癌症就是由易位引起的，易位将一个负责细胞增殖的基因 *c-myc* 移到了一个很活跃的启动子的下游（图 9.18）。

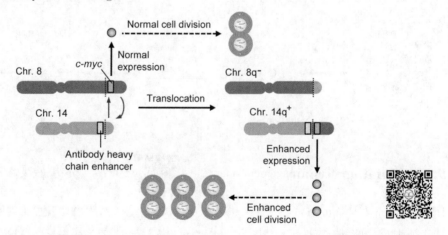

Figure 9.18 An example of chromosomal translocation

图 9.18 染色体易位的一个实例

9.5.4 Mutation Hot Spots

We have already seen that strand slippage occurs more frequently in sites of the genome that have many short sequence repeats. Sites such as these, which accumulate mutations more frequently than the rest of the genome, are called **mutation hot spots**. Another important example of mutation hot spots are CG sequences in which the C nucleotide has been modified with a methyl group.

9.5.4 突变热点

我们已经看到，链滑动在含有许多短序列重复的基因组位置发生得更为频繁。像这样一些比基因组其他位置更频繁地积累突变的位置称为**突变热点**。突变热点另一个重要例子是含有甲基化了的 C 的 CG 序列。

9.5 Consequences of DNA Mutations | *221*

Methylation of DNA is often used by cells to silence transcription. However, methylation of CG sequences provides a unique problem. When methylated C's undergo deamination, a fairly common occurrence in cells, they become indistinguishable from thymine [Figure 9.19(a)]. A mismatch occurs, but the cell's mismatch repair machinery has no way to know which nucleotide in the mismatch is correct—the T or the G. As a result, the mismatch may be incorrectly repaired. Base substitutions frequently occur at these methylated CG sequences [Figure 9.19(b)].

细胞通常采用 DNA 甲基化的方式来沉默转录。然而，CG 的甲基化产生了一个独特的问题。那就是，当甲基化了的 C 经历脱氨基反应（这一反应在细胞中会经常发生）时，它们转变成了与胸腺嘧啶不能加以区分的产物［图 9.19(a)］。这样，就会发生碱基错配，而细胞的错配修复系统没有办法知道错配碱基对中到底哪一个是正确的，即不知道 T 还是 G 是正确的。结果，错配处可能会被错误地修复，导致在甲基化的 CG 序列处频繁地发生碱基替换［图 9.19(b)］。

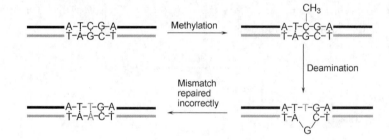

Figure 9.19 Mutation hot spots

图 9.19 突变热点

9.6 Mutation Repair

9.6 突变修复

If the long list of mutations we have covered in this chapter could not be repaired, life would be in deep trouble. Although small rates of mutation drive evolution, high rates of mutation make it impossible to faithfully pass genetic information from generation to generation. Even within the lifetime of one organism, such a slew of mutations would be disastrous. In animals, a small set of mutations in certain genes can put cells on the path to explosive growth and cancer.

Fortunately, all cells contain mechanisms to deal with mutations. Not all mutations can be repaired; translo-

如果本章中我们讲到的这一长串突变类型都不能被修复的话，那么生命会陷于极度的麻烦之中。虽然低的突变率推动了进化，但高的突变率会使遗传信息在世代之间不可能得到忠实传递。即使在一种生物的生活时期内，这么多的突变也将是灾难性的。在动物中，一些基因只要出现一小部分突变就能将细胞置于爆发性生长和产生癌症的境地。

幸运的是，所有细胞都具备一些机理来应对这些突变。不是所有的突变都能被

cations and transpositions, for example, are practically impossible to reverse. However, there are generally solutions for the more common mutations.

9.6.1 Direct Reversal

In the simplest cases, damage to a base can be reversed without having to remove the base. Pyrimidine dimers are one kind of lesion that can be directly repaired. The enzyme **photolyase** is used to break the covalent bonds between the pyrimidines, restoring the bases to their original states (Figure 9.20). Another example of direct reversal is the repair of a harmful alkylation to guanine. In this case, **methyltransferase** transfers the mutagenic alkyl (methyl) group onto one of its own amino acids. Methyltransferase is only good for one repair, and then is discarded by the cell (Figure 9.21).

9.6.1 直接回复

在最简单的情形下，一个碱基的损伤可以在不一定要移去这一碱基的情况下得到逆转。嘧啶二聚体就是这样一种能够被直接修复的损伤。**光解酶**用来断开嘧啶二聚体之间的共价键，从而将这两个碱基恢复到它们原来的状态（图9.20）。另一个直接逆转的例子是对鸟嘌呤烷基化损伤的修复。在这一例子中，**甲基转移酶**将诱变性的烷基（甲基）转移到它自身的氨基酸上。一个甲基转移酶只能用于一次修复，用过之后细胞就将它丢弃了（图9.21）。

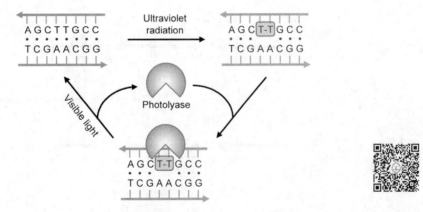

Figure 9.20　Photoreactivation by photolyase

图9.20　光解酶的光复活作用

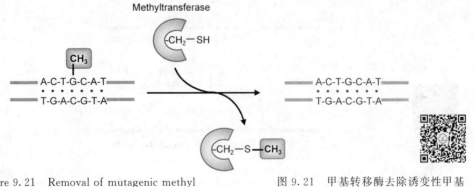

Figure 9.21　Removal of mutagenic methyl group by methyltransferase

图9.21　甲基转移酶去除诱变性甲基

9.6.2 Mismatch Repair

The two main errors that occur during DNA replication,

9.6.2 错配修复

错配的碱基和链滑动这两种主要的错误

mismatched bases and strand slippage, are both corrected by a mechanism called **mismatch repair (MMR)** (Figure 9.22). This process exists in both prokaryotes and eukaryotes. We discuss the method in *E. coli* here because it is simpler and better understood.

都是由**错配修复（MMR）**这一机理来矫正的（图9.22）。这一过程在原核生物和真核生物中都存在。在此我们讨论大肠杆菌中的方法，因为它更简单、更好理解。

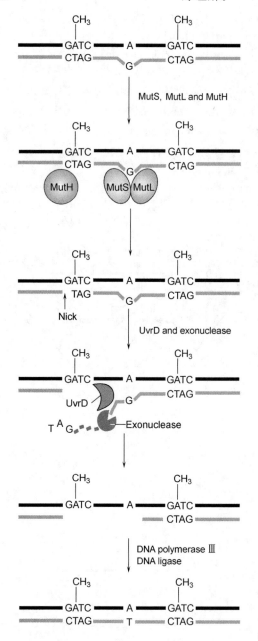

Figure 9.22　Mismatch repair in *E. coli*　　　　图9.22　大肠杆菌的错配修复

In a mismatched base pair, one base is correct and one base is abnormal. The incorrect base is almost always in the newly synthesized strand, as the composition of the template strand does not change during replica-

在一对错配的碱基中一个碱基是正确的而另一个碱基是不正确的。不正确的碱基几乎总是在新合成的链上，因为模板链在复制中不发生变化。因此MMR特

tion. MMR therefore specifically targets the mismatched base in the daughter strand for removal. GATC sequences within the *E. coli* genome provide a convenient way to distinguish parent strand from daughter strand. GATC sites are normally methylated on the adenine. However, for several minutes after synthesis of a new strand, the sequence remains unmethylated. MMR acts within this time frame, and is able to recognize the newly synthesized strand as the one with an unmethylated GATC sequence.

In the first step of MMR, the proteins MutS and MutL bind together to the mismatch. Simultaneously, a protein called MutH binds to a GATC sequence some distance away. MutS-MutL and MutH then form a complex that causes a single-stranded nick at the GATC site.

A helicase called UvrD then unwinds the double helix from the site of the nick to slightly past the site of the mismatch. This allows exonucleases to destroy a region of single-stranded DNA containing the error. Finally, DNA polymerase Ⅲ repairs the gap by synthesizing new DNA, presumably without error. The new DNA is then joined to the original strand by the enzyme ligase. Thus, although mismatches and IDLs generally affect only one or several bases, the cell's strategy is to remove a relatively large portion of DNA around the error and resynthesize it completely.

9.6.3 Nucleotide Excision Repair

Mismatch repair usually deals with normal bases that have been misaligned or incorrectly paired. A different repair system is used to deal with nucleotides and bases that have been chemically modified. This repair system is broadly termed **excision repair**, but there are two very distinct forms of this pathway, nucleotide excision repair and base excision repair. Nucleotide excision repair is used for more severe modifications, which may involve more than one nucleotide and may locally alter DNA's normal structure. Base excision repair is used for more common modifications, and is discussed in the next section.

异性地靶向子链中错配的碱基进行移除。基因组中的 GATC 序列提供了一种区别母链和子链的便利途径。GATC 位点中的 A 一般都是甲基化的。然而，在新链合成出来后的几分钟时间内 A 还保持着未甲基化的状态。MMR 正是在这一时间段里发挥作用，它能够识别新合成的链，因为在它上面有未甲基化的 GATC 序列。

在 MMR 的第一步，蛋白质 MutS 和 MutL 结合到错配位置。同时，一种称为 MutH 的蛋白质结合到一定距离以外的 GATC 序列上。MutS-MutL 和 MutH 之后形成一个复合体，在 GATC 位置产生一个单链缺口。

之后，称为 UvrD 的解旋酶从缺口的位置解开双螺旋到错配位点稍微后面一点。这使外切核酸酶得以摧毁含有错误的单链 DNA 区域。最后，DNA 聚合酶Ⅲ通过合成新的 DNA 修补这一空缺，一般来说不会再出错。新 DNA 片段然后由连接酶与原先的链连接起来。这样，虽然错配和 IDL 一般来说只是影响到一个或几个碱基，但细胞的策略是移去错配位置旁边相对较大一块 DNA，然后全部重新合成它。

9.6.3 核苷酸切除修复

错配修复处理的通常是被错排的或不正确配对的正常碱基。对那些发生了化学修饰的核苷酸和碱基，采用的是一种不同的修复系统。这种修复系统被统称为**切除修复**，但这一途径有两种互不相同的形式，即核苷酸切除修复和碱基切除修复。核苷酸切除修复用于更严重的修饰，这种更严重的修饰可能涉及不止一个核苷酸并且可能改变了 DNA 局部的正常结构。碱基切除修复用于较常见的修饰，我们将在下一节中进行讨论。

Nucleotide excision repair (NER) is superficially similar to mismatch repair (Figure 9.23). NER in *E. coli* begins when the proteins UvrA and UvrB bind to the DNA at the site of damage. UvrA soon unbinds, but UvrB stays attached. Next, UvrC binds to UvrB and makes single-stranded cuts on either side of the damaged base. The cuts are generally separated by about 12 nucleotides. The helicase UvrD then completely removes the single-stranded fragment containing the error. The gap is resynthesized by DNA polymerase.

核苷酸切除修复（NER）表面上看与错配修复相似（图 9.23）。大肠杆菌中 NER 开始于蛋白质 UvrA 和 UvrB 结合到 DNA 的损伤部位。UvrA 很快就离开了，但 UvrB 仍旧保持接触。之后，UvrC 结合到 UvrB 上并在损伤碱基的两边各产生一个单链切口。一般来说两个切口之间相距 12 个核苷酸。然后，解旋酶 UvrD 彻底去除含有错误的单链片段。空缺的部分由 DNA 聚合酶重新合成。

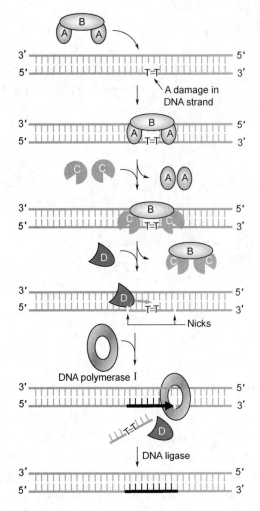

Figure 9.23 Nucleotide excision repair

图 9.23 核苷酸切除修复

NER, like MMR, involves the removal of a relatively large fragment of DNA around the error site, followed by resynthesis of DNA. It is important to remember, however, that NER and MMR recognize different kinds of errors, and use different proteins to fix them. Also, while MMR occurs chiefly during replication, NER is prevalent

像 MMR 一样，NER 也涉及将错误位置附近的一大块 DNA 片段去除并重新合成 DNA。然而，需要记住的重要一点是，NER 和 MMR 识别不同类型的错误并使用不同的蛋白质去进行修复。还有，MMR 主要在复制过程中进行，而 NER

throughout the cell cycle. The NER mechanism is not exactly the same in eukaryotes and prokaryotes, but the differences will not be discussed here.

9.6.4 Base Excision Repair

Base excision repair (BER) is used to replace bases that have undergone common and fairly minor modifications, such as deamination. BER is quite distinct mechanistically from NER and MMR (Figure 9.24). BER begins when a protein called DNA glycosylase binds to a damaged base and removes it by severing the bond that

在整个细胞周期中都能发挥作用。真核生物的 NER 机理与原核生物的不尽相同，但它们之间的不同点不在这儿讨论。

9.6.4 碱基切除修复

碱基切除修复（BER）用于取代经历了常见的和较小修饰（如脱氨基）的碱基。从机理上看，BER 与 NER 和 MMR 有很大不同（图 9.24）。BER 开始的时候，称为 DNA 糖基化酶的蛋白质结合到发生损伤的碱基上并切断它与脱氧核糖的

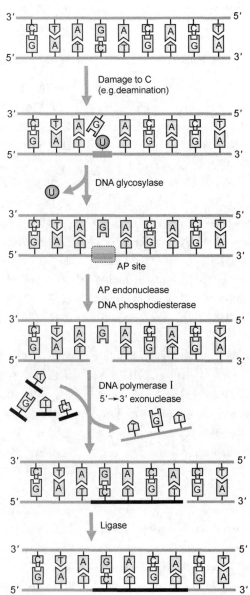

Figure 9.24　Base excision repair

图 9.24　碱基切除修复

connects it to deoxyribose. Cells contain a variety of DNA glycosylases, which recognize and remove different kinds of damaged bases. Next, the sugar and phosphate to which the base was formerly attached are also removed, leaving a small gap in the DNA. Proteins called AP endonuclease and DNA phosphodiesterase are responsible for this step. Finally, DNA polymerase I repairs the gap, adding a normal nucleotide where previously there was a damaged base. In reality, DNA polymerase I uses its unique $5'\rightarrow 3'$ exonuclease function to simulateously remove and resythesize several nucleotides downstream of the gap as well. BER is similar in eukaryotes and prokaryotes.

9.6.5 Double-Stranded Break Repair

Double-stranded breaks, a very serious form of DNA damage, can also be repaired by the cell. Two possible repair pathways are homologous recombination and **non-homologous end joining (NHEJ)**.

In NHEJ, the proteins Ku and DNA-PK bind to each broken end of DNA (Figure 9.25). The proteins on each end bind to the proteins on the other end, causing the ends to re-approach each other. Ku, a helicase, then unwinds each end of DNA, exposing single strands. If a strand on one end is even slightly complementary to a strand on the other end, these can now bind, rejoining the two ends. However, the rejoining is imperfect. Single strands that were not involved in rebinding are degraded, and as a result, non-homologous end joining often results in deletions. NHEJ is very common in mammals. This is probably because most DNA in mammalian genomes does not code for proteins; thus, the chance of a deletion occurring within a coding region and producing a mutation is slim.

Smaller eukaryotes like yeast tend to rely more on homologous recombination for repairing double-stranded breaks. This kind of repair is only possible in diploid (or polyploid) organisms, in which there are two nearly identical copies of each chromosome. After a break in a chromosome, its sister chromosome is used as a template to re-synthesize missing information. We will discuss this mechanism in

9.6.5 双链断裂修复

双链断裂是一种很严重的 DNA 损伤形式，它也能被细胞修复。有两种可能的修复途径，即同源重组和**非同源末端连接（NHEJ）**。

在 NHEJ 中，蛋白质 Ku 和 DNA-PK 结合到 DNA 的两个断头上（图 9.25）。在一个断头上的蛋白质与在另一个断头上的蛋白质结合，使两个断头重新靠拢在一起。Ku 是一种解旋酶，在这之后把每个 DNA 断头处的双链解开，暴露出单链。如果一个断头上的一条链与另一断头上的链有一点互补性，那么它们就可能结合在一起而重新将两个断头连接起来。然而，这样的连接是不完美的。在重新结合中没有起作用的单链被降解，结果，非同源末端连接常常带有缺失。NHEJ 在哺乳动物中十分常见。这可能是因为在哺乳动物基因组中大多数 DNA 并不编码蛋白质；因此，缺失发生在编码区内并产生一个突变的机会很少。

较小的真核生物如酵母更倾向于依赖同源重组来修复双链断裂。这种类型的修复只有在二倍体（或多倍体）生物中才有可能发生，因为它们的每条染色体有几乎完全相同的两个拷贝。在一条染色体发生断裂后，它的姐妹染色体被用来作为重新合成所丢失信息的模板。我们

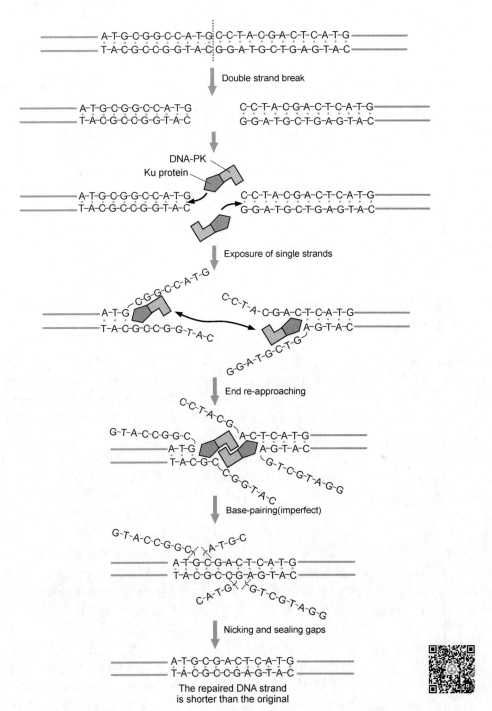

Figure 9.25 Non-homologous end joining to repair a double-stranded break

图 9.25 非同源末端连接用于双链断裂修复

detail in the following chapter. For now it is sufficient to know that the repaired chromosome is not exactly the same as before it was broken, but is quite close because it was repaired with sequences from a nearly identical chromosome.

将在下一章中详细讨论这一机理。现在只要知道，修复的染色体并不与断裂之前的完全一样，但也是非常接近的，因为修复时使用的是与它几乎相同的染色体上的序列。

9.7 Experiments

9.7.1 Nucleotide Excision Repair and Human Disease

Much has been learned about DNA repair pathways by studying humans who are deficient in certain aspects of DNA repair. Such patients have a tendency to accumulate mutations in their DNA much more rapidly than normal. A deficiency in nucleotide excision repair results in a disease called **xeroderma pigmentosum (XP)**. Patients with this disease develop a very high number of skin cancers if exposed to sunlight because they are unable to repair DNA damage from UV light.

By studying cells from XP patients it was found that mutations in several different genes can cause the disease. Many of these studies were performed by fusing cells from two different patients together. This technique is a kind of complementation analysis, similar in principle to the creation of partially diploid cells to study the genes involved in lactose metabolism, explained in chapter 3.

When two cells are fused together, one large cell is formed that has genes from both of the original cells. If these cells have mutations in different genes of the NER pathway, each cell will have a normal copy of the gene that the other cell lacks. Therefore, the fused cell should have at least one good copy of every gene in the NER pathway, and will respond normally when exposed to UV light.

By contrast, if the mutation in both cells that are fused together is in the same gene, the resulting cell will also be deficient for that gene. When it is exposed to UV light, it will be unable to repair the damage. By making many cell fusions between cells from many different patients, it was found that at least eight different genes were involved in XP.

This was an important jumping-off point for studying the disease. Once cells were grouped according to the genes that were mutated, each gene could in principal

9.7 实验研究

9.7.1 核苷酸切除修复与人类疾病

对缺失了某些 DNA 修复功能的患者进行研究可以帮助我们弄清不少 DNA 修复的途径。这样的患者具有一种倾向，他们的 DNA 比正常人的更容易积累突变。有一种核苷酸切除修复缺陷会导致称为**着色性干皮病（XP）**的疾病。这种病的患者如果暴露在阳光下的话会很容易得皮肤癌，因为他们不能够修复由紫外线导致的 DNA 损伤。

通过研究 XP 患者的细胞发现在几个不同的基因中发生突变能够引起这一疾病。这些研究多数是通过将两个不同患者的细胞融合在一起进行的。这一技术是一种互补分析，原理上类似于第 3 章中讲到的用创造出部分二倍体细胞的方法去研究乳糖代谢中涉及的基因。

当两个细胞被融合在一起时，得到的是一个具有两个原先细胞中基因的大细胞。如果两种细胞具有的突变发生在 NER 途径的不同基因上，则每个细胞会有另一个细胞所缺乏的基因的正常拷贝。因此，融合细胞至少具有 NER 途径各个基因的一个正常拷贝，将能够正常地对紫外线作出响应。

相反，如果融合在一起的两个细胞中发生的突变在同一个基因上，得到的细胞在那个基因上仍旧会有缺陷。当它被暴露在紫外线下时，将不能够修复产生的损伤。通过将来自许多不同患者的细胞制备许多融合细胞，发现至少有八个不同的基因与 XP 有关。

对研究这一疾病来说，这是一个很重要的起点。一旦根据它们的基因发生突变的情况对细胞完成了分组，则原理上可

be located and the protein it codes for could be determined. Any number of experiments could then be performed to determine the function of the protein. The fact that most of the names of genes and proteins in the human NER pathway begin with the letters XP indicates the importance of studying this disease for their discovery.

9.7.2 The Ames Test

The **Ames test** is not an experiment, it is a technique used to assess the mutagenic potential of a chemical. A strain of bacteria is used that has a mutation for a gene that synthesizes histidine, and therefore is unable to grow in medium without histidine [Figure 9.26(a)]. The bacteria are then mixed with the chemical to be tested, and the mixture is added to a plate without

以对每个基因进行定位，它们所编码的蛋白质也能够被分析出来。之后就可以开展任何其他实验去测定蛋白质的功能。在人类 NER 途径中的大多数基因和蛋白质的名称都以字母 XP 开头，这一事实表明了这些基因和蛋白质是通过研究这一疾病发现的，由此可见这一研究的重要性。

9.7.2 埃姆斯测验法

埃姆斯测验法不是一个实验，它是用来评定一种化学物质是否具有潜在诱变性的技术。该法使用一个组氨酸合成基因有突变的细菌菌株，它不能够在没有组氨酸的培养基中生长［图 9.26(a)］。之后将细菌与待试的化学物质混合，得到的混合物涂到不含组氨酸的培养基平

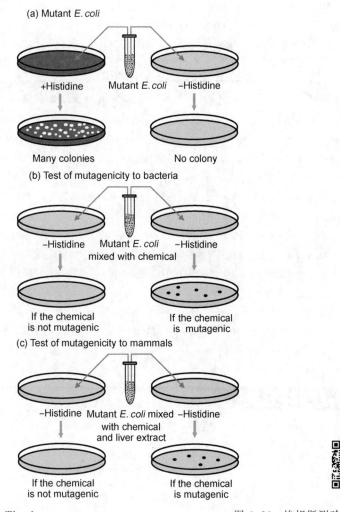

Figure 9.26　The Ames tests　　　　　　　图 9.26　埃姆斯测验法

histidine. If the chemical is mutagenic, there is a probability that in some cells it will cause a mutation that reverses the original mutation in the histidine synthesis gene. These cells will therefore be able to grow on the histidine-free medium and will appear as colonies [Figure 9.26(b)]. The number of colonies on the plate thus gives an indication of how mutagenic the chemical is.

The Ames test is quite convenient, as bacteria are easy to manipulate and grow. However, its relevance to humans and other organisms is often questionable. This is because some chemicals that are not mutagenic are converted to mutagenic molecules by the body's metabolism. *E. coli* does not have the same metabolism as humans, and therefore the real mutagenic potential of the chemical for a human would not be tested by the simple Ames test. To overcome this defect, a liver extract is often added along with the test chemical to mimic the body's natural metabolism [Figure 9.26(c)].

板上。如果这一化学物质是诱变性的，那么它就有可能在一些细胞中引起突变而逆转原来发生在组氨酸合成中的突变。这些细胞因此将能够在不含组氨酸的培养基上生长，出现菌落[图9.26(b)]。平板上的菌落数指示了该化学物质诱变性的强弱。

埃姆斯测验法是相当方便的，因为细菌易于操作和生长。然而，它与人类和其他生物之间的关联性常常被质疑。这是因为一些并不具有诱变性的化学物质会被生物体内的代谢功能转换成有诱变性的分子。大肠杆菌并没有跟人相同的代谢途径，因此对人类真正具有诱变性的化学物质进行试验不能采用简单的埃姆斯测验法。为了克服这一缺点，经常在待测化学物质中加入肝脏提取物来模拟人体内部的自然代谢功能[图9.26(c)]。

Summary   小结

In spite of the relative stability of the double helix, damage to the information held in the DNA occurs through a variety of mechanisms. Heritable damage to the DNA is called a mutation. Many mutations can be spontaneous or induced, depending on the source of the mutation. Mutations are also frequently caused by errors during DNA replication. Point mutations generally involve substitutions to individual bases. They can be quite damaging, but are usually less serious than insertion or deletions, which lead to frameshift mutations. Fortunately, much DNA damage can be repaired before becoming, or leading to, a mutation. Damaged bases can sometimes be directly repaired, but more commonly they are removed and resynthesized.

尽管双螺旋分子具有相对高的稳定性，对保存在DNA中信息的损害还是会以不同的方式发生。对DNA造成的可遗传的损伤称为突变。许多突变是自发的或诱发的，取决于突变的来源。在DNA复制过程中也会频繁地发生错误。点突变一般涉及单个碱基的替换。它们可能会造成很大危害，但通常不如插入和缺失带来的危害严重，后者导致移码突变。幸运的是，许多DNA损伤在变成突变或导致突变之前能够被修复。损伤的碱基有时能被直接修复，但更常见的是，它们被去除并重新合成。

Vocabulary   词汇

alkylating agent ['ælkileitiŋ] ['eidʒənt]	烷化剂	cross-linking	交联
		deamination [diːæmi'neiʃən]	脱氨基（作用）
Ames test	埃姆斯测验法	deletion [di'liːʃən]	缺失（突变）
base analogue ['ænəlɔg]	碱基类似物	depurination [diːpjuəri'neiʃən]	脱嘌呤（作用）
base excision repair (BER)	碱基切除修复	direct reversal [ri'vəːsəl]	直接回复
5-bromouracil [brəumə'juərəsil]	5-溴尿嘧啶	DNA damage	DNA损伤

DNA mismatch	DNA 错配	nonsense mutation	无义突变
enol ['i:nɔl]	烯醇	nucleotide excision repair (NER) [ek'siʒən]	核苷酸切除修复
ethylmethane sulfonate (EMS) ['sʌlfəneit]	乙基甲磺酸	photolyase ['fəutəuˌlaiəs]	光解酶
frameshift mutation	移码突变	point mutation	点突变
heritable alteration ['heritəbl] [ˌɔ:ltə'reiʃən]	可遗传改变	pyrimidine dimer	嘧啶二聚体
		reading frame [freim]	读码框
hypoxanthine [ˌhaipəu'zænθi:n]	次黄嘌呤	silent mutation	沉默突变
imino form ['iminəu]	亚氨基式	spontaneous mutation [spɒn'teiniəs]	自发突变
induced mutation	诱发突变	strand slippage [strænd] ['slipidʒ]	链滑动
insertion [in'sə:ʃən]	插入（突变）	tautomer ['tɔ:təmə]	互变异构体
intercalating agent [inˌtə:kə'leitiŋ]	嵌入剂	tautomerization [ˌtɔ:təmərai'zeiʃən]	互变异构化（作用）
inversion [in'və:ʃən]	倒位		
keto ['ki:təu]	酮（基）	transition [træn'siʒən]	转换
mismatch repair (MMR)	错配修复	translocation [ˌtrænsləu'keiʃən]	易位
missense mutation	错义突变	transversion [trænz'və:ʒən]	颠换
mustard gas ['mʌstəd]	芥子气	ultraviolet radiation [ˌʌltrə'vaiəlit]	紫外辐射
nitrous acid ['naitrəs]	亚硝酸	xeroderma pigmentation (XP) [ˌziərəu'də:mə]	着色性干皮病
non-homologous end joining (NHEJ)	非同源末端连接		

Review Questions / 习题

Ⅰ. True/False Questions（判断题）

1. Base excision repair is used to repair common damage to bases.
2. Alkylating agents add chemical groups to DNA bases.
3. The Ames test is used to check the mutagenic potential of a chemical.
4. Intercalating agents cause base insertions, but not deletions.
5. Mismatch repair only occurs in prokaryotes.
6. Methylated C's are converted to U's by deamination.
7. UV light can cause two purines to become covalently linked.
8. A, G, C, and T can all exist in tautomeric forms.
9. Chromosome translocations can cause two genes to be fused together.
10. An insertion of a nucleotide into a gene is more serious than the deletion of a nucleotide.

Ⅱ. Multiple Choice Questions（选择题）

1. What usually happens if 3 nucleotides are added to a site in the coding region of a gene?
   a. A frameshift mutation occurs and the gene becomes completely nonfunctional.
   b. When translated, an extra amino acid is added but no other changes occur.
   c. When translated, an extra amino acid is added and changes may occur to neighboring amino acids.
   d. 3 nucleotides will be deleted in another part of the gene.
   e. None of the above.

2. Which of the following occurs as part of a chemical reaction?
   a. deamination   b. strand slippage
   c. tautomerization   d. base intercalation
   e. none of the above

3. How does MMR know which base in a mismatch is incorrect?
   a. It always removes the GATC sequence.
   b. It always removes the CTAG sequence.
   c. It always removes the methylated base.
   d. It removes the base on the strand with unmethylated GATC.
   e. It removes both bases and resynthesizes both.
4. Why is non-homologous end joining common in mammals?
   a. Mammals are not diploid, so they cannot use homologous recombination.
   b. The mammalian genome has a lot of non-coding sequences, so the chance of a harmful deletion occurring is small.
   c. The mammalian genome has a lot of non-coding sequences, so homologous recombination is impossible.
   d. Double stranded breaks are not common in mammals.
   e. None of the above.
5. If a point mutation causes an alanine codon to be replaced by a glycine codon, this is called a _____.
   a. missense mutation
   b. nonsense mutation
   c. silent Mutation
   d. frameshift mutation
   e. insertion

Exploration Questions 思考题

1. An important part of repairing DNA damage, is recognizing the presence of damage and specifically what kind of damage it is. What are some mechanisms the cell uses to recognize various kinds of damage?
2. Do you think that cells always transcribe DNA damage repair genes? Do you think these genes are expressed more strongly under certain conditions? Why or why not?
3. What kinds of DNA damage are hardest for a cell to detect and fix?
4. What are some differences between mismatch repair and nucleotide excision repair?

Chapter 10　Recombination

In addition to being copied, cellular DNA can be rearranged, producing new molecules with different organization and even new genes. This property is broadly known as **recombination**. There are several different kinds of recombination, and they are divided into two main categories: homologous recombination and non-homologous recombination. These are also termed general recombination and site-specific recombination, respectively. **Homologous recombination** involves the swapping of DNA at sites that have very similar or identical DNA sequences. **Non-homologous recombination**, by contrast, involves the insertion or deletion of pieces of DNA at sites within a chromosome.

Non-homologous recombination is quite distinct from homologous recombination. Indeed, these occur by completely different mechanisms and at very different times and settings. The two processes share part of their names because they both cause rearrangements, and thus new combinations of DNA regions in a chromosome. We begin by discussing homologous recombination.

10.1　Homologous Recombination

Most animals, including humans and many laboratory species, are **diploid**, meaning their cells have two copies of each chromosome. Equivalent chromosomes in a diploid organism are called **homologues** or **homologous chromosomes**. Homologous chromosomes contain the same genes in the same order. However these genes are not necessarily identical. They may vary slightly in sequence and function. Each functional variation of a gene is called an **allele**.

Homologous recombination is a mechanism by which homologous regions of DNA can be interchanged. This mechanism is an important source of genetic variation in populations. In diploid organisms, each offspring receives one chromosome from the mother and one from the father. If these chromosomes were left unchanged, the offspring's offspring would receive either an exact copy of the grandfather's chromosome or of the grandmother's. Homologous recombination, which occurs during meiosis, exchanges DNA between the two chromosomes of each individual before reproduction. As a result, the chromosome passed to the grandchild by each parent is a unique combination of two chromosomes from different individuals.

Homologous recombination occurs in other contexts as well. It can be used to repair certain kinds of DNA damage, as information from one chromosome provides the basis for reconstruction of the other. Sometimes homologous recombination does not even occur between two chromosomes. Recombination may occur between a chromosome and a non-chromosomal piece of DNA with which it shares some sequence similarity. Also, homologous recombination may occur solely within one molecule if two regions in the molecule have homologous sequences.

10.1.1 Mechanism for Crossing-Over

Homologous recombination in meiosis causes large regions of DNA to be exchanged between homologous chromosomes. This is called **crossing-over**. When the result of crossing-over is viewed at the level of entire chromosomes, it appears as though a perfect break is made in each chromosome, and the severed regions are neatly reattached. However, a closer look at the border between the newly joined DNA regions shows that this is not the case. A **heteroduplex joint** is in fact produced, involving short regions in which a single strand from one chromosome has been paired with a single strand from the other chromosome (Figure 10.1). This joint hints that the process of homologous recombination is much more intricate than simply cutting and pasting.

We now explain homologous recombination as it occurs in yeast (Figure 10.2). The process begins with a double-stranded cut in one of two homologous chromosomes—call it chromosome A, for clarity. In the case of meiosis in yeast, this cut is made intentionally by the cell using an enzyme called spo11. Each new end produced by the cut

任何变化，则这一后代的后代得到的将是一条来自他祖父或祖母的完全拷贝染色体。发生在减数分裂时期的同源重组使每个个体在繁殖前能够互换两条染色体上的 DNA。结果，每个亲本传递给其孙辈的染色体是从不同个体来的两条染色体的独特组合。

同源重组在其他情况下也会发生。它能被用来修复一些类型的 DNA 损伤，因为从一条染色体来的信息提供了重建另一条染色体的基础。有时同源重组甚至不在两条染色体之间发生。它可以发生在一条染色体和一段非染色体成分的DNA 片段之间，只要它们之间具有一些相似序列。还有，如果这一分子的两个区域具有同源序列，同源重组也可能在单独的一个分子中进行。

10.1.1 交换机理

在减数分裂期间发生的同源重组引起同源染色体之间发生 DNA 大片段的交换，特称为**交换**。当从整个染色体水平来看交换的结果时，它看起来像在每条染色体上产生了一个完美的断裂，并且被割断的区域又很整齐地被重新连接在了一起。然而，更靠近一点看两个新连接起来的 DNA 区域交界处会让我们知道，它并不是那么简单。实际上，它产生了一个**异源双链接头**，涉及来自一条染色体的单链与来自另一条染色体的单链之间配对形成的小区域（图 10.1）。这一接头的存在暗示我们，同源重组过程比简单地切断和粘贴要复杂得多。

现在我们来解释在酵母中发生的同源重组（图 10.2）。这一过程开始于在两条同源染色体的一条上出现一个双链切口，为简单起见把这条染色体叫作染色体 A。在酵母减数分裂情形中，这一切口是细胞使用一种称为 spo11 的酶特地产生的。

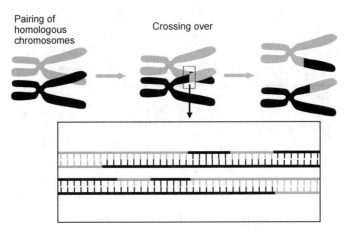

Figure 10.1 Crossing-over is not as simple as cutting and pasting. Chromosomal regions are connected by a heteroduplex joint (bottom) after crossing over

图 10.1 交换并不像剪与贴那么简单。交换发生后染色体上的区域由异源双链连在一起（图的下半部分）

is then digested by exonuclease to produce ends in which one single-strand dangles beyond its complementary strand. This occurs with the help of a protein complex called Mre11-Rad50-Xrs1, or MRX for short. The complex is also responsible for holding the two homologous chromosomes near each other during recombination.

之后，外切核酸酶开始消化因切割而产生出的每个新末端，使其中的一条单链悬挂在它的互补链上方。这是在一种称为 Mre11-Rad50-Xrs1（或简称为 MRX）的蛋白质复合体的帮助下实现的。这一复合体也负责在重组过程中将两条同源染色体保持在互相靠近的位置。

One of the two dangling single-strands is coated with a protein called Rad51 (The equivalent of Rad51 in prokaryotes is called RecA). The protein-coated strand is brought into close contact with the intact double helix of the homologous chromosome, call it chromosome B. In a process called **synapsis**, the single-stranded DNA molecule from chromosome A runs along the double helix of chromosome B, forming temporary base pairs. This occurs until a region that is highly complementary to the single strand is found. Rad51, which binds to single-stranded and double-stranded DNA, is the central player in this process.

两条悬着的单链中的一条被一种称为 Rad51 的蛋白质包裹起来（Rad51 在原核生物中的对等物是 RecA）。被蛋白质包裹的链靠近同源染色体（染色体 B）的完整双螺旋产生近距离接触。在一个称为**联会**的过程中，从染色体 A 来的单链 DNA 分子沿着染色体 B 的双螺旋移动，形成暂时的碱基对。这样的移动会一直持续下去直到找到与这条单链高度互补的区域。与单链和双链 DNA 结合的 Rad51 在这一过程中起着主导作用。

By the end of synapsis the single strand from chromosome A has opened up a small loop in chromosome B called a **D-loop**. The single strand has also hybridized to one strand in the D-loop. Small amounts of DNA synthesis now occur to lengthen the single strand from chromosome A using the strand from the D-loop as

到了联会的结束阶段，染色体 A 的单链已经在染色体 B 上打开了一个小环，称为 **D 环**。这段单链也已经与 D 环上的一条链杂交。现在，以 D 环中的链为模板合成一小部分 DNA 来延长染色体 A 上的单链。同时，D 环上的另一

10.1 Homologous Recombination | 237

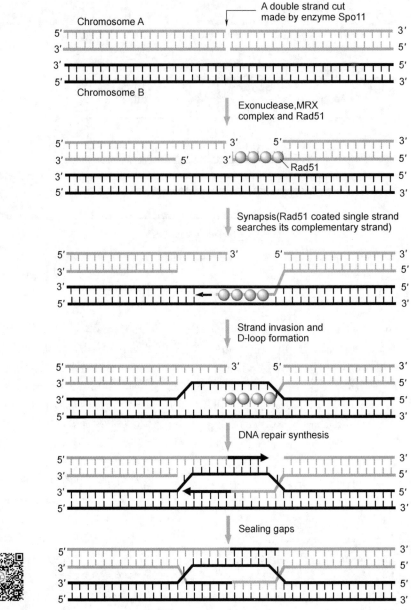

Figure 10.2　Mechanism of recombination in eukaryotes

图 10.2　真核生物重组机理

template. Meanwhile, the other strand of the D-loop is used as template to repair a different region of chromosome A that had been degraded by exonuclease.

After synthesis, both of the extended strands from chromosome A are covalently rejoined with the other half of chromosome A, sealing the double stranded cuts made at the beginning of replication. Chromosome A is not the same as it was originally, however. It now contains small regions that were directly derived from

条链用作模板去修复染色体 A 上另一个被外切核酸酶降解掉的区域。

合成之后，染色体上两条延伸出来的链均与染色体 A 的其他部分再次以共价键连接在一起，从而封闭掉复制开始时产生的双链切口。但是，染色体 A 已经与原来的不一样了。它现在含有直接来源于它的同源染色体的小区域。因此

its homologous chromosome. A small amount of recombination has therefore already occurred. Also, chromosome A is bound to chromosome B through hybridization of single strands from each chromosome.

The single strands cross each other, producing structures called **Holliday junctions.** Through a process called **branch migration**, these junctions can move and increase the amount of hybridization between the two chromosomes (Figure 10.3). The distance traveled by the Holliday junctions determines how large the heteroduplex joint will be when recombination is finished.

一个小范围的重组已经发生了。还有，染色体 A 通过不同染色体上的单链之间的杂交而结合到了染色体 B 上。

单链之间互相交叉，产生的结构称为 **Holliday 交叉**。通过一种称为**分支迁移**的过程，这些交叉能够移动并增加两条染色体之间的杂交区域（图 10.3）。Holliday 交叉的移动距离决定了当重组结束时异源双链接头的大小。

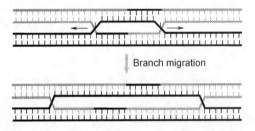

Figure 10.3　Branch migration of Holliday junctions

图 10.3　Holliday 交叉的分支迁移

After branch migration is completed, four DNA strands are tightly associated with each other. Somehow, these four strands must be separated into two double-stranded DNA molecules. This process is called **resolution** of the Holliday junction, and it varies depending on what recombination is being used for. To create a crossover for meiosis, four single-stranded breaks are made (Figure 10.4). Two nicks are made where strands cross at one Holliday junction. Two more nicks are made near the other Holliday junction, but in the strands that do not

在分支迁移完成后，四条 DNA 链互相紧密地联合在一起。不管怎样，这四条链必须被分开成两个双链 DNA 分子。这一过程称为 Holliday 交叉的**拆分**，不同的重组方式会出现不同的拆分结果。为了在减数分裂中创造出一个交换，产生了四个单链切口（图 10.4）。有两个切口产生在一个 Holliday 交叉的两条链上。另外两个切口产生在靠近另一个 Holliday 交叉的地方，但切口是在没有

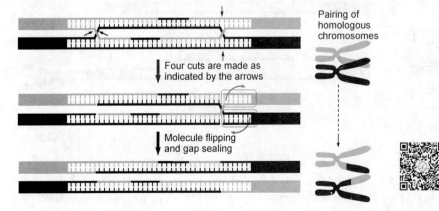

Figure 10.4　Resolution of the Holliday junction to create a crossover

图 10.4　产生交换的 Holliday 交叉拆分

participate in the junction. These nicks allow the two chromosomes to rotate around each other. The ends then reattach in such a way that crossing-over occurs.

10.1.2 Mechanism for Double-Stranded Break Repair

Homologous recombination can also be used to repair double-stranded breaks in DNA. When a double-stranded break occurs, there is no enzyme able to take both ends and stick them back together. Instead, degradation by exonucleases occurs at each end, producing staggered ends with dangling single strands. This situation is remarkably similar to the first step of meiotic recombination, and indeed, can be fixed by recombination.

Homologous recombination during **double-stranded break repair(DSBR)** is essentially the same as during meiosis. All steps of strand invasion, D-loop formation and Holliday junction formation, as well as strand migration are equivalent. The main difference is the resolution of Holliday junctions at the end of the recombination. Because DSBR does not require crossing-over, resolution is much simpler. Two nicks are made in each Holliday junction (Figure 10.5). These loose ends are then reattached in a combination that removes pairing between the chromosomes.

10.1.2 双链断裂修复机理

同源重组也能够用来修复 DNA 中的双链断裂。当发生了双链断裂时，没有酶能够抓住两个断头并把它们一起连接回去。这时，每个末端开始被外切核酸酶降解，产生具有单链悬着的交错末端。这一情形与减数分裂重组的第一步很相似，它也的确能通过重组作用得到修复。

在**双链断裂修复（DSBR）**中发生的同源重组基本上与减数分裂中的相同。所有单链入侵、D 环形成和 Holliday 交叉的形成以及链迁移步骤都是相同的。主要区别在于重组结束时 Holliday 交叉的拆分。由于 DSBR 不需要发生交换，所以这一拆分要简单得多。每个 Holliday 交叉处各产生两个切口（图 10.5）。之后，这些松开的末端重新连接在一起，这样的组合方式消除了两条染色体之间的配对。

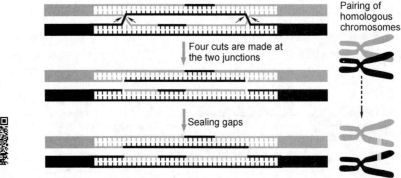

Figure 10.5 Resolution of Holliday junctions for non-crossover. Such resolution would be used in double-stranded break repair

图 10.5 不产生交换的 Holliday 交叉拆分。这种拆分方式也在双链断裂修复中得到应用

The importance of homologous recombination in preventing mutations from double-stranded breaks is apparent in some human pathology. Human females with mutated version of Brca1 and Brca2, proteins that help

同源重组在防止双链断裂产生突变上的作用在一些人类病理学方面的重要性是显而易见的。Brca1 和 Brca2 这两种蛋白质帮助 Rad51 行使正常的功能，带

Rad51 perform its function, accumulate mutations more quickly than normal and have much higher rates of breast cancer.

10.1.3 The RecBCD Pathway

Homologous recombination in prokaryotes is quite similar as in eukaryotes. One well-understood example of homologous recombination in *E. coli* is known as the **RecBCD pathway** (Figure 10.6). The protein RecBCD is made of three subunits, RecB, RecC, and RecD. The entire protein binds to double-stranded breaks in the DNA, and degrades the DNA until a site in the DNA, called a chi site (χ-site), with sequence 5′-GCTGGTGG-3′, is reached. At this point, the protein stops degrading the strand with a 3′ end at the site, but continues to degrade the complementary strand. This creates a single-strand with a 3′ end that is coated by single-stranded DNA binding proteins. With the help of a protein called RecA, this single strand invades a double-stranded DNA molecule and searches for a homologous DNA sequence. After homology is encountered, a D-loop is formed. A single-stranded nick is made in the double-stranded DNA and strands are exchanged and covalently linked to form a single Holliday junction. After branch migration, resolution of the junction can lead to either a crossover or a non-crossover recombination product.

10.1.4 Gene Conversion

Homologous recombination usually occurs with great precision. Even if the initial double-stranded break (during meiosis or accidental) is made within a gene, there is little risk of insertions or deletions being produced. There is, however, a very good chance of creating numerous mismatched bases. Remember that the heteroduplex joint involves hybridization between single-strands from homologous chromosomes. Although similar, these chromosomes may not have identical sequences, and the single strands probably will not pair to each other perfectly. Mismatches occur where one sequence differs from the other.

Generally, these mismatches are not a serious threat to the genome. However, when mismatches occur within a gene there can be interesting consequences. Even though there is only one DNA molecule, two genes are represented in it

有这两个基因突变的人类女性比正常人群更容易积累突变，也具有更高的乳腺癌患病比例。

10.1.3 RecBCD 途径

原核生物中的同源重组与真核生物中的有一定的相似之处。大肠杆菌同源重组研究得较为详尽的一个例子是 **RecBCD 途径**（图10.6）。RecBCD 蛋白由三个亚基组成，即 RecB、RecC 和 RecD。整个蛋白结合到 DNA 中的双链断裂处，然后开始降解 DNA 直到到达一个称为 χ 的位点（χ-位点），其特征序列是 5′-GCTGGTGG-3′。在这一位置上，RecBCD 蛋白停止降解在此位点处具有 3′ 末端的那条链，但会继续降解它的互补链。这样就产生了一条 3′ 末端突出的单链，该单链由 DNA 单链结合蛋白包裹。在 RecA 蛋白的帮助下，该单链 DNA 侵入一条双链 DNA 分子并且搜索与其同源的 DNA 序列。当遇到了同源序列后，会形成一个 D 环。之后在双链 DNA 上产生一个单链缺口，随之发生链的交换与链的共价连接，形成单一的 Holliday 交叉。在分支迁移之后，Holliday 交叉的拆分会得到发生了交换的重组产物，或得到没有发生交换的重组产物。

10.1.4 基因转换

一般来说同源重组过程的准确性很高。即使在最初的双链断裂（在减数分裂中发生或意外发生）出现在基因内部的情况下也几乎不会存在插入或缺失的危险。然而，它却会创造出很多错配碱基出现的机会。请记住，异源双链接头涉及来自同源染色体单链之间的杂交。虽然这些染色体是相似的，但它们的序列不一定完全相同，因此两条单链之间可能不会完全互相配对。在一条序列与另一条序列不同的地方会发生错配。

一般来说，这样的错配对基因组来说不会带来严重的威胁。而当错配发生在基因内部时，会出现有趣的结果。即使在只有一条 DNA 分子的情况下，它也会出

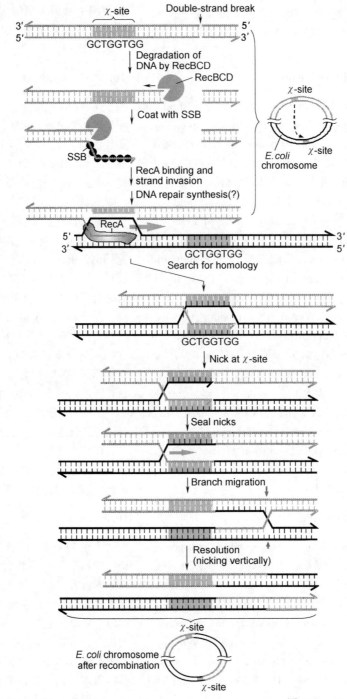

Figure 10.6　The RecBCD pathway

图 10.6　RecBCD 途径

(Figure 10.7). If replication proceeds with each slightly different strand as template, two different daughter molecules will be produced. One double-stranded molecule will have the version of the gene carried by one template strand; the other double-stranded molecule will have the gene carried by the other template strand.

现两个基因的情况（图 10.7）。如果以这两条序列稍有不同的链为模板进行复制，就会产生两条不同的子链分子。一条双链分子将具有一条模板链所携带的基因版本；另一条双链分子将具有另一条模板链携带的另一个基因版本。

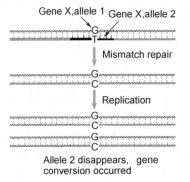
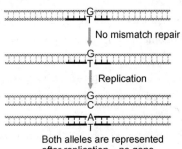

Figure 10.7 After recombination, two alleles can be represented on one piece of DNA. Mismatch repair leads to conversion of one of these alleles into the other. This is called gene conversion

图 10.7 重组后，两个等位基因序列会出现在一段 DNA 上。错配修复导致其中的一个等位基因转换成另一个等位基因。这称为基因转换

In some instances, however, the mismatched DNA undergoes mismatch repair (chapter 9) before replication. In this case, one of the strands is altered to look exactly like the other. When this 'repaired' molecule undergoes replication, both double-stranded daughter molecules will be identical; they will both carry only one version of the original two genes represented at the site. This is called **gene conversion**, because one of the genes was converted by mismatch repair and replication into a copy of the other gene.

然而，在有些实例中，错配的 DNA 在复制之前经历了错配修复（第 9 章）。这时，其中的一条链被改变后看起来与另一条链完全相同。当这一"修复的"分子进行复制的时候，两条双链子代分子将是完全相同的；两者都只携带在那一位点是原来两个基因中的一个的版本。这称为**基因转换**，因为其中的一个基因通过错配修复已经被转换并复制成了另一个基因的拷贝。

10.2 Non-homologous Recombination

10.2 非同源重组

Homologous recombination is not the only instance in which DNA is rearranged within a genome. In certain cases, a segment of DNA can be inserted or removed from a site with which it has no homology. This is called non-homologous recombination. The most common kind of non-homologous recombination involves mobile genetic elements called **transposons** or transposable elements.

同源重组并不是唯一可使基因组 DNA 发生重排的实例。在一些情况下，一段 DNA 能够插入一个与它没有任何同源性的位点，或从这样的位点移出，这称为非同源重组。非同源重组最常见的类型是涉及称为**转座子**或转座元件的可移动遗传元件。

10.2.1 Transposons

10.2.1 转座子

An insertion sequence is a simple type of transposable element found in prokaryotes (Figure 10.8). **Insertion sequences (ISs)** are defined as mobile elements that only encode for enzymes required for their transposition (**transposase**). Mobile DNA elements that encode additional genes are typically termed **transposons**. Insertion sequences are a diverse group of DNA elements. There

插入序列是在原核生物中发现的简单转座元件（图 10.8）。**插入序列（IS）**就是那些只编码了它们自身转座所需要的酶（**转座酶**）的可移动元件。具有额外编码基因的可移动 DNA 元件通常称为**转座子**。插入序列是一类差别很大的 DNA 元件。根据转座过程中转座酶切

are three major classes of bacterial insertion sequences, and their categorization is based on the enzymatic mechanisms by which their encoded transposase breaks and rejoins DNA elements during transposition (Table 10.1). The first category to be described, and the most common, includes ISs that express DDE transposases. These are so-called because of the critical function of two aspartate (D) residues and one glutamate (E) residue that coordinate metal ions in the active site for nucleophilic attack. These ISs commonly display two striking sequence properties. First, the two ends of the transposon are **inverted repeats**. In other words, both ends read the same if you turn one upside down and read it backwards. Second, the exact same sequence occurs in the genomic DNA immediately flanking an integrated insertion sequence. These are called **direct repeats**.

断以及重新连接 DNA 元件的作用机理不同，将细菌插入序列分为三大类（表 10.1）。第一类最为常见，是一些能表达 DDE 转座酶的插入序列；如此命名的原因是转座酶中的两个天冬氨酸（D）和一个谷氨酸（E）残基在协调活性位点金属离子发动亲核攻击中具有重要作用。这些插入序列通常具有两种突出的序列特性。首先，转座子的两端是**反向重复序列**；换句话说，如果把一个末端倒过来从后向前读，得到的序列是一样的。其次，在基因组中紧靠整合进这一插入序列的两侧 DNA 序列完全相同，这样的序列称为**同向重复序列**。

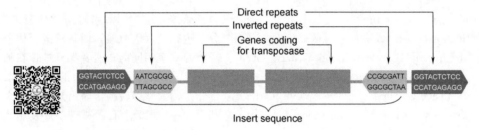

Figure 10.8 General structure of insertion sequences（transposon）

图 10.8 插入序列（转座子）的一般结构

Table 10.1 Examples of bacterial insertion sequences　　表 10.1 细菌插入序列范例

IS	Size (bps)	Direct repeats	Inverted repeats	Transposase type	Type of transposition
IS1	740~1180	8~9bp	Yes	DDE	Replicative
IS4	1400~1600	10~13bp	Yes	DDE	Non-replicative (conservative)
IS110	1200~1550	none	No	DEDD	Replicative
IS91	1500~2000	none	No	HUH	Replicative

A related family is the DEDD family of ISs. These do not display inverted repeats and do not always generate direct repeats; the mechanism of transposition for these ISs is not yet well-understood. Another family of ISs, the HUH family, employ transposases with a distinct catalytic mechanism. These have active sites with two histidine (H) residues separated by a large hydrophobic side chain (U), and use tyrosine a nucleophile in generating DNA breaks. Instead of inverted repeats, HUH ISs display stem-loop structures on either side

与上述插入序列有关的是 DEDD 类插入序列。这类插入序列没有反向重复序列，也不会每次都产生同向重复序列；目前还不清楚这类插入序列的转座机理。另一类插入序列，HUH 家族使用具有独特催化机制的转座酶。其活性位点包含由一个大的疏水侧链（U）连接在一起的两个组氨酸（H）；另外，在切断 DNA 链时使用酪氨酸作为亲核试剂。HUH 类插入序列两端没有反向重

of the element that direct specific binding by the transposase.

Transposases catalyze the movement of the insertion sequence into a new site in the genome. This sometimes involves making a copy of the transposon that inserts into the new location, while leaving the original transposon in its original location (Figure 10.9). In other instances the transposon is removed from its original site and inserted at the new site. The former method is called **replicative transposition**, the latter is termed **conservative transposition**.

转座酶催化插入序列进入基因组中的新位点。这样的移动有时涉及产生一个转座子的拷贝并将它插入新的位置，而原先的转座子还留在原来的位置（图10.9）。有时转座子会离开它原来的位置并插入到新的位点中。前者称为**复制型转座**，后者称为**保守型转座**。

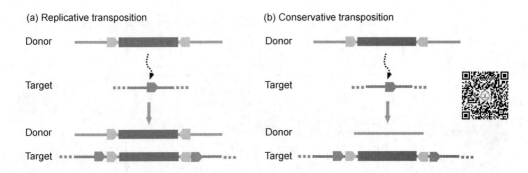

Figure 10.9 (a) In replicative transposition, a copy of the transposon is inserted into a target; (b) In conservative transposition, the transposon is removed from original site and then inserted into target

图 10.9 （a）在复制型转座中，一个转座子的拷贝插入了目标位置；（b）在保守型转座中，转座子离开原来的位置并插入目标位置

The mechanism of transposition is often depicted as shown in Figure 10.10. Staggered single-stranded cuts are made at the site of insertion in the target DNA. The strands are then separated, allowing the IS to enter the insertion site. Finally, the single strands are used as template to make double-stranded DNA. Because both single strands were initially complementary to each other, synthesis on each side produces two DNA sequences that are identical. This simple model is useful for demonstrating why ISs are often flanked by direct repeats in the target molecule. However it is a simplification of the actual process of transposition.

转座机理常常被描绘成图10.10所示的过程。首先，在目标DNA的插入位点处产生交错的单链切割。之后，DNA链被分开让转座子进入插入位点。最后，单链用作模板合成双链DNA。由于两条单链原来就是互补的，所以每一边的合成产生的是两条完全相同的DNA序列。这一简单的模型清楚地说明了为什么目标分子中转座子的两边是同向重复序列。然而，实际转座过程并不是如此简单。

Mechanisms of transposition are diverse, and largely dependent on sequence elements of the IS and the specific transposase employed. One characterized mechanism

转座机理是多种多样的，通常取决于插入序列具有什么结构特性和使用什么样的转座酶。一种已经弄清了的DDE类

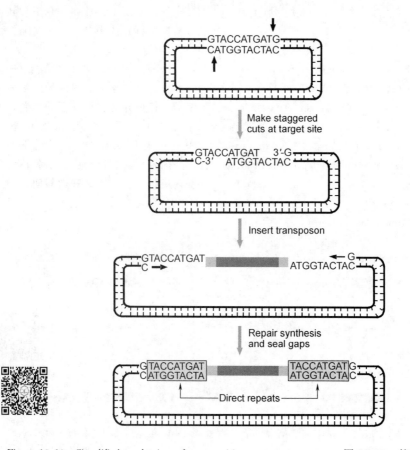

Figure 10.10 Simplified mechanism of transposition

图 10.10 简化的转座机理

for some replicative transposition of DDE-type ISs is shown in Figure 10.11. The result of transposition here is that the transposon moves from its site in a circular DNA molecule to a site in the cell's circular chromosome. First, single-stranded staggered cuts are made on each side of the insertion site, and on each side of the transposon. The single stranded ends at the transposon site are then separated and covalently linked to the single-stranded ends at the insertion site. This fuses the transposon and the chromosome.

During this process, the original double-stranded transposon has been denatured to exposed two, separated single strands. Each strand is used as template for DNA synthesis. Because the two strands are complementary, they produce two identical double-stranded copies of the original transposon. After synthesis, any breaks in the DNA are sealed, producing one large circular DNA with two transposons.

插入序列复制型转座机理如图 10.11 所示。这里，转座的结果是转座子从一个环状 DNA 分子上移到了细胞的环状染色体上。首先，分别在插入位点和转座子的两边产生单链交错切口。然后，在转座子处的单链末端被分开并共价连接到插入位点的单链上。这导致了转座子与染色体的融合。

在这一过程中，原来的双链转座子发生变性暴露出两条分开的单链，每条链均用作 DNA 合成的模板。因为这两条链是互补的，所以它们产生的是与原来的转座子完全相同的两个双链拷贝。合成完成后，DNA 中的所有裂口均被封闭，产生一个含有两个转座子的环状 DNA 大分子。

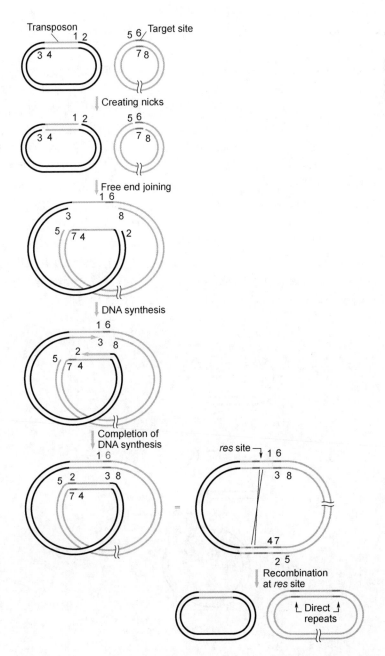

Figure 10.11　Mechanism for replicative transposition. Transposition is represented as it might occur in prokaryotes, between a circular piece of DNA (plasmid) and a circular chromosome

图 10.11　复制型转座机理。按原核生物中的可能机理表示，转座发生在环状 DNA 片段（质粒）和环状染色体之间

This structure must be resolved to produce two separate molecules: one chromosome with a new transposon, and one DNA molecule that still has a transposon in its original location. To achieve this, homologous recombination at special sequences within these transposons is used,

这一结构必须被拆分以产生两个分开的分子：一条含有新转座子的染色体和一个在原来位置仍然带有转座子的 DNA 分子。为达成此目的，需要在这些转座子的特殊序列上发生同源重组，从而将

joining one half of one transposon to the other half of the other transposon, and vice-versa.

If the same transposon and target were used for conservative transposition, the initial steps of transposition would be the same. Staggered cuts would be made, and the ends of the insertion site would be linked to the ends of the transposon. At this stage, however, the mechanisms diverge (Figure 10.12). For conservative transposition, two more cuts are made on either side of the transposon, completely severing its attachment to its original site in the circular DNA. The transposon remains linked at its new site by two covalent bonds to single-strands of the insertion site, and it is not denatured. All that remains is to use replication to make double-stranded DNA from the short single-stranded regions, and the conservative transposition is complete.

一个转座子的一半连接到另一个转座子的另一半上，反之亦然。

如果保守型转座使用的是相同的转座子和目标，转座的初始步骤是相同的。也会产生交错切口，插入位点的末端也与转座子的末端连接在一起。然而，从这个阶段开始，两种机理开始出现不同（图10.12）。对保守型转座而言，将在转座子的两边再产生两个切口，完全割断转座子与环形 DNA 在原来位置的连接。转座子在新位点上保持与插入位点单链的共价连接，并且不发生变性。所有剩下的工作就是采用复制方法从短的单链区域产生双链 DNA，整个保守型转座过程结束。

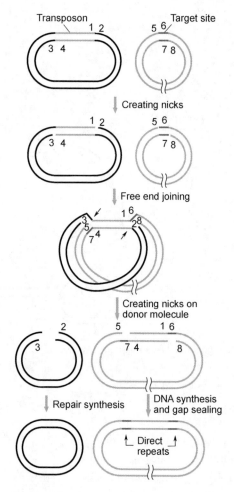

Figure 10.12 Mechanism of conservative transposition, as it might occur in prokaryotes

图 10.12 保守型转座，按原核生物中可能的机理表示

10.2.2 Retrotransposons

The mechanism of replicative transposition described above only applies to certain kinds of transposons. Mobile genetic elements called retrotransposons move around the genome using a completely different mechanism. There are two classes of **retrotransposons**, LTR-retrotransposons and non-LTR retrotransposons. Both classes replicate by first making an RNA copy of the transposon DNA.

For **LTR-retrotransposons**, the RNA copy is translated and among the proteins produced are the enzymes **reverse-transcriptase** and **integrase** (Figure 10.13). Reverse transcriptase is a very unique enzyme that can perform transcription in reverse, making DNA from RNA. It uses the single-stranded RNA, originally transcribed from the transposon DNA, to make a double-stranded DNA copy of the transposon DNA. Integrase, which is quite similar to transposase, then inserts the DNA copy into a target site in the chromosome. LTR-retrotransposons are highly similar to retroviruses, a class of viruses that make a DNA copy of their RNA genome and insert it into the host genome.

10.2.2 反转录转座子

上边描述的复制型转座机理只适用于一部分转座子。称为反转录转座子的可移动遗传元件采用完全不同的机理在基因组中移动。存在两种**反转录转座子**，即 LTR 反转录转座子和非 LTR 反转录转座子。两种类型都首先产生转座子 DNA 的 RNA 拷贝。

对 **LTR 反转录转座子**而言，RNA 拷贝被翻译产生**反转录酶**和**整合酶**等蛋白质（图 10.13）。反转录酶是一种很特别的酶，它能催化以反方向进行的转录，从 RNA 产生 DNA。它使用单链 RNA（从原来的转座子 DNA 上转录而来）产生反转录转座子 DNA 的双链 DNA 拷贝。整合酶与转座酶很相似，之后将这个 DNA 拷贝插入染色体的目标位点。LTR 反转录转座子与反转录病毒很相似，这一类病毒从它们的 RNA 基因组产生 DNA 拷贝并插入宿主基因组中。

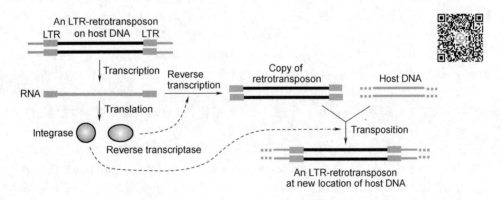

Figure 10.13 Mechanism of transposition by an LTR-retrotransposon

图 10.13 LTR 反转录转座子的转座机理

Non-LTR retrotransposons transpose via a different mechanism. A single stranded nick is made in the target site of transposition. The loose strand of DNA produced by this nick is then used in the reverse transcription reaction involving an RNA copy of the retrotransposon. By a mechanism not well understood, the DNA copy produced from the reaction is then integrated

非 **LTR 反转录转座子**通过不同的机理进行转座。它首先在转座的目标位点产生一个单链切口。之后，这一切割产生的 DNA 松散链被用在反转录反应中，后者又涉及反转录转座子的 RNA 拷贝。然后，通过一种还不清楚的机理，从上述反应产生的 DNA 拷贝整合到目

into the DNA at the target site.

Non-LTR retrotransposons are extremely relevant to human genetics. More than 500000 copies of the one called **LINE** (**long interspersed nuclear element**) exist in the human genome. This constitutes more than 15% of the total DNA in a human! Many of these, however, are not fully intact and are no longer able to transpose. All the same, they have been an important factor in causing mutations during human evolution, and are still the source of some disease-causing mutations. LINE is not the only significant retrotransposon in humans; no less than 42% of the human genome consists of DNA from retrotransposons. An additional 3% comes from DNA transposons, which have no RNA intermediate.

10.2.3 Bacteriophage λ Integration

Non-homologous recombination does not always involve random integration of DNA into a genome. Bacteriophage lambda, a virus that infects *E. coli*, provides an example of integration into specific target sites of the genome. Lambda phage contains a small, circular DNA genome. After infection of a cell, the virus proceeds along one of two possible pathways; it may initiate lysis, whereby the genome is rapidly replicated, transcribed and translated to make hundreds of viruses that burst open the cell. Alternatively, the phage may integrate its genome into the host's chromosome and lay dormant, sometimes over the course of several generations. This is called **lysogeny**.

A lysogenic phage DNA generally integrates at sites within the *E. coli* chromosome, called *att* B (**attachment site of bacterium**), that shares homology to a site in its genome called *att* P (**attachment site of phage**). This insertion is aided by the viral protein integrase, which recognizes both the phage and bacterial attachment sites and joins them to each other (Figure 10.14). Although the lambda phage DNA contains some homology to the *E. coli* chromosome, its integration is considered an example of non-homologous recombination. Lambda DNA integration is often considered to be an example of **site-specific recombination**.

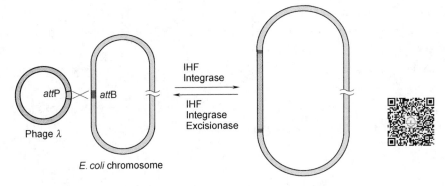

Figure 10.14 Site-specific integration of lambda phage DNA into *E. coli* chromosome

图 10.14 λ 噬菌体 DNA 向大肠杆菌基因组的位点特异性整合

10.3 Gene Editing

Gene editing is an emerging genetic engineering technology. It enables genetic modification at the target site in genome of an organism, including deletion or insertion of a specific gene and deletion or substitution of specific bases. Traditional genetic engineering technology only allows random insertion of DNA fragment into target genome, while gene editing is capable of precise modification at a specific genomic locus.

To review contents in Chapter 9 and this chapter: DNA **double-stranded break** can be repaired through **non-homologous end joining** and **homologous recombination**. Therefore, if a double-stranded break is artificially created at a specific site, gene editing can be achieved by DNA repair using either non-homologous end joining or homologous recombination. Thus, it can be seen that artificially creating a double-stranded break at specific genomic locus is crucial for gene editing. This process is completed through two steps: firstly recognizing and then cutting the target DNA. The development of gene editing technology has experienced the following three stages, because different target DNA recognizing and/or cutting mechanisms are used.

10.3.1 ZFNs System

Zinc finger nucleases (ZFNs) system is an artificially designed gene editing system with **endonuclease** activity on target DNA. It is composed of two ZFNs, each of which has a DNA recognition domain and a non-specific

10.3 基因编辑

基因编辑是一种新兴的基因工程技术。它可以对生物基因组中目标位点做遗传修饰，包括特定基因的删除或插入、特定碱基的缺失或替换。早期的基因工程技术只能将 DNA 片段随机插入目标基因组里，而基因编辑则能实现对特定基因组位置的精准修饰。

回顾第 9 章和本章内容：DNA **双链断裂**可以通过**非同源末端连接**和**同源重组**进行修复。因此，如果在特定位置人为创建双链断裂，通过非同源末端连接或同源重组来修复双链断裂，即可实现基因编辑。由此可见，实现基因编辑的关键是在基因组中特定位置人为创建双链断裂。这一过程分两步完成，即：先识别，然后切断目标 DNA。由于使用了不同的目标 DNA 识别和/或切割机制，基因编辑技术的发展经历了以下三个阶段。

10.3.1 ZFN 系统

锌指核酸酶（ZFN）系统是一种人工设计的基因编辑系统，对目标 DNA 具有**核酸内切酶**活性。该系统由两个 ZFN 组成，每个 ZFN 含有一个 DNA 识别结

endonuclease domain. Its DNA recognition domain is responsible for locating target DNA, which endows the system with specificity. And its non-specific endonuclease domain acts as "molecular scissors", which endows the system with cleavage activity. The association of both domains enables the system to cleave target DNA sequence at a specific genomic locus (Figure 10.15).

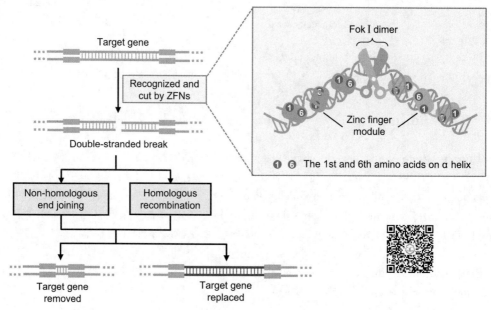

Figure 10.15 Structure and functioning mechanism of ZFNs system

The DNA recognition domain of ZFN is usually composed of 3~4 tandemly linked **zinc finger** structures, each of which has the "finger" like conserved structural module. The 1st and 6th amino acids in **α helix** of the zinc finger module determine triple base recognition preference of ZFN. Sequence-specific recognition of DNA can be achieved by tandemly linking multiple zinc fingers. The DNA endonuclease domain of ZFN is a non-specific **Fok Ⅰ monomer**. Fok Ⅰ monomer has no cleavage activity. Only after two Fok Ⅰ monomers dimerize, they become catalytically active in cutting DNA strand.

Each Fok Ⅰ monomer is connected to a DNA recognition domain through linker peptide, generating the ZFN holoenzyme. When two ZFNs simultaneously recognize and bind to two specific sequences spaced 6~8 bases on two separate DNA strands (one in 5′→3′ direction and the other in 3′→5′ direction), two Fok Ⅰ monomers interact and generate active **dimer**. Next, the Fok Ⅰ dimer mediates a double-strand cleavage at this site. Afterwards, target gene can be removed or replaced through repairing the double-stranded break by non-homologous end joining or homologous recombination (Figure 10.15).

The emergence of ZFNs system has greatly promoted development of gene editing technology. However, for a given DNA sequence, it cannot strictly correspond to a specific zinc finger recognition domain, which makes the design of ZFN extremely complicated. Therefore, for each given target sequence, a huge **gene expression library** needs to be constructed for zinc finger proteins, and repeated experimental screening is required to obtain the zinc finger structure that specifically binds to target sequence. This disadvantage has greatly limited the promotion and application of ZFNs system.

10.3.2 TALENs System

In order to overcome the disadvantage of ZFNs system, **transcription activator-like effector nucleases**（TALENs）system was designed to achieve simpler and more efficient gene editing. Actually, these two systems work in a very similar way. In TALENs system, a **transcription activator-like effector**（TALE）structural module is responsible for specific recognition of target DNA.

The DNA recognition domain of TALE module consists of 5~30 tandemly linked repetitive structural units (Figure 10.16). Each unit is composed of 34 amino acids, of which 32 amino acids are fixed. The 12th and 13th amino acids are different in different units, which are known as **repeat variable di-residues**（RVDs）. RVDs determine base recognition preference, constituting the core recognition region of TALEN. It is worth noting that specific DNA bases have their one-to-one corresponding RVDs. In general, A, T, C and G bases are recognized by di-residues NI, NG, HD and NN, respectively. Thus, a TALE that recognizes a given DNA can be obtained by tandemly linking repetitive units with corresponding RVDs. One TALEN is formed by fusing multiple TALE modules with Fok Ⅰ

序列后，两个 Fok Ⅰ 单体相互作用形成有活性的**二聚体**，进而在该位点对 DNA 双链进行切割。之后，借助非同源末端连接或同源重组对双链断裂进行修复，即可剔除或替换目的基因（图 10.15）。

ZFN 系统的出现大大促进了基因编辑技术的发展。但对于一段给定的 DNA 序列，并不能将其严格对应到特定的锌指识别结构域，这使得 ZFN 的设计变得异常繁杂。因此，对于每一个给定的靶序列，都需要构建庞大的锌指蛋白**基因表达文库**，经过反复的实验筛选才能获得特异性结合目标序列的锌指结构。这个弊端严重限制了 ZFN 系统的推广与应用。

10.3.2 TALEN 系统

为了克服 ZFN 系统的弊端，设计出了**转录激活因子样效应物核酸酶**（TALEN）系统，以实现更加简单高效的基因编辑。事实上，两个系统起作用的模式非常相似。在 TALEN 系统中，由**转录激活因子样效应物**（TALE）结构模块负责对目标 DNA 的特异性识别。

TALE 模块的 DNA 识别结构域由 5~30 个重复结构单元串联而成（图 10.16）。每个单元都由 34 个氨基酸构成，其中 32 个氨基酸是固定的。第 12、13 位氨基酸在不同的单元中存在差异，被称为**重复序列可变的双氨基酸残基**（RVD）。RVD 决定碱基识别的倾向性，构成了 TALEN 的核心识别区域。值得注意的是，特定 DNA 碱基具有与其一一对应的 RVD。一般来说，碱基 A、T、C 和 G 分别被双氨基酸残基 NI、NG、HD 和 NN 识别。因此，对于给定的一段 DNA，只需按照其序列将具有相应 RVD 的重复单元进行串联，即可获得识别该序列的 TALE，将其与 Fok Ⅰ 蛋

monomer. Similar to the use of Fok I domain in ZFNs system, formation of Fok I dimer in TALENs system is also required for gene editing.

白单体连接就形成了 TALEN。因为使用与 ZFNs 系统相同的 Fok I 结构域，所以在 TALENs 系统中也需要形成 Fok I 二聚体才能实现基因编辑。

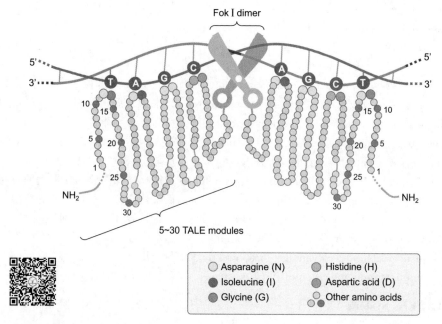

Figure 10.16　Structure and functioning mechanism of TALENs system

图 10.16　TALEN 系统结构及其作用机理

In TALENs system, the one-to-one correspondence between target DNA sequence and recognition unit makes the design of TALE traceable, which represents a great advancement in the field of gene editing.

在 TALENs 系统中，DNA 识别单元与目标 DNA 的一一对应特性，使得 TALE 的设计有迹可循，这是基因编辑领域的巨大进步。

10.3.3　CRISPR/Cas System

As the need for gene editing continued to grow, the traditional ZFNs and TALENs systems have been continuously improved and perfected. Meanwhile, a novel gene editing system named **CRISPR/Cas** came into play. This system was originally observed in bacteria and archaea as an adaptive immune mechanism against phage or exogenous DNA infection. It is composed of **clustered regularly interspaced short palindromic repeats** (CRISPR) and CRISPR-associated (Cas) nucleases (Figure 10.17).

Upon the first invasion of an exogenous DNA, bacteria integrate a segment of the invading DNA as **spacer sequence** into its own CRISPR array, separated by **repeat**

10.3.3　CRISPR/Cas 系统

随着对基因编辑需求的持续攀升，传统 ZFN 和 TALEN 系统得到了不断改进与完善。与此同时，一种称为 **CRISPR/Cas** 的新型基因编辑系统横空出世。该系统起初是在细菌和古菌中观察到的一种用于抵御噬菌体或外源 DNA 侵染的适应性免疫机制，它由**成簇规律间隔短回文重复**序列（CRISPR）以及 CRISPR 相关（Cas）的核酸酶组成（图 10.17）。

当外源 DNA 初次入侵时，细菌会将该 DNA 的一段作为**间隔序列**整合到自身的 CRISPR 阵列中，并由**重复序列**隔开

sequences [Figure 10.17 (a)]. When the same exogenous DNA invades again, CRISPR array is transcribed into pre-CRIPSR RNA (pre-crRNA), which contains the segment corresponding to previously invaded DNA. Additionally, the transactivating crRNA (tracrRNA) with **hairpin structure** is transcribed at the same time. Pre-crRNA assembles with tracrRNA through base pairing, which is then cleaved, modified and processed to generate mature crRNA-tracrRNA by RNase Ⅲ. Finally, crRNA-tracrRNA combines with Cas protein to form functional crRNA-tracrRNA-Cas complex. And the activated Cas nuclease cleaves invading DNA at the recognition site, thereby accomplishing the purpose of eliminating exogenous nucleic acids [Figure 10.17 (b)].

［图 10.17（a）］。当相同外源 DNA 再次入侵时，细菌从 CRISPR 阵列中转录生成前体 CRISPR RNA（pre-crRNA），这些 pre-crRNA 包含一段与先前入侵 DNA 相对应的序列。此外，具有**发夹结构**的反式激活 crRNA（tracrRNA）也会被转录出来。pre-crRNA 和 tracrRNA 随后通过碱基互补进行组装，经 RNase Ⅲ 剪切、修饰、加工后产生成熟的 crRNA-tracrRNA。最终，与 Cas 蛋白结合形成功能性 crRNA-tracrRNA-Cas 复合体。被激活的 Cas 核酸酶在序列识别处切割入侵 DNA，从而达到消灭外源核酸的目的［图 10.17（b）］。

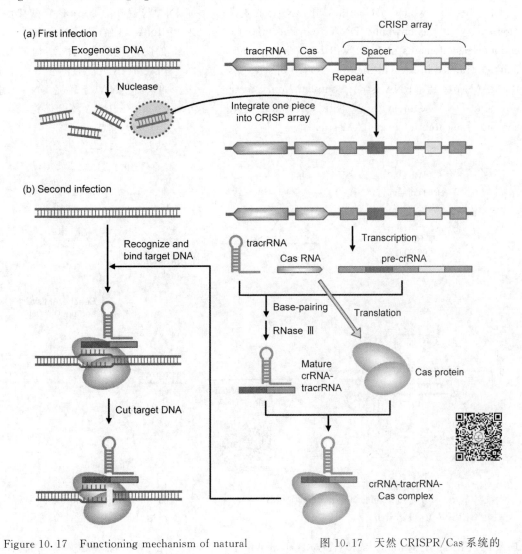

Figure 10.17 Functioning mechanism of natural CRISPR/Cas system

图 10.17 天然 CRISPR/Cas 系统的作用机理

To date, a variety of natural CRISPR/Cas systems have been identified and engineered for gene editing, among which CRISPR/Cas9 system is the most widely used. It is derived from the simplest type Ⅱ CRISPR system, which merely consists of a **single-guide RNA** (sgRNA) and an endonuclease Cas9. The sgRNA is artificially designed according to tertiary structure of crRNA-tracrRNA. Cas9 contains two endonuclease domains named HNH and RuvC (Figure 10.18). After Cas9/sgRNA complex is formed, it will search for **protospacer adjacent motif** (PAM) site with 5′-NGG (N represents any base) in the genome. When a PAM site is reached, the double-stranded DNA upstream of the PAM site will be unwound, allowing sgRNA to contact the 20 unwound nucleotides. If sgRNA is complementary to the unwound DNA segment, the two endonuclease domains of Cas9 will be activated and a double-stranded break will then be created upstream of the PAM site. If sgRNA is not complementary to the unwound DNA segment, Cas9/sgRNA complex will continue to search for next PAM site.

迄今为止，已有多种天然 CRISPR/Cas 系统得到鉴定与改造用于基因编辑，其中属 CRISPR/Cas9 系统应用最为广泛。该系统由最简单的Ⅱ型 CRISPR 系统改造而来，只包含一条**单链引导 RNA**（sgRNA）和 Cas9 核酸酶。其中 sgRNA 是根据 crRNA-tracrRNA 的高级结构人工设计的，Cas9 含两个称为 HNH 和 RuvC 的核酸内切酶结构域（图 10.18）。当 Cas9/sgRNA 复合体形成后，会在基因组范围内搜索序列为 5′-NGG（N 为任何碱基）的**前间区序列邻近基序**（PAM）位点。到达 PAM 位点后，会将 PAM 位点上游的 DNA 双链解旋，从而允许 sgRNA 与解旋的 20 个核苷酸接触。如果 sgRNA 与解旋 DNA 片段互补，Cas9 的两个核酸内切酶结构域会被激活，并在 PAM 位点上游产生双链断裂。如果 sgRNA 与解旋 DNA 片段不互补，Cas9/sgRNA 复合体将继续搜索下一个 PAM 位点。

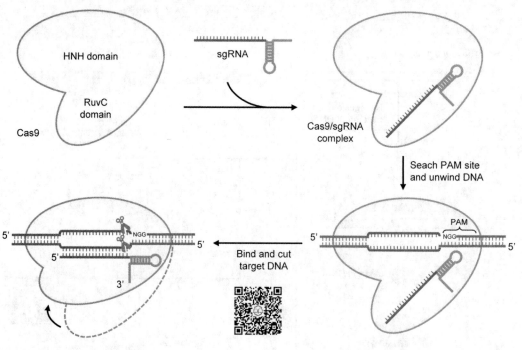

Figure 10.18　The mechanism of action of CRISPR/Cas9 system

图 10.18　CRISPR/Cas9 系统的作用机理

In CRISPR/Cas9 system, sgRNA is responsible for precise positing of gene sequences that need to be edited.

在 CRISPR/Cas9 系统中，sgRNA 负责对需要编辑的基因片段进行精准定位。

Thus, target DNA sequence can be located by only designing a 20 nucleotide sequence-specific sgRNA. Compared with ZFNs and TALENs systems that need to design recognition domains for each given DNA sequence, it is particularly simple to design the sgRNAs in this system. Besides, Cas9 itself contains two endonuclease domains. Thus, a single Cas9/sgRNA is sufficient for gene editing. Such advantage is unmatched by ZFNs and TALENs systems.

因此，只需设计 20 个核苷酸的序列特异性 sgRNA，即可定位目标 DNA 序列。相比 ZFN 和 TALEN 系统需要为每条给定的 DNA 序列设计识别结构域，为该系统设计 sgRNA 显得尤为简便。此外，Cas9 蛋白本身具备两个核酸内切酶结构域，因而仅需单个 Cas9/sgRNA 即可实现基因编辑，这也是 ZFN 和 TALEN 系统不能比拟的。

Summary

Homologous recombination is the exchange of DNA or genetic information between two homologous sites. It occurs in many contexts, including meiosis and the repair of double-strand breaks in the DNA. The mechanism of homologous recombination involves hybridization between single-strands from different DNA molecules and eventually the production of a Holliday junction. The mechanism used to resolve the junction determines the final products of recombination. DNA can also be recombined using non-homologous recombination. This often involves the insertion of mobile DNA elements, or transposons, into chromosomes. Various kinds of transposons exist, and some are very common in the human genome, especially retrotransposons.

小结

同源重组是在两个同源位点之间 DNA 或遗传信息的交换。它在许多情景下都会发生，包括减数分裂和 DNA 中双链断裂的修复。同源重组机理涉及来自不同 DNA 分子的单链之间的杂交和最终 Holliday 交叉的产生。用来拆分这一交叉的机理决定了重组的最终产物。DNA 也能以非同源重组的方式发生重组。这通常涉及可移动 DNA 元件（或转座子）的插入。存在着不同种类的转座子，有些在人类基因组中很普遍，特别是反转录转座子。

Vocabulary 词汇

allele [ə'li:l]	等位基因	integrase ['intiˌgreis]	整合酶
attachment site [ə'tætʃmənt]	附着位点	integration host factor (IHF) [ˌinti'greiʃən]	整合宿主因子
branch migration	分支迁移		
chromatid ['krəumətid]	染色单体	replicative transposition ['replikeitiv]	复制型转座
conservative transposition [kən'sə:vətiv]	保守型转座	resolution [ˌrezə'lu:ʃən]	拆分
crossing-over	交换	resolvase [ˌrezəl'veis]	解离酶
diploid ['diploid]	二倍体	retrotransposon ['retrəu'trænspəusən]	反转录转座子
direct repeat	同向重复（序列）		
excisionase [ek'siʒəneis]	切除酶	site-specific recombination	位点特异性重组
gene conversion [kən'və:ʃən]	基因转换	staggered ends ['stægə:d]	交错末端
heteroduplex [ˌhetərəu'dju:pleks]	异源双链	synapsis [si'næpsis]	联会
Holliday junction ['hɔlidei 'dʒʌŋkʃən]	Holliday 交叉	transposase [ˌtrænspə'seis]	转座酶
		transposition [ˌtrænspə'ziʃən]	转座
homologous recombination ['ri:kəmbi'neiʃən]	同源重组	transposon [træns'pəusən]	转座子

Review Questions

I. True/False Questions (判断题)

1. Gene conversion is when one gene is deleted from the genome.
2. Holliday junctions may be resolved in more than one way.
3. Retrotransposons require reverse transcriptase to make copies of themselves.
4. Bacteriophage lambda usually integrates into a specific site in the genome of *E. coli*.
5. Spo11 makes a double-stranded cut in the DNA.
6. All organisms have two copies of each chromosome.
7. Crossing over occurs during meiosis in *E. coli*.
8. Double-stranded break repair with homologous recombination involves formation of Holliday junctions.
9. Homologous recombination may result in a DNA mismatch.
10. Proteins are usually required for transposons to insert into a chromosome.

II. Multiple Choice Questions (选择题)

1. The RecBCD mechanism operates in which organism?
   a. Yeast
   b. All eukaryotes
   c. Mammals
   d. *E. coli*
   e. None of the above

2. The homologue of Rad51 in prokaryotes is _____.
   a. RecA
   b. RecBCD
   c. Spo11
   d. Topoisomerase I
   e. Topoisomerase II

3. Which of the following statements is true of insertion sequences?
   a. They contain inverted repeats.
   b. They are surrounded by direct repeats.
   c. They code for transposase.
   d. They are inserted into DNA by making staggered cuts at the insertion site.
   e. All of the above.

4. Approximately how much of the human genome is made of transposons and retrotransposons?
   a. 2%
   b. 10%
   c. 25%
   d. 33%
   e. 45%

5. What is true of homologous chromosomes?
   a. They have exactly the same DNA sequence.
   b. They pair to each other during meiosis.
   c. Each cell only has one of the two homologous chromosomes.
   d. They are circular.
   e. All of the above.

Exploration Questions (思考题)

1. What is gene conversion, and how does it occur?
2. How does the distance that a Holliday junction migrates during branch migration affect the characteristics of the heteroduplex joint?
3. Do you think transposons have mostly negative or positive effects on organisms? Why?
4. What are the differences between double-stranded break repair and crossing-over?

Answers to Review Questions 习题答案

Chapter 1 (第1章)

Ⅰ. True/False Questions (判断题)

1. False, proteins can be different lengths.
2. False, there are thousands of proteins in the cell.
3. False, arginine is positively charged, so it is more likely to be found on the surface of the protein. The protein core is usually hydrophobic.
4. False.
5. False, it occurs because the separation of hydrophobic molecules from water creates more disorder (entropy).
6. True.
7. False, many sequences, but certainly not every sequence, can form an alpha helix.
8. False, proteins can bind to many molecules, including DNA.
9. False, kinesin is one kind of motor protein. There are many other kinds.
10. False, some proteins require modifications after translation in order to function.

Ⅱ. Multiple Choice Questions (选择题)

1. c 2. e 3. e 4. a 5. a

Chapter 2 (第2章)

Ⅰ. True/False Questions (判断题)

1. False, they have one less oxygen atom.
2. False, DNA can usually be renatured by restoring normal conditions.
3. False, RNA molecules can be different lengths, as can DNA molecules.
4. False, the strongest bonds are covalent bonds. The bonds between two bases are hydrogen bonds, not covalent bonds.
5. True.
6. False, purines always form base pairs with pyrimidines.
7. True.
8. False, RNA uses U instead of T.
9. False, bases are connected to the 1' carbon of ribose.
10. False, RNA is always used by the cell, and has different functions from DNA.

Ⅱ. Multiple Choice Questions (选择题)

1. b 2. e 3. b 4. c 5. b

Chapter 3 (第3章)

Ⅰ. True/False Questions (判断题)

1. False, the sequence of the promoter can vary.
2. False, RNA polymerase can transcribe DNA even without sigma subunit. Sigma subunit is required to bind to specific DNA sequences.
3. False, cells are almost always transcribing genes.
4. True.
5. False, mRNA is used to bring information about protein sequence to the ribosome.
6. False, the proteins are transcribed at the same rate, but may be translated and degraded at different rates.
7. True.
8. False, these proteins are transcribed together, but are translated as separate proteins, and have different locations in the cell.
9. True.
10. True.

Ⅱ. Multiple Choice Questions (选择题)

1. d 2. a 3. e 4. c 5. a

Chapter 4（第 4 章）

Ⅰ. True/False Questions（判断题）

1. True.
2. False，enhancers are DNA elements.
3. True.
4. False，many class Ⅱ genes do not contain a TATA box in the promoter.
5. True.
6. False，prokaryotes only have one kind of RNA polymerase.
7. True.
8. False，helix-loop-helix motifs are completely different from helix-turn-helix motifs.
9. True.
10. True.

Ⅱ. Multiple Choice Questions（选择题）

1. b 2. e 3. d 4. e 5. a

Chapter 5（第 5 章）

Ⅰ. True/False Questions（判断题）

1. False，polyadenylation does not occur on proteins.
2. True.
3. True.
4. True.
5. False，genes often have a large number of introns and exons.
6. True.
7. True.
8. False，mRNA is still degraded，but usually much more slowly.
9. False，genes are required to code for the protein component（s）and for the RNA component（s）.
10. False，RNA polymerase Ⅱ does not add the poly（A）tail. This is added by poly（A）polymerase.

Ⅱ. Multiple Choice Questions（选择题）

1. e 2. b 3. a 4. c 5. a

Chapter 6（第 6 章）

Ⅰ. True/False Questions（判断题）

1. False，there is only one start codon.
2. False，tRNA can be charged or uncharged.
3. True.
4. True.
5. False，tRNA is not T-shaped in 3-dimensional structure.
6. False，tRNA does not bind to stop codons.
7. True.
8. False，ribosomes are composed of many proteins and RNAs.
9. True.
10. False，some amino acids are only coded for by one codon.

Ⅱ. Multiple Choice Questions（选择题）

1. d 2. e 3. c 4. d 5. c

Chapter 7（第 7 章）

Ⅰ. True/False Questions（判断题）

1. True.
2. False，it is composed of 8 histones，but only 4 different kinds of histones（2 of each）.
3. False，there are many thousands of nucleosomes on each chromosome.
4. True.
5. True.
6. False，miRNA can cause mRNA to be degraded and can affect translation.
7. True.
8. False，dephosphorylated 4E-BP blocks translation.
9. True.
10. False，prokaryotic mRNAs are not polyadenylated.

Ⅱ. Multiple Choice Questions（选择题）

1. e 2. e 3. b 4. e 5. e

Chapter 8 (第8章)

Ⅰ. True/False Questions (判断题)

1. False, there is one origin of replication in prokaryotes, but eukaryotes have many origins.
2. False, the sliding clamp holds the polymerase on the DNA.
3. True.
4. False, only DNA polymerase Ⅰ has this ability.
5. True.
6. False, DNA polymerase Ⅰ is used to replace RNA primers with DNA.
7. True.
8. True.
9. False, they are joined together by ligase.
10. False, DnaA is a protein.

Ⅱ. Multiple Choice Questions (选择题)

1. d 2. c 3. a 4. b 5. b

Chapter 9 (第9章)

Ⅰ. True/False Questions (判断题)

1. True.
2. True.
3. True.
4. False, intercalating agents can cause base insertions and deletions.
5. False, it occurs in eukaryotes as well.
6. False, methylated C's are converted to T's by deamination.
7. False, UV light causes two pyrimidines to become linked.
8. True.
9. True.
10. False, both can lead to frameshift mutations depending where they occur.

Ⅱ. Multiple Choice Questions (选择题)

1. c 2. a 3. d 4. b 5. a

Chapter 10 Questions (第10章)

Ⅰ. True/False Questions (判断题)

1. False, gene conversion does not involve deletion of DNA.
2. True.
3. True.
4. True.
5. True.
6. False, many organisms have only one copy of each chromosome, and some have more than two copies of each chromosome.
7. False, *E. coli* does not undergo meiosis.
8. True.
9. True.
10. True.

Ⅱ. Multiple Choice Questions (选择题)

1. d 2. a 3. e 4. e 5. b

Index（Chinese） 中文索引

A

阿拉伯糖	70
埃姆斯测验法	231
安芬森实验	21
氨基	5
2-氨基嘌呤	210
氨基酸	5
氨酰-tRNA 合成酶	138
氨酰基（A）位	128

B

半保留复制	178
半不连续复制	183
半乳糖	71
β-半乳糖苷酶	61
半乳糖苷转乙酰基酶	61
胞嘧啶	27
保留复制	200
保守型转座	245
报告基因	98
闭合启动子复合体	53
变性	21，32
别构调节	20
病毒载量	35
部分二倍体	72
部分双键	6

C

操纵基因	59
操纵子	59
插入	213
插入/缺失环（IDL）	213
插入序列（IS）	243
拆分	239
长散布核元件（LINE）	250
常染色质	148
超螺旋	193
沉默突变	217
沉默子	91
成簇规律间隔短回文重复	254
持家基因	83
持续合成能力	186
重复序列	254
重复序列可变的双氨基酸残基	253
重组	235
初级转录本	102
次黄苷	139
错配碱基	206
错配修复（MMR）	223
错义突变	218

D

大沟	32
代谢物激活蛋白（CAP）	63
单链 DNA	30
单链 DNA 结合蛋白（SSB）	185
单链引导 RNA	256
蛋白质	5
蛋白质域	12
倒位	216
等位基因	235
颠换	206
点突变	206
电子显微术	172
读码框	219
端粒	199
端粒酶	199
多聚核糖体	126
多顺反子 mRNA	60
多肽	8

E

α-鹅膏蕈碱	97
二倍体	235
二级结构	9
二聚体	94，253
二硫键	14

F

发夹环	57
发夹结构	255
翻译	2
翻译起始因子（IF）	127
反密码子	137
反式剪接	113
反向平行	31
反向重复序列	244
反转录酶	249
反转录转座子	249
范德瓦耳斯力	15
方向舵	54
非翻译区（UTR）	126
非 LTR 反转录转座子	249
非同源末端连接（NHEJ）	228, 251
非同源重组	235
分解代谢	65
分支迁移	239
辅阻遏物	67
负调控	62
复性	32
复制叉	181
复制泡	180
复制起点	179
复制型转座	245
富含 AU 元件（ARE）	161

G

冈崎片段	184
共价键	14
共有序列	51
构象	8
光解酶	223

H

含氮碱基	25, 26
合成代谢	65
核苷酸	25
核苷酸切除修复（NER）	226
核孔	158
核内小核糖核蛋白	108
核酸	24
核酸检测	35
核酸内切酶	251
核糖核苷酸	26
核糖核酸	27
核糖核酸酶	103
核糖体	125
核小体	147
核心组蛋白	147
赫尔希-蔡斯实验	44
后随链	184
互变异构体	208
互补	30
互补 DNA	35
互补分析	73
滑行夹	187
滑行夹加载器	187
活性位点	17

J

肌动蛋白	17
肌动蛋白丝	17
基础转录	88
基序	94
基因	38
基因编辑	251
基因表达	59
基因表达文库	253
基因转换	243
激活蛋白	89
激酶	154
加帽	102
N-甲酰甲硫氨酸	126
间隔序列	254
剪接	106
剪接前导序列（SL）	112
剪接体	108
剪接位点	107
碱基堆积	32
碱基对	30
碱基类似物	209
碱基切除修复（BER）	227
交换	236
交联	213

焦磷酸解作用	55	尿嘧啶	27
解旋酶	185	**O**	
聚合物	5	偶联转录-翻译作用	68
聚腺苷酸化	104	**P**	
K		配体	67
开放启动子复合体	53	嘌呤	26
可变剪接	114	**Q**	
可变聚腺苷酸化	157	启动子	50
−35框	50	启动子清空	54
−10框	50	起点识别复合体（ORC）	198
L		起始tRNA	126
Ⅰ类内含子	110	起始	53
Ⅱ类内含子	110	30S起始复合体	126
离子键	15	起始密码子	124
联会	237	起始子（Inr）	83
链滑动	213	前导子	67
亮氨酸拉链	96	前复制复合体（pre-RC）	198
磷酸二酯键	28	前间区序列邻近基序	256
侣伴蛋白	14	前起始复合体	84
α螺旋	9,252	前体mRNA	102
螺旋-环-螺旋（HLH）	97	嵌入剂	215
螺旋-转角-螺旋（HTH）	94	切割激活因子（CstF）	104
M		切割因子Ⅰ和Ⅱ（CFⅠ和CFⅡ）	104
麦塞尔逊-斯托尔实验	200	切割与聚腺苷酸化特异因子（CPSF）	104
酶	16	切片	163
密度超速离心	201	氢键	15
密码子	124	驱动蛋白	18
嘧啶	26	全局控制	166
嘧啶二聚体	212	缺失	213
模块	13	**R**	
模块化	98	染色体	38
3′末端	28	染色质重塑蛋白	153
5′末端	29	热休克蛋白	159
N		冗余的	124
内部核糖体进入序列（IRES）	133	溶原性	250
内共生学说	40	融合蛋白	220
内含子	106	乳糖	61
内含子晚期论	113	乳糖操纵子	61
内含子早期论	113	乳糖渗透酶	61
内在型终止	57	**S**	
鸟嘌呤	27	三级结构	9

散乱复制	200
扫描	133
上游区域	50
渗漏突变	218
释放因子	132
疏水相互作用	15
衰减子	67
衰减作用	67
双链 DNA	30
双链断裂	251
双链断裂修复（DSBR）	240
双螺旋	32
双向复制	180
水解	55
顺乌头酸酶	161
四级结构	9
羧基	5
锁-钥机理	19

T

肽	8
肽基（P）位	127
肽基转移酶	128
肽键	6
探针	33
糖-磷酸骨架	29
套索	107
特异转录因子	88
铁蛋白	169
通用转录因子	84
同向重复序列	244
同源染色体	235
同源体	235
同源异形域	95
同源重组	235,251
突变	205
突变热点	221
脱氨基	207
脱嘌呤	207
脱氧核糖核苷酸	26,27
脱氧核糖核酸酶（DNase）	173

W

外来体	160
$3'\rightarrow 5'$ 外切核酸酶活性	187
$5'\rightarrow 3'$ 外切核酸酶活性	192
外显子	106
烷化剂	210
微卫星	214
微小 RNA（miRNA）	163
χ-位点	241
位点特异性重组	250
无义突变	218

X

细胞核	39
细胞周期	197
细菌附着位点（*att* B）	250
下游启动子元件（DPE）	83
先导链	184
30nm 纤维	148
线珠模型	147
腺苷酸环化酶	64
腺嘌呤	27
小干涉 RNA（siRNA）	163
小沟	32
小卫星	33
校正	55,187
协同调控	59
锌指	95,252
锌指核酸酶	251
Ⅰ型拓扑异构酶	193
Ⅱ型拓扑异构酶	193
胸腺嘧啶	27
5-溴尿嘧啶（BU）	209

Y

30S 亚基	125
50S 亚基	125
亚基	12
亚硝酸	211
延伸因子（EF）	130
摇摆	139
一级结构	9
ρ 依赖型终止	57
移码突变	219
移位	128

遗传密码	124
遗传物质	24
乙基甲磺酸（EMS）	211
乙酰化	151
异染色质	148
异源二聚化作用	96
异源双链接头	236
易位	216
引发酶	191
引物	191
诱导-契合	19
诱导物	67
运铁蛋白	161

Z

杂交	33,49
载脂蛋白 B（APOB）	115
增强子	91
增殖细胞核抗原（PCNA）	196
β折叠	9
真核起始因子	135
整合酶	249
正调控	63
指导 RNA（gRNA）	117
中心法则	2
终止	57
终止密码子	124
转换	206
转录	2
转录激活因子样效应物	253
转录激活因子样效应物核酸酶	253
转录泡	54
转座酶	243
转座子	214,243
着色性干皮病（XP）	230
紫外辐射（UV）	212
自发突变	207
自我剪接	110
自主复制序列（ARS）	197
阻遏蛋白	91
组成型突变体	73
组蛋白	146
组蛋白甲基转移酶（HMT）	154
组蛋白密码	152
组蛋白脱乙酰基转移酶（HDAC）	154
组蛋白尾	151
组蛋白乙酰基转移酶（HAT）	154

其他

A（氨酰基）位	128
A 框	87
A→I 编辑	115
AraC 蛋白	71
ara 操纵子	70
att B（细菌附着位点）	250
att P（噬菌体附着位点）	250
B 框	87
C-末端	8
CRISPR/Cas	254
Ct 值	35
CTD	103
C→U 编辑	115
D 环	237
DNA 错配	206
DNA 分型术	34
DNA 聚合酶 I	191
DNA 聚合酶Ⅲ	185
DNA 聚合酶Ⅲ核心酶	185
DNA 聚合酶Ⅲ全酶	185
DNA 类型	34
DNA 链	30
DNA 旋转酶	193
DNA 指纹	33
DNA 指纹分析	34
E 位	128
Fok Ⅰ单体	252
gal 操纵子	71
gal 阻遏蛋白	72
Holliday 交叉	239
INF-β 基因	154
lac 阻遏蛋白	62
LTR 反转录转座子	249
mRNA 特异性控制	166
N-末端	8
P（肽基）位	128
poly(A) 聚合酶	105

poly(A) 尾	104	RNAi 沉默复合体（RISC）	163
R 基团	5	SD 序列	126
RecBCD 途径	241	Southern 转印	34
RNA 编辑	115	Sxl 蛋白	156
RNA 干涉（RNAi）	163	TATA 结合蛋白（TBP）	84
RNA 聚合酶 I	81	TATA 框	82
RNA 聚合酶	49,51	TBP 相关因子 II（TAF II）	85
RNA 聚合酶 II	81	TF II B 识别元件（BRE）	83
RNA 聚合酶 III	81	*trp* 操纵子	64
RNA 聚合酶核心	51	trp 阻遏蛋白	65
RNA 聚合酶全酶	52		

Index (English) 英文索引

A

A (aminoacyl) site	128
acetylation	151
aconitase	161
Actin	17
actin filaments	17
activators	89
active site	17
Adenine	27
adenylyl cyclase	64
A→I editing	115
alkylating agents	210
allele	235
allosteric regulation	20
Alternative polyadenylation	157
alternative splicing	114
α-amanitin	97
Ames test	231
amino acids	5
amino group	5
aminoacyl-tRNA synthetase	138
2-aminopurine	210
anabolic	65
Anfinsen experiment	21
anticodon	137
anti-parallel	31
APOB (apolipoprotein B)	115
apolipoprotein B (APOB)	115
ara operon	70
arabinose	70
AraC protein	71
ARE (AU-rich element)	160
ARS (autonomously replicating sequence)	197
attachment site of bacterium (*att*B)	250
*att*B (attachment site of bacterium)	250
attenuation	67
attenuator	67
attachment site of phage (*att*P)	250
*att*P (attachment site of phage)	250
AU-rich element (ARE)	160
autonomously replicating sequence (ARS)	197

B

basal transcription	88
base analogues	209
Base excision repair (BER)	227
base pair	30
base-stacking	32
beads-on-a-string model	147
BER (Base excision repair)	227
bidirectional replication	180
−35 box	50
−10 box	50
box A	87
box B	87
branch migration	239
BRE (TFⅡB recognition element)	83
5-bromouracil (BU)	209
BU (5-bromouracil)	209

C

CAP (catabolite activator protein)	63
capping	102
carboxyl group	5
catabolic	65
catabolite activator protein (CAP)	63
cell cycle	197
central dogma	2
chaperones	14
chromatin remodeling proteins	153
chromosome	38
clamp loader	187
cleavage and polyadenylation specificity factor (CPSF)	104
cleavage factors Ⅰ and Ⅱ (CF Ⅰ and CF Ⅱ)	104

cleavage stimulation factor protein (CstF)	104
closed promoter complex	53
clustered regularly interspaced short palindromic repeats	254
codon	124
complementary	30
complementary DNA	35
complementation analysis	73
conformations	8
consensus sequences	51
conservative replication	200
conservative transposition	245
constitutive mutants	73
coordinate regulation	59
core histones	147
co-repressor	67
coupled transcription-translation	68
covalent bond	14
CPSF (cleavage and polyadenylation specificity factor)	104
CRISPR/Cas	254
crossing-over	236
cross-linking	213
CstF (cleavage stimulation factor protein)	104
CTD	103
Ct value	35
C-terminus	8
C→U editing	115
Cytosine	27

D

deamination	207
deletions	213
denaturation	32
denaturing	21
density ultracentrifugation	201
deoxyribonuclease (DNase)	173
deoxyribonucleic acid	27
deoxyribonucleotides	26
ρ-dependent termination	57
depurination	207
Dicer	164
dimer	94, 253

diploid	235
direct repeats	244
dispersive replication	200
disulfide bond	14
D-loop	237
DNA fingerprinting	34
DNA fingerprints	33
DNA gyrase	193
DNA mismatch	206
DNA polymerase I	191
DNA polymerase III	185
DNA polymerase III core	185
DNA polymerase III holoenzyme	185
DNA strand	30
DNA type	34
DNA typing	34
DnaA	180
DNase (deoxyribonuclease)	173
double helix	32
double-stranded break	251
double-stranded break repair (DSBR)	240
double-stranded DNA	30
downstream promoter element (DPE)	83
DPE (downstream promoter element)	83
Drosha	164
DSBR (double-stranded break repair)	240

E

E (exit) site	128
EF (elongation factors)	130
eIFs	135
electron microscopy	172
elongation factors (EF)	130
EMS (ethylmethane sulfonate)	211
endonuclease	251
endosymbiotic hypothesis	40
enhancers	91
enzymes	16
ethylmethane sulfonate (EMS)	211
euchromatin	148
eukaryotic initiation factors	135
excision repair	225
exit (E) site	128

exons	106
3′→5′ exonuclease activity	187
5′→3′ exonuclease activity	192
exosome	160

F

ferritin	169
five prime end	29
Fok I monomer	252
N-formylmethionine	126
frameshift	219
fusion protein	220

G

gal operon	71
gal repressor	72
GAL4	98
galactose	71
β-galactosidase	61
galactoside transacetylase	61
gene	38
gene conversion	243
Gene editing	251
gene expression	59
gene expression library	253
General transcription factors	84
genetic code	124
genetic material	24
global control	166
gRNAs (guide RNAs)	117
group I introns	110
group II introns	110
Guanine	26
guide RNAs (gRNAs)	117

H

hairpin loop	57
hairpin structure	254
HATs (Histone acetyltransferases)	154
HDACs (histone deacetylases)	154
heat-shock proteins	159
helicase	185
α helix	9,252
helix-loop-helix (HLH)	97
helix-turn-helix (HTH)	94

Hershey-Chase experiment	44
heterochromatin	148
heterodimerization	96
heteroduplex joint	236
histone acetyltransferases (HATs)	154
histone code	152
histone deacetylases (HDACs)	154
histone methyltransferases (HMTs)	154
histones	146
histone tails	151
HLH (helix-loop-helix)	97
HMTs (Histone methyltransferases)	154
Holliday junctions	239
homeodomain	95
homologous chromosomes	235
homologous recombination	235,251
homologues	235
housekeeping genes	83
HTH (helix-turn-helix)	94
hybridization	33
hybridize	49
Hydrogen bonds	15
hydrolysis	55
Hydrophobic interaction	15

I

IDL (insertion/deletion loop)	213
IF (initiation factors)	127
induced mutations	209
induced-fit mechanism	19
inducer	67
INF-β gene	154
Initiation	53
initiation factors (IF)	127
initiator (Inr)	83
initiator tRNA	126
inosine	139
Inr (initiator)	83
insertion/deletion loop (IDL)	213
insertions	213
insertion sequences (ISs)	243
integrase	249
intercalating agents	215

intrinsic termination	57
introns	106
introns-early theory	113
introns-late theory	113
inversion	216
inverted repeats	244
Ionic bonds	15
ISs (Insertion sequences)	243

K

kinases	154
kinesin	18

L

lac operon	61
lac repressor	62
lactose	61
lactose permease	61
lagging strand	184
lariat	107
leader	67
leading strand	184
leaky mutation	218
leucine zipper	96
LexA	99
ligand	67
LINE (long interspersed nuclear element)	250
long interspersed nuclear element (LINE)	250
lock-and-key mechanism	19
LTR-retrotransposons	249
lysogeny	250

M

major grooves	32
Meselson-Stahl experiment	200
microRNAs (miRNA)	163
microsatellites	214
minisatellites	33
minor grooves	32
miRNA (microRNAs)	163
mismatched base	206
mismatch repair (MMR)	224
missense mutation	218
MMR (mismatch repair)	224
modularity	98
modules	13
motif	94
mRNA-specific control	166
mutation	205
mutation hot spots	221

N

negative regulation	62
NER (Nucleotide excision repair)	226
NHEJ (non-homologous end joining)	228
nitrogenous base	25,26
nitrous acid	211
30nm fiber	148
non-homologous end joining (NHEJ)	228,251
Non-homologous recombination	235
Non-LTR retrotransposons	249
nonsense mutation	218
N-terminus	8
nuclear pores	158
nucleic acid testing	35
nucleic acids	24
nucleosome	147
nucleotide excision repair (NER)	226
nucleotides	25
nucleus	39

O

Okazaki fragments	184
open promoter complex	53
operator	59
operons	59
ORC (origin recognition complex)	198
OriC	179
origin of replication	179
origin recognition complex (ORC)	198

P

P (peptidyl) site	128
partial diploid	72
partial double-bond	6
PCNA (proliferating cell nuclear antigen)	196
peptide	8
peptide bond	6
peptidyl (P) site	128
peptidyl transferase	128

phosphodiester bond	28
photolyase	223
point mutations	206
poly(A) polymerase	105
poly(A) tail	104
polyadenylation	104
polycistronic mRNA	60
polymers	5
polypeptide	8
polysome	126
positive regulation	63
pre-initiation complex	84
pre-mRNA	102
pre-RC (pre-replicative complex)	198
pre-replicative complex (pre-RC)	198
primary structure	9
primary transcript	102
primase	191
primers	191
probe	33
processivity	186
proliferating cell nuclear antigen (PCNA)	196
promoter	49
promoter clearance	54
proofread	55
proofreading	187
protein domain	12
proteins	5
protospacer adjacent motif	256
purines	26
pyrimidine dimer	212
pyrimidines	26
pyrophosphorolysis	55

Q

quaternary structure	9

R

R group	5
reading frame	219
RecBCD pathway	241
recombination	235
redundant	124
release factor	132

renaturation	33
repeat sequences	254
repeat variable di-residues	253
replication bubble	180
replication fork	181
replicative transposition	245
reporter gene	98
repressor	91
resolution	239
retrotransposons	249
reverse-transcriptase	249
ribonucleases	104
ribonucleic acid	27
ribonucleotides	26
ribosome	125
RISC (RNAi silencing complex)	163
RNA editing	115
RNA interference (RNAi)	163
RNAi silencing complex (RISC)	163
RNA polymerase	49, 51
RNA polymerase I	81
RNA polymerase III	81
RNA polymerase II	81
RNA polymerase core	51
RNA polymerase holoenzyme	52
rudder	54

S

30S subunit	125
50S subunit	125
scanning	133
secondary structures	9
self-splicing	110
semi-conservative replication	178
semi-discontinuous replication	183
β sheet	9
Shine-Dalgarno sequence	126
silencers	91
silent mutations	217
single-guide RNA	256
single-strand DNA binding proteins (SSBs)	185
single-stranded DNA	30
30S initiation complex	126

siRNA (small interfering RNAs)	163	TFⅡD	84
χ-site	241	three prime end	28
site-specific recombination	250	thymine	27
SL (spliced leader)	112	transcription	2
SL1	86	transcription activator-like effector	253
slicing	163	transcription activator-like effector nucleases	253
sliding clamp	187	transcription bubble	54
small interfering RNAs (siRNA)	163	transferrin	161
small nuclear ribonucleoproteins	108	transition	206
snRNPs	108	translation	2
Southern blotting	34	translocation	128, 216
spacer sequence	254	transposase	243
specific transcription factors	88	transposons	243
spliced leader (SL)	112	*trans*-splicing	113
spliceosome	108	transversion	206
splice sites	107	*trp* operon	64
splicing	106	trp repressor	65
spontaneous mutations	207	type Ⅰ topoisomerases	193
SSBs (single-strand DNA binding proteins)	185	type Ⅱ topoisomerases	193
start codon	124		
stop codons	124	**U**	
strand slippage	213	UAS	98
subunit	12	UBF	86
sugar-phosphate backbone	29	Ultraviolet radiation (UV)	212
supercoil	193	untranslated regions (UTRs)	126
Swi/Snf	153	upstream region	50
Sxl protein	156	uracil	27
synapsis	237	UTRs (untranslated regions)	126
		UV (Ultraviolet radiation)	212
T		**V**	
TATA-binding protein (TBP)	84	Van der Waals forces	15
TAFⅡ (TBP-Associated Factors Ⅱ)	85	virus load	35
TATA box	82	**W**	
tautomers	208	wobble	139
TBP (TATA-binding protein)	84	**X**	
TBP-Associated Factors Ⅱ (TAFⅡ)	85	xeroderma pigmentosum (XP)	230
telomerase	199	XP (xeroderma pigmentosum)	230
telomere	199	**Z**	
termination	57	zinc finger	95, 252
tertiary structure	9	zinc finger nucleases	251
TFⅢB	88		
TFⅡB recognition element (BRE)	83		
TFⅢC	88		

Glossary

A (aminoacyl) site
First site on the ribosome to which tRNAs bind, bringing new amino acids. Named after the acyl bond that attaches amino acids to tRNA.

Acetylation
The addition of an acetyl group to a molecule.

Aconitase
Protein that regulates iron metabolism. Controls mRNA stability and translatability for transferrin and ferritin, respectively.

Activators
Proteins that increase transcription of a gene.

Active site
Site on an enzyme that is directly responsible for catalyzing reactions.

Affinity chromatography
A kind of column chromatography technique. The column is packed with molecules that bind to a specific subset of proteins. Affinity = attraction.

A→I editing
A form of post-transcriptional modification to mRNA in eukaryotes in which adenine is deaminated to form inosine, an unusual base.

Alkyl group
A class of chemical groups composed of carbon and hydrogen. Examples: methyl group ($—CH_3$) and ethyl group ($—CH_2CH_3$).

Alleles
Versions of the same gene that differ slightly in function and sequence.

Allosteric regulation
Form of regulation in which a small molecule binds to a regulatory site on a protein, causing a structural and functional change at the active site. 'Steric' is related to the word 'structure'.

Alternative polyadenylation
The ability to make mRNAs of varying sizes from one coding region, by altering the site of pre-mRNA cleavage and polyadenylation. Can alter protein size and properties of the mRNA, such as stability.

名词解释[1]

A（氨酰基）位
带有新氨基酸的 tRNA 与核糖体结合的第一个位置。根据氨基酸连接到 tRNA 上的酰基键命名。

乙酰化作用
将一个乙酰基加到某个分子上的过程。

顺乌头酸酶
调控铁代谢的蛋白质。分别控制运铁蛋白 mRNA 的稳定性和铁蛋白 mRNA 的可翻译性。

激活蛋白
促进基因转录的蛋白质。

活性位点
酶分子中直接负责催化反应的位点。

亲和层析
柱层析技术的一种。柱中用能与一组特殊的蛋白质结合的分子填充。亲和 = 吸引。

A→I 编辑
真核生物中对 mRNA 进行转录后修饰的一种形式，修饰时将腺嘌呤脱氨基形成次黄苷，次黄苷是一种非普通碱基。

烷基
由碳和氢组成的一类化学基团。例如：甲基（$—CH_3$）和乙基（$—CH_2CH_3$）。

等位基因
相同基因的不同版本，它们之间在功能和序列上稍有不同。

别构调节
调节的一种形式，通过一个小分子结合到蛋白质的调节位点而引起其活性位点结构和功能的改变。"steric"（空间的）一词与"structure"（结构）有联系。

可变聚腺苷酸化
通过改变前体 mRNA 的切割位点和聚腺苷酸化位点而从一个编码区产生不同大小 mRNA 的能力。能改变蛋白质分子的大小和 mRNA 的性质（如稳定性）。

---

[1] 按英文名词字母先后顺序排列；希腊字母（α、β、σ等）、数字及英文单位均不参加排序。

Alternative splicing

The ability to make various proteins from one coding region by choosing between the inclusion/exclusion of certain introns and exons.

α-amanitin

A toxin that inhibits the three eukaryotic RNA polymerases to different extents. Name derives from mushroom of genus *Amanita* in which toxin is found.

Ames test

Technique to assess the mutagenic potential of a chemical.

Amino acids

Small molecules that can be polymerized to form proteins. Name derives from the presence of an amino group as well as an acidic carboxyl group.

Amino group

A chemical group comprising nitrogen bound to two hydrogen atoms.

Aminoacyl-tRNA synthetase

A protein that matches tRNAs with the correct amino acid. Name origin: enzyme synthesizes the acyl bond that joins amino acids to tRNAs.

Anticodon

Three base sequence in a tRNA that binds to one or more codons. The prefix 'anti-' here means 'opposite' or 'complementary.'

Anti-parallel

Two strands that are parallel but oriented in the opposite direction. Often used to describe the orientation of strands in a DNA molecule relative to each other.

***ara* operon**

Operon containing genes that metabolize the sugar arabinose.

Attenuation

A mechanism in the *trp* operon to ensure that genes are not transcribed in the presence of tryptophan.

Attenuator

Transcribed sequence just downstream of the *trp* operon promoter that is central to the attenuation mechanism. Can form a terminating hairpin structure that stops transcription of the operon.

AU-rich element (ARE)

Sequence in the 3′-UTR of certain eukaryotic mRNAs that is involved in regulation of mRNA stability. Named for prevalence of A and U bases in the sequence.

Autonomously replicating sequence (ARS)

An origin of replication in yeast. Name origin: pieces of DNA containing this sequence are able to replicate autonomously, meaning even if they are not part of a chromosome.

可变剪接

通过选择包含或不包含某些内含子和外显子而从一个编码区产生不同蛋白质的能力。

α-鹅膏蕈碱

一种能不同程度地抑制三种真核生物 RNA 聚合酶的毒素。名称来自产生此毒素的 *Amanita* 属蘑菇。

埃姆斯测验法

用来评价化学物质是否具有潜在诱变性的技术。

氨基酸

能聚合形成蛋白质的小分子。名称来自于它们所带的氨基以及酸性的羧基。

氨基

由氮原子与两个氢原子结合形成的化学基团。

氨酰-tRNA 合成酶

一种将 tRNA 与正确的氨基酸匹配的蛋白质。名称来源：酶合成了将氨基酸连接到 tRNA 上去的酰基。

反密码子

tRNA 上能与一个或多个密码子结合的三碱基序列。前缀"anti-"在这儿的意思是"相反的"或"互补的"。

反向平行的

两条平行但方向相反的链。常用来描述 DNA 分子中的链相对于各自的方向。

***ara* 操纵子**

含有阿拉伯糖代谢基因的操纵子。

衰减作用

trp 操纵子中用于确保当色氨酸存在时基因不被转录的一种机理。

衰减子

紧接在 *trp* 操纵子的启动子之后转录出来的序列，对衰减作用机理很重要。能形成终止型发夹结构而使操纵子的转录停止。

富含 AU 元件（ARE）

在调节 mRNA 稳定性中起作用的一些真核生物 mRNA 3′-UTR 序列。因序列中含有许多 A 和 U 而得名。

自主复制序列（ARS）

酵母中的一种复制起点。名称来源：含有此序列的 DNA 片段即使不是染色体的一部分也能够自主地复制。

Basal transcription

The low rate of transcription that occurs in eukaryotes if only the pre-initiation complex is present (without activators). Basal ∼ base = a low part.

Base analogues

Molecules that resemble DNA bases in structure and can be incorporated into DNA. 'Analogues' = two things that resemble each other. Analogue ∼ analogy.

Base excision repair (BER)

DNA damage repair pathway usually used to fix common damage to DNA bases.

Base pair

Two complementary nucleotides bound by hydrogen bonds. Often used to measure the length for DNA.

Base-stacking

Term to describe the placement of base pairs in a DNA molecule. Base pairs lie in parallel planes one above the other, giving the impression that they are 'stacked'.

Beads-on-a-string

Term to describe the lowest level of DNA organization. Nucleosomes are separated by stretches of naked DNA.

Bi-directional replication

Term to describe DNA replication that proceeds in two opposite directions from an origin of replication. The prefix 'bi-' = 'two'.

−10 box

Common promoter element in *E. coli*. Named for its location approximately 10 bases upstream of the transcription start site.

−35 box

Common promoter element in *E. coli*. Named for its location approximately 35 bases upstream of the transcription start site.

Branch migration

The movement of a Holliday junction that causes different hybridizations between the homologous chromosomes.

5-bromouracil (BU)

Mutagenic base analogue. Resembles thymine, but easily interconverts to tautomeric form that pairs with guanine. Structure resembles uracil with an attached bromine group.

Capping

Post-transcriptional modification in which a derivative of guanosine is attached to the 5′ end of the pre-mRNA. In normal English, a 'cap' is anything that goes on the head of something. For example, a hat is often called a 'cap'.

基础转录

真核生物中如果只有前起始复合体存在（没有激活蛋白）的情况下所发生的低速率的转录。基础 ∼ 基底 = 少量。

碱基类似物

结构上与 DNA 碱基类似的分子，能被整合到 DNA 中。"类似物" = 两种互相相像的事物。类似物 ∼ 类似。

碱基切除修复（BER）

DNA 损伤修复途径，通常用来修复 DNA 碱基的普通损伤。

碱基对

通过氢键结合在一起的互补的核苷酸。常用来度量 DNA 的长度。

碱基堆积

用来描述 DNA 分子中碱基排布位置的术语。碱基对一个接一个地平放在平行的平面上，给人的印象是它们是"堆放"在那里的。

线珠结构

用于描述 DNA 组织最低一级水平的术语。核小体由裸露的 DNA 片段连接在一起。

双向复制

用来描述 DNA 复制从复制起点开始沿两个相反的方向进行的术语。前缀 "bi-" = "二"。

−10 框

大肠杆菌中常见的启动子元件。由于它位于转录起始位点上游约 10 个碱基处而得名。

−35 框

大肠杆菌中常见的启动子元件。由于它位于转录起始位点上游约 35 个碱基处而得名。

分支迁移

指 Holliday 交叉的移动，它会引起同源染色体不同区段之间发生杂交。

5-溴尿嘧啶（BU）

诱变性的碱基类似物。与胸腺嘧啶相似，但很容易转换成与鸟嘌呤配对的互变异构形式。带有溴基团的结构与尿嘧啶相似。

加帽

将鸟嘌呤核苷的衍生物加到前体 mRNA 5′端上去的转录后修饰作用。在日常英语中，"cap"（帽、盖）指放到某些事物头上的任何东西。例如，有沿帽常被叫作 "cap"。

Carboxyl group

Acidic chemical group in which a carbon atom is bound to two oxygen atoms. Name origin: the group contains carbon and oxygen.

Catabolite activator protein (CAP)

Protein in prokaryotes that responds to glucose/cAMP concentration to activate transcription. Used to enhance transcription of the *lac* and *ara* operon genes, which produce proteins involved in catabolism.

Chaperones

Proteins that help other proteins to fold into the correct structure. In normal English, a chaperone is a person who accompanies somebody else.

Chromatin immunoprecipitation (ChIP)

A co-immunoprecipitation technique that checks for binding of specific DNA sequence to a specific protein.

Chromatin remodeling proteins

Proteins that alter the association of histones with DNA as a way of regulating transcription. In normal English, to 'model' = 'to build' or 'to organize'. The prefix 're-' = 'again' or 'new'. Chromatin remodeling causes a new organization of histones with DNA.

Clamp loader

A subset of prokaryotic DNA polymerase subunits that load the sliding clamp onto DNA.

Co-immunoprecipitation (CoIP)

A technique to check for binding of a protein to other proteins or molecules. Immunoprecipitation of a protein is performed using antibodies. Any proteins that bind to that protein will also be precipitated and identified. The prefix 'co-' = 'with'. Used here because proteins that immunoprecipitate *with* the protein that binds to the antibody are identified.

Colonies

Isolated populations of cells on a plate. Theoretically, all cells in a colony are identical because they are descended from one cell. In normal English, a 'colony' is a group of people that settles in a foreign land.

Column chromatography

A general set of techniques used to purify proteins by passing them through a column. Proteins exit the column at different time depending on their properties and the properties of the column.

Complementary

Term used to describe bases that can pair with each other. In normal English, things that 'complement' each other are things that go well together. 'Complementary' should not be confused with 'complimentary'.

羧基

一个碳原子连接到两个氧原子上形成的酸性化学基团。名称来源：该基团含有 carbon（碳）和 oxygen（氧）。

代谢物激活蛋白（CAP）

原核生物中对葡萄糖/cAMP 浓度进行响应而激活转录的蛋白质。用于增强 *lac* 和 *ara* 操纵子基因的转录，产生的蛋白质在分解代谢中起作用。

伴侣蛋白

帮助其他蛋白质折叠成正确结构的蛋白质。在日常英语中，chaperone 指一个陪伴另一个人的人。

染色质免疫沉淀法（ChIP）

一种免疫共沉淀技术，用于检查特异 DNA 序列与特异蛋白质之间的结合。

染色质重塑蛋白

改变组蛋白与 DNA 之间结合方式的蛋白质，是一种调控转录的方式。在日常英语中，to "model" = "建造" 或 "组织"。前缀 "re-" = "再次" 或 "新的"。染色质重塑使组蛋白与 DNA 产生新的组织方式。

滑行夹加载器

由原核生物 DNA 聚合酶的一部分亚基组成的将滑行夹套到 DNA 上去的装置。

免疫共沉淀法（CoIP）

一种检查某种蛋白质是否与其他蛋白质或其他分子结合的技术。蛋白质的免疫沉淀采用抗体进行。任何与该蛋白质结合的蛋白质也将被沉淀和鉴定出来。前缀 "co-" = "与"。用在此处的原因是，它鉴定出了那些与该蛋白质（结合在抗体上）一起免疫沉淀出来的蛋白质。

集落

平板上分开的细胞群体。理论上说，一个集落中的所有细胞是完全相同的，因为它们都是一个细胞的后代。在日常英语中，"colony" 指定居在外国土地上的一群人。

柱层析

使蛋白质穿过层析柱而对它们进行纯化的一套综合技术。蛋白质的性质和层析柱的性质决定了蛋白质流出柱子的时间不同。

互补的

用来描述能互相配对的碱基的术语。在日常英语中，能互相 "complement" 的事物是那些可以很好地在一起的事物。请不要将 "complementary"（互补的）与 "complimentary"（赞美的）混淆。

Conformations
Used in relation to proteins as synonym for 'structure'.

Consensus sequence
The most probable sequence of a sequence element. In normal English, a 'consensus' is when everybody agrees about something.

Conservative replication
A model of replication in which an entirely new DNA molecule is produced, and the parental DNA molecule is conserved.

Conservative transposition
A type of transposition in which the transposon is removed from its original location to be inserted into a new location.

Constitutive
A commonly used word in biology meaning 'always' or 'nonstop'. Constitutive *lac* operon mutants are those that transcribe *lac* operon genes always, under any condition.

Coordinate regulation
Transcriptional regulation in which a set of genes are regulated together.

Core histones
Histones that come together to form the core of the nucleosome.

Co-repressor
A small molecule that binds to a repressor protein to allow repressive activity. Prefix 'co-' = 'with'. In this case the co-repressor works with the repressor to cause repression.

Coupled transcription-translation
Translation that occurs while transcription is still in progress.

Covalent bond
A bond between two atoms in which electrons are shared.

CRISPR/Cas
Clustered regularly interspaced short palindromic repeats (CRISPR) and CRISPR-associated (Cas) nucleases, an artificially designed gene editing system by forming single-guide RNA and Cas nuclease complex to recognize and cleave target DNA sequence.

Crossing-over
A recombinational event in which regions of DNA are exchanged between homologous chromosomes.

Cross-linking
The covalent linkage of bases opposite each other in a DNA molecule.

构象
作为"结构"的同义词在与蛋白质有关的表述中使用。

共有序列
一个序列元件的最有可能的序列。在日常英语中，"consensus"指每个人都同意某件事。

保留复制
一种复制模型，认为复制产生一个全新的 DNA 分子，而亲本 DNA 分子保持不变。

保守型转座
一种转座类型，转座时转座子从它的原始位置离开并插入一个新的位置中。

组成型的
是生物学常用词，表示"总是"或"不停的"。组成型 *lac* 操纵子突变体在任何条件下都一直转录 *lac* 操纵子基因。

协同调控
一组基因在一起进行调控的转录调控方式。

核心组蛋白
组合在一起形成核小体核心的组蛋白。

辅阻遏物
结合到阻遏蛋白上使其产生阻遏活性的小分子。前缀 "co-" = "与"。在此辅阻遏物与阻遏蛋白一起引发阻遏作用。

偶联转录-转译作用
当转录还在进行的时候发生的翻译作用。

共价键
两个原子共享电子形成的键。

成簇规律间隔短回文重复序列以及相关核酸酶
成簇规律间隔短回文重复序列（CRISPR）以及 CRISPR 相关（Cas）核酸酶，一种人工设计的基因编辑系统，通过形成单链引导 RNA 与 Cas 核酸酶复合体来识别并切割目标 DNA 序列。

交换
DNA 区域在同源染色体之间发生交换的重组事件。

交联
在 DNA 分子中位于相对链上的碱基之间形成的共价连接。

C-terminus

The end of a polypeptide containing a carboxyl group. Name origin: C refers to carboxyl, and terminus = end. *Pl.*: C-termini.

Ct value

Cycle threshold value, defined to measure quantity of target DNA in a test sample. The lower the value, the higher the quantity of target DNA.

C→U editing

A form of post-transcriptional modification to mRNA in eukaryotes in which cytosine is deaminated to form uracil.

Deamination

The removal of an amino group from a base. The prefix 'de-' = 'undo' or 'remove' in this case.

Deletion

The removal of base pairs from a DNA molecule.

Denaturation

The disruption of non-covalent interactions in a macromolecule that alters its three-dimensional structure. With respect to DNA, denaturation involves separations of single-strands. With respect to protein, denaturation involves unfolding of the polypeptide.

Density ultracentrifugation

A technique that uses a centrifugation (fast spinning) to separate molecules according to density. Prefix 'ultra-' = very. In this technique, centrifugation is very very fast.

Deoxyribonucleases (DNase)

Enzymes that cut deoxyribonucleic acid, DNA.

Deoxyribonucleic acid (DNA)

A nucleic acid made by polymerization of deoxyribonucleotides.

Deoxyribonucleotides

Nucleotides that are similar to ribonucleotides but are missing an —OH group. Name origin: The prefix 'de-' = 'undo' or 'remove' here. Deoxyribonucleotides are ribonucleotides with the oxygen removed.

ρ-dependent termination

A form of transcription termination in prokaryotes that depends on the protein ρ as well as on sequences in the DNA/RNA.

Depurination

The removal of a purine base from the DNA backbone. Prefix 'de-' = 'undo' or 'remove' here.

Dideoxy method

A technique for sequencing DNA that relies on the use of dideoxyribonucleotides.

Dideoxyribonucleotides

Nucleotides that resemble ribonucleotides but are

C-末端

多肽上含有羧基的末端。名称来源：C 指 carboxyl（羧基），terminus＝末尾。复数：C-termini。

Ct 值

循环临界值，用来定义试样中目标 DNA 的含量。该值越低，表示目标 DNA 含量越高。

C→U 编辑

一种形式的真核生物转录后修饰作用，修饰时将胞嘧啶脱氨基形成尿嘧啶。

脱氨基

从一个碱基上去除一个氨基的过程。在这里，前缀 "de-" ＝ "消除" 或 "去除"。

缺失

从 DNA 分子中去除碱基对的过程。

变性

大分子中非共价相互作用力的破坏导致三维结构改变的过程。对 DNA 来说，变性涉及单链的分离。对蛋白质来说，变性涉及多肽链的解折叠。

密度超速离心

根据密度应用离心（快速旋转）对分子进行分离的技术。前缀 "ultra-" ＝ 非常。在此技术中，离心速度是非常非常快的。

脱氧核糖核酸酶（DNase）

切割脱氧核糖核酸（DNA）的酶。

脱氧核糖核酸（DNA）

由脱氧核糖核苷酸聚合而来的核酸。

脱氧核糖核苷酸

与核糖核苷酸类似的核苷酸，但它们缺少—OH 基团。名称来源：在这里，前缀 "de-" ＝ "去掉" 或 "去除"。脱氧核糖核苷酸就是 oxygen（氧）被 removed（去除）了的核糖核苷酸。

ρ 依赖型终止

原核生物中的一种转录终止方式，依靠 ρ 蛋白和 DNA/RNA 序列引起转录终止。

脱嘌呤

从 DNA 骨架上去除嘌呤碱基的过程。在这里，前缀 "de-" ＝ "去掉" 或 "去除"。

双脱氧法

依赖于使用双脱氧核糖核苷酸进行 DNA 序列测定的技术。

双脱氧核糖核苷酸

与核糖核苷酸类似的核苷酸，但它们缺少两个

missing two —OH groups. Name origin: Prefix 'di-' = 'two'. Prefix 'de-' = 'undo' or 'remove'. Dideoxyribonucleotides are ribonucleotides that have had two oxygens removed.

Dimer
A molecule composed of two smaller molecules. Lactose is an example of a dimer, as it is made of glucose and galactose. A protein made of two subunits can also be called a dimer. Prefix 'di-' = 'two'.

Diploid
Organisms with two copies of each gene. Prefix 'di-' = two.

Direct repeats
Sequences that are exactly the same, and present in the same orientation.

Dispersive replication
A model for replication in which parental DNA is fragmented. The fragments are dispersed, and daughter DNA molecules are made by connecting the fragments with newly synthesized DNA.

Disulfide bond
A covalent bond formed between the R groups (—SH) of two cysteine amino acids. Prefix 'di-' = 'two'. Disulfide bonds are formed by two chemical groups that contain sulfur.

D-loop
A structure formed during homologous recombination after synapsis. Two DNA strands in one chromosome are separated, forming an opening that resembles a loop.

DNA fingerprint
A distribution pattern of DNA usually with dozens of bands, obtained from DNA fingerprinting.

DNA fingerprinting
An analytical method of displaying DNA fingerprint of an organism by using one probe that hybridizes to 'minisatellite' at all loci.

DNA gyrase
A common topoisomerase Ⅱ in prokaryotes. Undoes supercoiling by gyrating one part of the chromosome relative to another.

DNA microarray
A technique that is used to test the difference in transcription of thousands of genes between two different conditions. Name origin: prefix 'micro-' = 'small'. A DNA microarray is a very small piece of glass in which an array of very small spots of DNA are placed.

—OH基团。名称来源：前缀"di-"="二"。前缀"de-"="去掉"或"去除"。双脱氧核糖核苷酸就是 two oxygen（两个氧）被 removed（去除）了的核糖核苷酸。

二聚体
由两个较小的分子组成的分子。乳糖就是一个二聚体的实例，它是由葡萄糖和半乳糖组成的。由两个亚基组成的蛋白质也可称为二聚体。前缀"di-"="二"。

二倍体
每个基因有两个拷贝的生物。前缀"di-"="二"。

同向重复序列
完全相同并以相同方向出现的序列。

散乱型复制
一种复制模型，认为亲本 DNA 呈短片段状。这些 DNA 片段是分散的，子代 DNA 分子是由这些片段与新合成的片段连接而成的。

二硫键
在两个半胱氨酸的 R 基团（—SH）之间形成的共价键。前缀"di-"="二"。二硫键就是由 two（两个）含 sulfur（硫）的化学基团所形成的键。

D 环
联会之后的同源重组过程中形成的结构。一条染色体上的两条 DNA 链被分开，形成一个环状开口。

DNA 指纹
一种 DNA 分布样式，通常含几十条 DNA，由 DNA 指纹分析获得。

DNA 指纹分析
一种显示生物 DNA 指纹的分析方法，使用一个探针与所有位点上的"小卫星"杂交。

DNA 旋转酶
原核生物中一种常见的拓扑异构酶Ⅱ。通过将染色体的一部分相对另一部分进行旋转（gyrating）而消除超螺旋。

DNA 芯片
用来试验两种不同条件下几千个基因不同转录情况的一种技术。名称来源：前缀"micro-"="小"。DNA 芯片就是一块被放上了一个很小的 DNA 阵列的很小的玻璃片。

DNA mismatch
When two or more bases in a DNA molecule are not matched correctly. Prefix 'mis-' = 'incorrect'.

DNA polymerase
Enzyme that polymerizes deoxyribonucleotides to make DNA.

DNA polymerase Ⅰ
A prokaryotic DNA polymerase with a special $5'\rightarrow 3'$ exonuclease activity, used to remove primers.

DNA polymerase Ⅲ
A prokaryotic DNA polymerase with high processivity that performs most DNA replication.

DNA polymerase Ⅲ core
The smallest collection of subunits required for DNA polymerase Ⅲ to make DNA.

DNA polymerase Ⅲ holoenzyme
The full collection of DNA polymerase Ⅲ subunits. Necessary for replication to occur with high processivity.

DNA strand
A covalently linked chain of deoxyribonucleotides. The double helix is composed of two DNA strands.

DNA type
A distribution pattern of DNA usually with only a few bands, obtained from DNA typing.

DNA typing
An analytical method of displaying DNA types for identifying a particular individual by using three to five probes, each of which hybridizes to 'minisatellite' at one locus.

DnaA
A protein involved in initiation of DNA replication in *E. coli*. Binds to DNA 9-mers in the *OriC*.

Double helix
The structure of DNA, consisting of two DNA strands that join together and form a helical shape.

Double-stranded break repair (DSBR)
General term for the repair of double-stranded breaks in the DNA.

Downstream promoter element (DPE)
A promoter element that occurs downstream of the transcription start site.

E (exit) site
The site on a ribosome to which tRNAs bind before exiting the ribosome.

Electron microscope
A microscope that uses electron beams to visualize subcellular components.

DNA 错配
指 DNA 分子中两个或多个碱基没有互相正确匹配。前缀 "mis-" = "不正确的"。

DNA 聚合酶
将脱氧核糖核苷酸聚合形成 DNA 的酶。

DNA 聚合酶 Ⅰ
一种原核生物的 DNA 聚合酶,具有特殊的 $5'\rightarrow 3'$ 核酸外切酶活性,可用于去除引物。

DNA 聚合酶 Ⅲ
一种原核生物的 DNA 聚合酶,具有很强的持续合成能力,执行大多数 DNA 的复制任务。

DNA 聚合酶 Ⅲ 核心
DNA 聚合酶 Ⅲ 能产生 DNA 的最少亚基组合。

DNA 聚合酶 Ⅲ 全酶
DNA 聚合酶 Ⅲ 的完整亚基组合。是持续进行复制所必需的。

DNA 链
脱氧核糖核苷酸以共价键连接起来的链。双螺旋由两条 DNA 链组成。

DNA 类型
一种 DNA 分布样式,通常只含几条带,由 DNA 分型术获得。

DNA 分型术
一种显示 DNA 类型的分析方法,用于鉴定特定个体,使用三至五个探针,每个探针只与一个位点上的"小卫星"杂交。

DnaA
在大肠杆菌 DNA 复制起始中起作用的一种蛋白质。在 *OriC* 的位置与 DNA 的 9-mer 部分结合。

双螺旋
DNA 的一种结构,由两条链组成,两条链结合在一起产生螺旋形。

双链断裂修复(DSBR)
修复 DNA 中双链断裂的常用术语。

下游启动子元件(DPE)
出现在转录起始位点下游的启动子元件。

E(退出)位
核糖体上的位点,是 tRNA 在退出核糖体之前所在的位置。

电子显微镜
应用电子束对亚细胞成分进行成像的显微镜。

Elongation factors
Proteins involved in the elongation phase of translation.

Endonuclease
An enzyme that cleaves DNA or RNA at site inside the molecule (as opposed to cutting at ends of the molecule). Prefix 'endo-' = 'inside'.

Enhancers
Regulatory DNA elements to which activators bind to enhance the rate of transcription.

Enzymes
Proteins that catalyse reactions.

Ethylmethane sulfonate (EMS)
An alkylating agent that adds an ethyl groups (a kind of alkyl group) to target molecules.

Euchromatin
Chromatin regions that are less tightly packaged than heterochromatin and contain transcribed genes.

Excision repair
A general term for mechanisms that repair damaged bases. Excise = remove a piece from something. In excision repair, the damaged base (and some surrounding bases) are removed from the DNA and resynthesized.

Exons
Parts of a gene that are expressed as protein. Exons are formed by the interruption of coding regions by introns.

Exonuclease
Enzymes that degrade DNA or RNA from the ends of the molecule; in other words, from the exterior.

$3' \rightarrow 5'$ exonuclease activity
Function of DNA polymerases. Allows $3' \rightarrow 5'$ removal of incorrect nucleotides after polymerization. See also exonuclease.

$5' \rightarrow 3'$ exonuclease activity
Function of DNA polymerase I in prokaryotes. Allows for removal of nucleotides in the direction of synthesis. Often used for removal of RNA primers.

Exosome
A protein complex containing exonucleases that degrades mRNA in the $3' \rightarrow 5'$ direction.

Ferritin
Protein that binds to iron in the cytoplasm. 'Ferr' comes from the Latin word for iron.

30nm fiber
An higher-level structure of chromatin. The elongated structure has a width of approximately 30nm.

Five prime end
The end of a DNA strand that terminates with the five

延伸因子
在翻译延伸阶段发挥作用的蛋白质。

核酸内切酶
在分子的内部切割 DNA 或 RNA 的酶（与在分子的末端切割相反）。前缀"endo-"="内部"。

增强子
具有调控作用的 DNA 元件，激活蛋白结合上去后可以增强转录速率。

酶
催化反应的蛋白质。

乙基甲磺酸（EMS）
一种烷化剂，能将一个乙基（烷基的一种）加到目标分子上。

常染色质
比异染色质包装松散并含有已转录基因的染色质区域。

切除修复
受损碱基修复机理的通用名称。Excise = 从某事物中去除。在切除修复中，受损的碱基（以及一些邻近碱基）被从 DNA 中去除并重新合成。

外显子
基因中被表达（expressed）成蛋白质的部分。外显子由内含子打断编码区域而形成。

核酸外切酶
从分子的末端降解 DNA 或 RNA 的酶；换句话说，是从外面（exterior）进行切割。

$3' \rightarrow 5'$ 核酸外切酶活性
DNA 聚合酶的活性。允许在聚合反应后以 $3' \rightarrow 5'$ 方向去除不正确的核苷酸。请参照"exonuclease"。

$5' \rightarrow 3'$ 核酸外切酶活性
原核生物中 DNA 聚合酶 I 的一种功能。允许在 DNA 合成方向上去除核苷酸。常在去除 RNA 引物中使用。

外来体
一种含有核酸外切酶、以 $3' \rightarrow 5'$ 方向降解 mRNA 的蛋白质复合体。

铁蛋白
在细胞质中与铁结合的蛋白质。"Ferr"来自"铁"的拉丁文。

30nm 纤维
一种染色质的高级结构。其延长的结构宽度大约是 30nm。

5'-末端
DNA 链的一个末端，以脱氧核糖核苷酸中核糖的

prime (5′) carbon (which has a phosphate attached) of the ribose in a deoxyribnoucleotide. 'Prime' means nothing here; it simply denotes that the label 'five' comes from arbitrary numbering of the ribose carbons.

Forward genetics
Genetic/molecular biology research in which a phenotype is first considered, and later the gene responsible is isolated.

Frameshift
A mutation that causes a shifting of the reading frame of an mRNA. Caused by insertions and deletions.

Fusion protein
A protein that is made by fusing together two or more different proteins.

β-galactosidase
Enzyme that cleaves lactose into galactose and glucose. Name origin: the bond cut by this enzyme is called a β-galactosidic bond.

Gel electrophoresis
A technique for separating DNA according to length. The DNA is pulled through a piece of gel by a voltage because of its electric charge.

Gel-filtration chromatography
A form of column chromatography that separates proteins according to size. Uses beads made gel, with small tunnels that trap smaller proteins more easily than larger proteins. Filtration = separation of components in a mixture.

Gene conversion
Consequence of homologous recombination, in which hybridization between different alleles causes one allele to be converted into the other. This occurs by mismatch repair.

Gene editing
A novel genetic engineering technology capable of modifying gene structure at target site in genome of an organism.

Gene expression
The process of making a protein from a gene. In normal English, to 'express' yourself = to say what you are thinking. In gene expression, the cell makes something from the information that the gene is holding.

Genetic material
A general term to describe a material that can pass traits from generation to generation. The term was mainly used in the past, before DNA was discovered to be the genetic material.

Global control
Regulation of all translation in the cell at once. In nor-

五一撇（5′）碳（上面连接着磷酸）结尾。"一撇"在这儿没有含义，它只是简单地表示"五"这一记号来自对核糖上碳的人为编号。

正向遗传学
一种遗传学/分子生物学研究，先考虑表型，之后再分离相关的基因。

移码
一种引起 mRNA 读码框位置发生变化的突变。由插入和缺失引起。

融合蛋白
通过融合两个或多个不同的蛋白质产生的蛋白质。

β-半乳糖苷酶
将乳糖分解为半乳糖和葡萄糖的酶。名称来源：该酶切割的键称为 β-半乳糖苷键。

凝胶电泳
根据长度分离 DNA 的技术。由于 DNA 上带有电荷（electric charge），它们在电压作用下被牵引着在凝胶中移动。

凝胶过滤层析
一种根据分子大小分离蛋白质的柱层析。采用珠状凝胶，凝胶中带有小的通道，小分子比大分子更容易被通道捕捉。过滤 = 从混合物中分离组分。

基因转换
同源重组的后果，在不同等位基因之间的杂交使其中的一个等位基因被转换成了另一个等位基因。这是由错配修复引起的。

基因编辑
一种新型基因工程技术，能在生物基因组中目标位点修饰基因结构。

基因表达
从基因产生蛋白质的过程。在日常英语中，to "express" yourself = 说出你的想法。在基因表达中，表示细胞从基因持有的信息中产生某种物质。

遗传物质
用来描述可以将性状在世代间传递的物质的常用术语。该术语主要在过去使用，即在 DNA 被发现是遗传物质之前。

全局控制
细胞中一次就对所有翻译过程进行的调控。在日

mal English, global = something that happens everywhere.

Group I introns
Self-splicing introns that use a free guanylate nucleotide in the splicing mechanism.

Group II introns
Self-splicing introns that follow nearly the same basic splicing reaction as normally occurs with the spliceosome.

Guide RNAs (gRNAs)
RNAs that guide mRNA editing in trypanosomes. Direct addition and deletion of uracils from the mRNA.

Hairpin loop
An RNA structure caused by hybridization between neighboring regions of RNA. Somewhat resembles a hairpin.

Helicase
Enzyme that separates the two strands of the double helix by breaking hydrogen bonds between the two strands.

α helix
A helical secondary structure in proteins. $Pl.$: α helices.

Helix-loop-helix (HLH)
A DNA-binding domain in proteins. Each half of the domain consists of two α helices connected by a peptide loop. The two halves, usually present on different proteins, come together at the binding site on DNA.

Helix-turn-helix (HTH)
A common DNA-binding motif in prokaryotic proteins. Consists of two α helices connected by a short peptide turn.

Heterochromatin
Regions of DNA tightly packaged with proteins, usually does not contain genes to be expressed.

Heterodimerization
Formation of dimers between two different subunits. The prefix 'hetero-' = different.

Heteroduplex joint
Site on a chromosome, produced after homologous recombination, where a region of DNA from one chromosome meets a region from a different chromosome. At the joint, single-stranded DNAs from the different chromosomes are hybridized. The prefix 'hetero-' = different.

Histone code
Combinations of covalent modifications on histones that have a functional significance and are recognized by proteins.

Histone tails
Long unstructured ends of histones that stick out from the nucleosome. Can bind to other DNA and to other

常英语中，global = 在任何地方都发生。

I 类内含子
自我剪接型内含子，在剪接过程中使用游离鸟苷酸。

II 类内含子
自我剪接型内含子，剪接过程几乎与有剪接体参与的剪接反应相同。

指导 RNA（gRNA）
在锥虫中指导 mRNA 编辑的 RNA。指导向 mRNA 添加或从中删去尿嘧啶。

发夹环
一种 RNA 结构，由 RNA 邻近区域之间杂交产生。看起来有些像发夹。

解旋酶
通过打断两条链之间的氢键而将双螺旋（helix）的两条链分开的酶。

α 螺旋
蛋白质中一种螺旋形的二级结构。复数：α helices。

螺旋-环-螺旋（HLH）
蛋白质的一种 DNA 结合域。该结合域的每一半由两个 α 螺旋通过一个肽环连接在一起。该结合域的两半通常来自不同的蛋白质，它们会聚集到目标 DNA 位置并发生结合。

螺旋-转角-螺旋（HTH）
原核生物中常见的 DNA 结合基序。由两个 α 螺旋通过一个肽转角连接在一起。

异染色质
与蛋白质紧密包装在一起的 DNA 区域，通常没有需要表达的基因。

异源二聚化作用
两个不同亚基之间形成二聚体的过程。前缀"hetero-" = 不同的。

异源双链接头
染色体上由同源重组产生的位点，其中来自一条染色体的 DNA 区域与来自另一条染色体的区域相遇。在接头处，来自不同染色体的单链 DNA 杂交在一起。前缀"hetero-" = 不同的。

组蛋白密码
在组蛋白上发生的共价修饰组合，具有功能意义，能被蛋白质识别。

组蛋白尾
从核小体中伸出的组蛋白非结构化长末端。能与其他 DNA 和其他组蛋白结合，能被共价修饰。

histones, and can be covalently modified.

Histones
Proteins around which DNA is wrapped to organize it and regulate transcription.

Holliday junctions
Structure formed during homologous recombination connection of single strands of one chromosome to the other chromosome. Named after a scientist called Holliday.

Homeodomain
DNA-binding domain, found in many eukaryotic proteins involved in organismal development.

Homologous chromosomes
Chromosomes that are not identical, but have the same set of genes. The prefix 'homo-' = 'same'.

Homologous recombination
Recombination between DNA molecules with similar DNA sequences. The prefix 'homo-' = 'same'.

Housekeeping genes
Genes that must be transcribed in all cells, because they have a function that is basic to cell survival. In normal English, 'housekeeping' refers to the daily activities required in a home, like cleaning and cooking. Housekeeping genes are genes perform the daily activities required in a cell.

Hybridization
The binding of two strands of nucleic acid. Hybridization can occur between two different DNA molecules, and between DNA and RNA.

Hydrogen bonds
Relatively weak interatomic bonds involving hydrogen atoms that are partially positively charged.

Hydrophobic interaction
Interaction that occurs between non-polar molecules when in the presence of water. Prefix ' hydro-' = 'water', and suffix '-phobic' = 'afraid of'.

Immunoprecipitation
Technique in which proteins are isolated by antibodies. Name origin: antibodies are sometimes called immunoglobulins, because they are active in the immune system. The technique creates a clump of proteins called a precipitate.

Induced mutations
Mutations caused by factors not normally present in the cell.

Induced-fit
A change in the structure of an enzyme's active site during catalysis.

组蛋白
用于缠绕 DNA 以组织 DNA 并调控转录的蛋白质。

Holliday 交叉
在同源重组中一条染色体与另一条染色体单链连接形成的结构。根据科学家的名字 Holliday 命名。

同源异形域
DNA 结合域，在许多真核生物中发现，在生物发育中起作用。

同源染色体
不是完全相同但具有同样一套基因的染色体。前缀 "homo-" = "相同的"。

同源重组
在具有相似 DNA 序列的 DNA 分子之间发生的重组。前缀 "homo-" = "相同的"。

持家基因
在所有细胞中必须被转录的基因，因为它们具有的功能对细胞生存来说是最基本的。在日常英语中，"housekeeping"指在家里需要做的日常事务，比如打扫和做饭。持家基因是那些需要在细胞中执行日常事务的基因。

杂交
核酸的两条链之间的结合。杂交可以发生在两条不同的 DNA 分子之间，以及 DNA 和 RNA 之间。

氢键
相对较弱的原子之间形成的键，由氢原子上带有部分正电荷而引起。

疏水相互作用
当非极性分子被置于水中发生的相互作用。前缀 "hydro-" = "水"，后缀 "-phobic" = "害怕"。

免疫沉淀法
通过抗体对蛋白质进行分离的技术。名称来源：抗体有时称为**免疫球蛋白**（immunoglobulins），因为它们在免疫系统中很活跃。该技术产生一种蛋白质聚集形成的沉淀。

诱导突变
由通常不是细胞的正常成分引起的突变。

诱导契合
在催化过程中酶的活性位点结构所发生的改变。

Inducer

A small molecule that binds to a transcription factor, leading to activation of transcription. May act by preventing a repressor protein from repressing transcription.

30S initiation complex

In prokaryotes, complex of mRNA, 30S ribosomal subunit, and initiator tRNA placed at the start codon.

Initiator tRNA

A specialized tRNA that functions in translation initiation.

Initiator (Inr) promoter element

An element of eukaryotic promoters, involved in initiation of transcription.

Inosine

A nitrogenous base. Sometimes found in tRNA anticodons, where it has the ability to recognize three different bases at the third position in the codon.

Insertion sequences

Simple prokaryotic transposons. They are sequences that insert themselves into random sites on a piece of DNA.

Insertion/deletion loop (IDL)

Loops caused by strand slippage during DNA replication. Leading to deletions or insertions, depending on which strand the loop occurs.

Insertions

With reference to DNA, the addition of bases into a coding region.

Integrase

Protein coded by LTR-retrotransposons, similar in function to transposase. Helps to integrate the transposon into a site on the DNA.

Intercalating agents

Molecules that resemble base pairs in structure and insert between base pairs in DNA, causing insertions and deletions. The prefix 'inter-' = 'between'.

Internal ribosome entry sequence (IRES)

A sequence in eukaryotic mRNAs that allows ribosomes to begin translation downstream of the normal start codon.

Intrinsic termination

A transcription termination mechanism in prokaryotes in which the transcribed RNA forms a hairpin loop followed by a string of A's. Called 'intrinsic' because the RNA alone is able to stop transcription, without help from proteins.

Introns

Sequences that do not code for protein that intervene (or interrupt) among coding regions.

诱导物

与转录因子结合从而激活转录的小分子。可以通过防止阻遏蛋白对转录的阻遏而发挥作用。

30S 起始复合体

是原核生物中由 mRNA、30S 核糖体亚基和起始 tRNA 在起始密码子位置组成的复合体。

起始 tRNA

在翻译起始中起作用的一种特殊 tRNA。

起始子（Inr）启动子元件

在转录起始中起作用的真核启动子元件。

次黄苷

一种含氮碱基。有时出现在 tRNA 的反密码子中，具有识别位于密码子第三个位置的三个不同碱基的能力。

插入序列

简单的原核生物转座子。它们是一些能使自身插入一段 DNA 随机位置中的序列。

插入/缺失环（IDL）

在 DNA 复制过程中由于链滑动而形成的环。能导致缺失或插入突变，依环出现在哪条链而定。

插入

指在 DNA 编码区增加了碱基。

整合酶

LTR 反转录转座子编码的蛋白质，与转座酶功能相似。帮助将转座子整合（integrate）到 DNA 的位点中。

嵌入剂

结构与碱基对相似并能插入 DNA 的碱基对之间的分子，能引起插入与缺失突变。前缀"inter-" = "在……之间"。

内部核糖体进入序列（IRES）

真核 mRNA 序列中位于正常起始密码子下游、允许核糖体开始翻译的区域。

内在型终止

原核生物转录终止的一种机理，终止时转录出来的 RNA 形成发夹环结构，紧接其后的是一连串 A。称之为"内在型"是因为 RNA 自身即能终止转录，不需要蛋白质的帮助。

内含子

不编码蛋白质并干涉（intervene）或打断（interrupt）编码区的序列。

Introns-early theory
Theory that introns were present in early life forms and were retained by eukaryotes, but lost by prokaryotes.

Introns-late theory
Theory that introns were introduced into eukaryotes after the lineage separated from prokaryotes.

Inversion
When a region of a chromosome becomes rotate, or inverted, in orientation.

Inverted repeats
Sequences that are the same if you take the complement of one and read it backwards.

Ion-exchange chromatography
A type of column chromatography in which proteins are separated according to charge. Column is loaded with ions.

Ionic bonds
An attraction between two ions of opposite charge.

Kinases
Proteins that add phosphate groups to other proteins.

Knockdown
The use of RNA interference to eliminate specific mRNAs in the cell. In colloquial English, 'knock down' means to throw something to the floor.

Knockout
A technique to eliminate the presence of specific gene in an entire organism. In colloquial English, a 'knockout' is when someone is defeated in a fight.

lac **operon**
An operon containing genes involved in lactose metabolism.

lac **repressor**
Repressor protein that regulates the *lac* operon.

Lactose
A sugar composed of glucose and galactose. Often found in milk. The root 'lact' is related to the Latin word for milk.

Lactose permease
A protein that brings lactose into the cell. In other words, it makes the cell permeable to lactose.

Lagging strand
The strand that is replicated discontinuously during DNA replication. In normal English, 'lagging' = 'slower'. Lagging strand replication is a little bit slower, and less direct, than leading strand replication.

Lariat
An intermediate, looped, structure in splicing. In nor-

内含子早期论
认为内含子在早期生命形式中即存在并被真核生物保留而被原核生物丢失的理论。

内含子后期论
认为内含子是真核生物在与原核生物进化谱系分开以后才开始有的理论。

倒位
一个染色体区域的方向发生旋转或颠倒的现象。

反向重复序列
互补序列从反方向读与它自身相同的序列。

离子交换层析
根据蛋白质所带电荷的不同而进行分离的一种柱层析技术。层析柱用离子装填。

离子键
带相反电荷的两个离子之间的引力。

激酶
将磷酸基团加到其他蛋白质上去的蛋白质。

基因敲低
应用 RNA 干涉从细胞中去除特殊 mRNA 的技术。在口头英语中，"knock down"的意思是：把……丢到地上。

基因敲除
将某个特殊的基因从整个生物中去除的技术。在口头英语中，"knockout"的意思是：某人被打败。

lac 操纵子
含有乳糖（lactose）代谢基因的操纵子。

lac 阻遏蛋白
调控 *lac* 操纵子的阻遏蛋白。

乳糖
一种由葡萄糖和半乳糖组成的糖。牛奶中常见。词根'lact'与拉丁语中的牛奶一词有关。

乳糖渗透酶
一种将乳糖送进细胞的蛋白质。也就是说，它使细胞对乳糖变成可渗透的（permeable）。

后随链
在 DNA 复制中以不连续方式复制的链。在日常英语中，"lagging" = "较慢的"。相对于先导链的复制而言，后随链的合成稍微有一点慢而且不那么直接。

套索
在剪接中产生的环状中间结构产物。在日常英语

mal English, a 'lariat' is a loop made out of rope that is used for catching animals.

Leading strand
The strand that is replicated continuously in DNA replication. In normal English, something that is 'leading' is ahead of other things. Indeed, the leading strand is replicated a little more quickly and more directly than the lagging strand.

Leaky mutation
A missense point mutation that has an affect on protein function. In normal English, something 'leaky' abnormally allows liquid to pass through. For example, a cup with a hole at the bottom is a 'leaky' cup. Leaky things are usually slightly broken or damaged. Likewise, proteins with leaky mutations are slightly damaged.

Leucine zipper
A DNA-binding domain form by two α helices, usually from two different proteins, that come together at the DNA. The helices are held to each other by hydrophobic interaction, as both are lined with leucine amino acids.

Ligand
A small molecule that binds to a protein and changes its function.

Ligase
An enzyme that repairs single-stranded cuts in DNA.

Lock-and-key mechanism
A mechanism in which the active site of the enzyme does not change shape during catalysis. Name origin: when a key fits into a lock, the shape of each component is perfectly suited to accommodate the other, and no change of shape is necessary.

LTR-retrotransposons
A class of retrotransposons that integrate into DNA using integrase. Name origin: LTR stands for 'long terminal repeats'. These transposons have long repeated sequences at each end. The term prefix 'retro-' = 'reverse' or 'backwards'. Retrotransposons use reverse transcriptase to make DNA from RNA. This is backwards from the normal progression of the central dogma.

Macromolecule
A large molecule made by covalently joining smaller molecules. Prefix 'macro-' = large.

Major grooves
Indentations on the side of the double helix. Some of these grooves are large, some are small. The large grooves are called major grooves.

中，'lariat'指一种用来捕捉动物的环形绳圈。

先导链
在DNA复制中连续复制的链。在日常英语中，something that is "leading"的意思是：它在其他事物之前。确实，先导链比后随链复制得稍微快一点而且也更直接。

渗漏突变
对蛋白质功能有影响的一种错义点突变。在日常英语中，something "leaky"的意思是允许液体以不正常的方式通过。例如，底部有一个洞的杯子是一只漏（leaky）的杯子。漏的东西一般来说是有轻微碎裂或损坏的东西。同样，具有渗漏突变的蛋白质也是被轻微损坏的。

亮氨酸拉链
由2个α螺旋组成的DNA结合域，通常来自2个不同的蛋白质，它们在DNA上聚集在一起。这两个螺旋通过疏水相互作用保持在一起，因为它们都有一排亮氨酸氨基酸。

配体
一种与蛋白质结合并改变蛋白质功能的小分子。

连接酶
一种修补DNA单链缺口的酶。

锁-钥机理
一种在催化反应中酶活性位点形状不发生改变的机理。名称来源：当一把钥匙插进锁中时，每个部分的形状都与其他部分完全匹配，不需要改变任何形状。

LTR反转录转座子
一类使用整合酶将自身整合进DNA的反转录转座子。名称来源：LTR表示"long terminal repeats"（长末端重复序列）。这些转座子在两头都有长的重复序列。前缀"retro-" = "相反的"或"向后"。反转录转座子使用反转录酶从RNA产生DNA。这与中心法则的正常顺序相反。

大分子
小分子以共价键连接起来产生的大的分子。前缀"macro-" = 大的。

大沟
双螺旋侧面的沟槽。这些沟槽有的大、有的小。大的沟槽叫作大沟。

Mass spectrometry

A technique that determines the mass of molecules using a machine.

Methyltransferase

An enzyme that transfers methyl groups from a damaged base onto itself, directly reversing damage to the base.

MicroRNA (miRNA)

Small hairpin shaped RNAs that regulate mRNA stability and translation. Prefix 'micro-' = very small.

Minisatellite

A short DNA sequence composed of many tandemly repeated elements each of which has a dozen or more base pairs.

Mismatch repair (MMR)

A mechanism for the repair of mismatched bases in a DNA molecule.

Mismatched base

A base pair in which the two bases are not complementary to each other. Prefix 'mis-' = incorrect. 'Match' = put two things together. Mismatched bases have been incorrectly put together, often by DNA polymerase.

Missense mutation

A point mutation that causes a change in one amino acid of a protein. Prefix 'mis-' = incorrect. 'Sense' = meaning (in this case). Missense mutations often give proteins an incorrect meaning, or function.

Modules

Protein domains that are found in many proteins and are well conserved in evolution.

Molecular cloning

A technique to isolate a gene and have many copies of it available. 'Cloning' = making identical copies of something.

Monomer

Single molecules that are joined with other molecules to make a larger molecule or polymer. Prefix 'mono-' = 'one' or 'single'.

Motif

A small, basic protein structure that is found in many different proteins.

mRNA

Short for 'messenger' RNA. Is the kind of RNA used to copy genetic information in DNA for use by the ribosome. In other words, it acts as a messenger of genetic information.

mRNA-specific control

A kind of translation control in which the translation of

质谱

一种应用仪器测定分子质量的技术。

甲基转移酶

从一个损伤碱基上将甲基（methyl）转移（transfers）到自己身上的一种酶，能直接逆转损伤碱基的状态。

微小 RNA（miRNA）

调控 mRNA 稳定性及其翻译的发夹形小 RNA 分子。前缀"micro-"=很小。

小卫星

小片段 DNA 序列，由许多重复元件串联而成，每个元件含有 12 个或更多个碱基对。

错配修复（MMR）

一种修复 DNA 分子中错配碱基的机理。

错配碱基

一种两个碱基互相不互补的碱基对。前缀"mis-"=不正确的。"Match"=将两样事物放到一起。错配碱基是被不正确地放到一起的碱基，通常由 DNA 聚合酶引起。

错义突变

引起蛋白质中一个氨基酸变化的点突变。前缀"mis-"=不正确的。"Sense"=含义（在这儿的意思）。错义突变常常赋予蛋白质一种错误的含义或功能。

模块

在许多蛋白质中存在并且在进化中相当保守的蛋白质域。

分子克隆

一种分离基因并获得许多它的拷贝的技术。"Cloning"=产生许多……的相同拷贝。

单体

连接在一起产生更大的分子或聚合体的小分子。前缀"mono-"="一个"或"单个"。

基序

在许多不同蛋白质中存在的小的、基本的蛋白质结构。

信使 RNA

信使（messenger）RNA 的缩写。是用来从 DNA 中拷贝遗传信息供核糖体使用的一种 RNA。换句话说，它作为遗传信息的信使。

mRNA 特异性控制

一种翻译控制的种类，它调控的是特殊的 mRNA，

specific mRNAs is regulated, as opposed to all mRNAs.

Mutation
DNA damage that causes a heritable change in the DNA.

Negative regulation
With respect to transcription, means that the binding of a protein causes repression of transcription.

Nitrogenous base
One or two-ringed molecules that are an important component of nucleic acids. Each ring has a number of nitrogen atoms.

Nitrous acid
A mutagenic chemical that causes deamination and conversion of several bases into abnormal bases.

Non-homologous end joining (NHEJ)
A form of double-stranded break repair in which broken ends are directly rejoined, without using homologous recombination.

Non-homologous recombination
Rearrangement of DNA regions that are not similar (homologous) to each other.

Non-LTR retrotransposons
see LTR-retrotransposons. Non-LTR retrotransposons do not have long terminal repeats, and integrate into DNA using a completely different mechanism.

Nonsense mutation
A point mutation that introduces a stop codon before the normal stop codon of the gene. In normal English, 'nonsense' = something that has no meaning. The introduction of a stop codon is a serious mutation, and often leads to mRNAs that cannot be transcribed or that make very damaged proteins.

Northern blotting
A technique to identify individual RNA molecules after gel electrophoresis. Name origin: the word 'Northern' here is a play on words. The first such technique to be developed was for DNA. It was called a Southern blotting because the scientist who developed the technique was named Southern. Southern is also a word that indicates direction. When a similar technique was developed for RNA, scientists wanted to give the technique a similar but different name, so they called it a Northern blotting. 'Northern' is not anybody's last name, but it is the opposite direction as 'southern'.

N-terminus
The end of a protein containing a free amino group. The 'N-' derives from the fact that amino groups contain nitrogen.

而不是全部 mRNA。

突变
使 DNA 发生可遗传变化的 DNA 损伤。

负调控
就转录而言，意思是当一种蛋白质结合上去以后引起转录的阻遏。

含氮碱基
一种一个环或两个环的分子，是核酸的重要组成部分。每个环有几个氮（nitrogen）原子。

亚硝酸
能引起脱氨基和将几种碱基转变成异常碱基的诱变性化学物质。

非同源末端连接（NHEJ）
一种双链断裂的修复形式，修复时将断头直接连接起来，不需要进行同源重组。

非同源重组
互相不相似的（即不同源的）DNA 区域之间的重排。

非 LTR 反转录转座子
参照"LTR 反转录转座子"。非 LTR 反转录转座子没有长末端重复序列，采用完全不同的机理整合到 DNA 中。

无义突变
在基因的正常终止密码子之前产生一个终止密码子的点突变。在日常英语中，"nonsense" = 没有意义的事物。产生终止密码子是一种严重的突变，常常导致 mRNA 不能被翻译或产生严重损坏的蛋白质。

Northern 印迹法
在凝胶电泳后对单个 RNA 分子进行鉴定的技术。名称来源："Northern" 在这儿是个双关语。第一种这样的技术是用在 DNA 上的。它被称为 Southern 印迹法，因为发展这一技术的科学家名叫 Southern。Southern（南方的）也是一个指方向的词。当在 RNA 上发展出类似的技术时，科学家想为它取一个类似的但又不同的名称，所以它被叫作 Northern 印迹法。"Northern" 不是任何人的姓，它指 "Southern" 的相反方向。

N-末端
蛋白质的含有自由氨基的末端。"N-" 来自氨基中含有氮（nitrogen）这一事实。

Nuclear pores

Large protein complexes in the nuclear membrane that allow molecules to pass between the nucleus and the cytoplasm. A 'pore' = a small hole in a surface.

Nucleic acids

Macromolecules that are polymers of nucleotides. They are acidic because of the phosphodiester bond, and are found in high concentrations in the nuclei of cells.

Nucleic acid testing

A technique of detecting existence of target DNA or RNA molecule, usually with the purpose of controlling epidemic disease if conducted to the public.

Nucleosome

A structure with DNA wrapped around a core of histones. In normal English, 'nuclear' = 'core'.

Nucleotide excision repair (NER)

A kind of excision repair usually used to fix nucleotides that have undergone large or unusual modifications.

Nucleotides

Small molecules that can be polymerized to form nucleic acids.

Nucleus

The compartment of eukaryotic cells that houses most of the DNA. In normal English, 'nucleus' = 'core' or 'center'. The nucleus of an atom is the small core of protons and neutrons.

Okazaki fragments

Individual pieces of newly synthesized DNA created during discontinuous synthesis. Okazaki is the name of the scientist who discovered these fragments.

Operator

DNA element in prokaryotes downstream of the promoter. Binding site for proteins that regulate transcription. In normal English, an 'operator' = somebody who controls a system.

Operons

An organization of related genes in which all genes are under the control of one regulatory region and are expressed on one mRNA transcript.

OriC

The origin of replication on an *E. coli* chromosome.

Overexpression

A technique in which a particular protein is expressed in a cell in large concentrations, over the normal concentration.

P (peptidyl) site

Site on a ribosome to which a tRNA moves after being

核孔

核膜上的大的蛋白质复合体，允许分子在细胞核与细胞质之间通行。"pore" = 表面上的小孔。

核酸

一种大分子，是核苷酸的聚合物。由于它们含有磷酸二酯键，所以它们是酸性（acidic）的，并且在细胞核（nuclei）中以很高的浓度存在。

核酸检测

一种检测目标 DNA 或 RNA 分子是否存在的技术，如果面向公众进行则通常是为了防控流行病。

核小体

DNA 包裹在组蛋白核心外面形成的结构。在日常英语中，"nuclear" = "核心"。

核苷酸切除修复（NER）

一种切除修复方法，通常用来修复经受了大的或不同寻常修饰的核苷酸。

核苷酸

能够聚合形成核酸的小分子。

细胞核

为大多数 DNA 提供停留场所的真核细胞内的隔离空间。在日常英语中，"nucleus" = "核心"或"中心"。原子的核是由质子和中子组成的小核心。

冈崎片段

在不连续合成中产生的新合成的单独 DNA 片段。冈崎是发现这些片段的科学家的姓名。

操纵基因

原核生物中位于启动子下游的 DNA 元件。是转录调控蛋白的结合位点。在日常英语中，"operator" = 控制某一系统的人。

操纵子

一种相关基因的组织方式，其中所有基因位于一个调控区域的控制之下并且被表达成一个 mRNA 转录本。

OriC

大肠杆菌染色体（chromosome）上复制的起点（origin）。

过量表达

一种在细胞中大量表达（expressed）某种特殊蛋白质的技术，其表达量超出了（over）正常的浓度。

P（肽基）位

核糖体上的一个位置，是 tRNA 从 A 位移出后所

in the A site. In the P site, the bond that joins the tRNA to a polypeptide is broken, and the polypeptide is rejoined to the tRNA (+ amino acid) in the A site by a peptide bond.

PAM site
Protospacer adjacent motif site that can be recognized by sgRNA-Cas complex in CRISPR/Cas9 gene editing system.

Partial diploids
Organisms to which an extra set of certain genes has been added (see Diploid).

Peptide
A term often used to denote a small polypeptide.

Peptide bond
The bond that connects amino acids in a polypeptide.

Peptidyl transferase
The enzyme functioning in ribosomes that transfers the polypeptide from the P site tRNA to the A site tRNA (+ amino acid), creating a new peptide bond.

Phosphodiester bond
A bond joining nucleotides in a nucleic acid. The bond contains one phosphorous atom, and two ester bonds. The prefix 'di-' = 'two'.

Photolyase
Enzyme that directly reverses pyrimidine dimers, which usually are caused by UV light. Prefix 'photo' = 'light'.

Point mutations
Mutations to individual bases in DNA, usually leading to substitution of a base with another base. In normal English, a 'point' = something very small and precisely localized. Likewise, point mutations only occur to single bases, a very small part of the whole DNA molecule.

Poly(A) polymerase
Specialized RNA polymerase that adds many adenine nucleotides to the end of pre-mRNAs to form the poly(A) tail.

Poly(A) tail
A post-transcriptional addition to mRNA in eukaryotes that involves addition of many adenine (A) nucleotides to the 3′ end of the transcript. Prefix 'poly-' = 'many'.

Polycistronic mRNA
mRNA in prokaryotes that contains more than one gene to be translated. Cistron = gene. Prefix 'poly-' = 'many' or 'more than one'.

Polymerase chain reaction (PCR)
A technique used to replicate specific regions of a DNA

处的位置。在 P 位上，tRNA 与多肽链之间的连接被打断，之后此多肽链与 A 位上的 tRNA （＋氨基酸）之间形成肽键。

PAM 位点
前间区序列邻近基序位点，在 CRISPR/Cas9 基因编辑系统中能被 sgRNA-Cas 复合体识别。

部分二倍体
指一些特殊的生物，其体内的一些基因有额外的一套拷贝（参照"二倍体"）。

肽
时常用来说明小多肽的术语。

肽键
在多肽中连接氨基酸的键。

肽基转移酶
在核糖体中发挥作用的一种酶，它将 P 位 tRNA 上的多肽转移（transfers）到 A 位的 tRNA （＋氨基酸）上，产生新的肽（peptide）键。

磷酸二酯键
在核酸中连接核苷酸的键。该键含有一个磷原子和两个酯键。前缀 "di-" = "二"。

光解酶
直接逆转嘧啶二聚体突变（通常由 UV 光线引起）的酶。前缀 "photo" = "光"。

点突变
DNA 中单个碱基的突变，通常造成某个碱基被另一个碱基替换。在日常英语中，"point" = 很小、很精确地定位的事物。同样，点突变只发生在单个的碱基上，在整个 DNA 分子中很小的部分上。

Poly(A) 聚合酶
在前体 mRNA 的末尾加上许多腺嘌呤核苷酸以形成 poly(A) 尾的特殊的酶。

Poly(A) 尾
真核生物在转录后加在 mRNA 的 3′末端后面的许多腺嘌呤（A）核苷酸。前缀 "poly-" = "许多"。

多顺反子 mRNA
含有不止一个需转译基因的原核生物 mRNA。cistron（顺反子）＝ 基因。前缀 "poly-" = "许多" 或 "不止一个"。

聚合酶链式反应（PCR）
一种用来对 DNA 模板上特殊区域进行许多次复

template many times. A chain reaction = reaction that grows larger and larger over time. With each cycle of PCR, the amount of DNA copies produced grows almost exponentially.

Polymer
Macromolecule created by linking many smaller molecules. Prefix 'poly-' = 'many'. Suffix '-mer' = 'subunits' or 'smaller components'.

Polypeptide
A polymer of amino acid connected by peptide bonds.

Polysome
A translation complex in which multiple ribosomes are translating one mRNA at the same time. Prefix 'poly-' = 'many' or 'more than one'.

Positive regulation
In reference to transcription, denotes the activation of transcription by binding of a protein.

Pre-initiation complex
The group of general transcription factors Ⅱ (TFⅡs) and RNA polymerase Ⅱ that assemble at the promoter of each Class Ⅱ gene before initiation of transcription. Prefix 'pre-' = 'before'.

Pre-mRNA
The precursor to a eukaryotic mRNA, which has been transcribed but has not yet undergone post-transcriptional modifications. Prefix 'pre-' = 'before'.

Pre-replicative complex (Pre-RC)
Complex of proteins that mark origins of replication in eukaryotes and initiate replication. Often forms long before replication begins. Prefix 'pre-' = 'before'.

Primary structure
Amino acid sequence of a protein. Perhaps termed a 'structure' because the sequence is often enough to determine the folded structure of the protein.

Primary transcript
see pre-mRNA. In normal English, a transcript = a copy of something. Primary = first. The primary transcript is the first RNA copy of the DNA, before any modifications have been made.

Primase
The enzyme that adds primers to DNA.

Primers
Short pieces of RNA that are hybridized to DNA so that DNA polymerase can initiate replication. Prefix 'prim-' = 'first'. Primers must be made first, before DNA synthesis can begin.

Probe
A short piece of nucleic acid with traceable label (e.g.

制的技术。链式反应 = 随时间推移变得越来越多的反应。经过每一循环的 PCR，产生出 DNA 量几乎以指数形式增长。

聚合物
通过连接许多小分子而产生出的大分子。前缀 "poly-" = "许多"。后缀 "-mer" = "亚基"或 "更小的组分"。

多肽
氨基酸由肽键连接在一起形成的聚合物。

多核糖体
一种转译复合体，其中有多个核糖体同时在翻译同一条 mRNA。前缀 "poly-" = "许多"或"不止一个"。

正调控
对于转录，表示通过一种蛋白质的结合而激活转录。

前起始复合体
在转录起始前一组通用转录因子Ⅱ（TFⅡs）和 RNA 聚合酶Ⅱ在每个Ⅱ类基因启动子位置组装产生的结构。前缀 "pre-" = "在……之前"。

前体 mRNA
真核 mRNA 的前体，已经被转录出来但还没有经过转录后修饰。前缀 "pre-" = "在……之前"。

前复制复合体（Pre-RC）
真核生物中标明复制起点并启动复制的蛋白质复合体。常常在复制开始前好久就形成了。前缀 "pre-" = "在……之前"。

初级结构
蛋白质的氨基酸序列。称之为"结构"也许是因为这样的序列常常足以决定该蛋白质折叠出的结构。

初级转录物
参照 "pre-mRNA"。在日常英语中，a transcript = 某种事物的拷贝。primary = 第一、首先。初级转录物是 DNA 的第一个 RNA 拷贝，是在任何修饰发生前的拷贝。

引发酶
将引物加到 DNA 上去的酶。

引物
短的 RNA 片段，与 DNA 杂交以便 DNA 聚合酶启动转录。前缀 "prim-" = "第一、首先"。在开始 DNA 合成前必须先有引物。

探针
小片段核酸序列，具有可追踪的标签（例：荧光

fluorescent dye) that enables identification of target DNA sequence complementary to it.

Processivity
The amount of DNA that DNA polymerase can replicate in one run, before falling off the template.

Promoter
DNA element responsible for binding to RNA polymerase (and general transcription factors, in eukaryotes). Often involved in regulation of transcription. In normal English, 'promote' = to encourage.

Proofread
The process of rechecking work and correcting errors. DNA polymerase and RNA polymerase both have some ability to proofread the strands they are synthesizing.

Proteases
Proteins that cut other proteins.

Protein domain
A region of protein structure that has a specific and isolated function.

Pull-down assay
A technique very similar to affinity chromatography. A column is packed with a specific protein, and proteins that bind to that protein, directly or via other proteins, are retained in the column. Called a 'pull-down' because all proteins that bind to the target are pulled out of the cell extract.

Purines
Nitrogenous bases containing two rings.

Pyrimidine dimer
A form of DNA damage in which two adjacent pyrimidines in a DNA strand become covalently bound to each other. Often caused by UV light.

Pyrimidines
Nitrogenous bases containing only one ring.

Quaternary structure
The structure of a protein with multiple subunits. Quaternary = fourth degree. This structure is one level or organization higher than tertiary (third degree) structure.

R group
The variable chemical group in an amino acid.

Reading frame
The organization of bases in a coding region into groups of three, marked at the beginning and end by start and stop codons.

Recombination
General term for a process that changes the order of pieces of DNA, creating new combinations of DNA re-

染料），能鉴定出与其互补的目标 DNA 序列。

持续合成能力
DNA 聚合酶在从模板上脱落前一次能够合成的 DNA 量的大小。

启动子
负责与 RNA 聚合酶结合的 DNA 元件（在真核生物中还负责与通用转录因子的结合）。常常涉及转录调控。在日常英语中，"promote" = 鼓励。

校正
再次检查合成出的产物并更正其中错误的过程。DNA 聚合酶和 RNA 聚合酶都具有一定的校正自身合成的链的能力。

蛋白酶
切割其他蛋白质的蛋白质。

蛋白质域
具有特殊、独立功能的蛋白质结构中的一个区域。

下拉分析
一种与亲和层析非常相似的技术。层析柱用一种特殊蛋白充填，能与这种蛋白质直接或间接结合的蛋白质被留在了层析柱里。称为"下拉"是因为所有与目标蛋白结合的蛋白质都从细胞提取物中被拉了出来。

嘌呤
具有两个环的含氮碱基。

嘧啶二聚体
一种 DNA 损伤形式，其中 DNA 链上两个相邻的嘧啶被共价连接到了一起。通常由紫外线引起。

嘧啶
只有一个环的含氮碱基。

四级结构
具有多个亚基的蛋白质的结构。quaternary = 第四个等级。这一结构的水平或组织比三级结构高一个等级。

R 基团
氨基酸中一种可变的化学基团。

读码框
编码区中碱基以三个一组形成的结构，以起始密码子和终止密码子为开始和结束的标志。

重组
用于描述 DNA 片段顺序发生改变、产生 DNA 区域新组合过程的通用术语。前缀"re-" = "又

gions. Prefix 're-' = 'again' or 'new'.

Redundant

In normal English, redundant = repetitive. The genetic code is called redundant because several codons may code for one amino acid.

Release factor

A protein that binds to stop codons, releasing ribosomes, tRNA, and mRNA from each other.

Renaturation

The recovery of non-covalent interactions in a macromolecule. With respect to DNA, renaturation involves formation of double-stranded molecule from the separated single-strands. With respect to protein, renaturation involves re-folding of the unfolded polypeptide chain.

Replication bubble

Separation of single-strands in DNA helix creates a somewhat circular opening where DNA replication can take place.

Replication fork

A separation of single strands that is the site for one direction of DNA replication. In normal English, a 'fork' = the site where a larger path splits into smaller paths. At the replication fork, the thick double-stranded DNA molecule splits into two single-stranded DNA molecules.

Replicative transposition

A transposition mechanism in which the original transposon is replicated, and the copy is inserted into a new site in the DNA.

Restriction endonucleases

Enzymes that cut DNA at specific sequences inside the molecule (*see* Endonuclease). The term 'restriction' comes from the fact that these enzymes were originally discovered because they restrict the entry of viruses into bacteria, by cutting viral DNA. These proteins are sometimes simply called 'restriction enzymes'.

Reverse genetics

Form of genetics/molecular biology research in which a gene is first identified, and then the function/phenotype associated with the gene is explored.

Reverse transcription

The process of making DNA from RNA. This is the reverse of transcription.

Reverse transcriptase

Enzyme that performs reverse transcription.

Ribonucleases

Enzymes that cleave ribonucleic acid (RNA).

或"新的"。

冗余的

在日常英语中，redundant = 重复的。遗传密码是冗余的，因为存在几个密码子编码同一种氨基酸的现象。

释放因子

一种结合到终止密码子上的蛋白质，能将核糖体、tRNA 和 mRNA 各自释放出来。

复性

大分子中非共价相互作用的恢复。对 DNA 而言，复性涉及已经分开的单链形成双链分子。对蛋白质而言，复性涉及已经解折叠的多肽链重新折叠。

复制泡

在 DNA 螺旋中单链分离产生的有点像环形的开口，在这儿可以发生 DNA 复制。

复制叉

单链分离产生的可以向一个方向复制 DNA 的位点。在日常英语中，"fork" = 一条大路分成几条小路的地点。在复制叉处，粗的双链 DNA 分子被分成两条单链 DNA 分子。

复制型转座

一种转座机理，其中原始的转座子被复制，转座子的拷贝被插入一个 DNA 的新位点中。

限制性核酸内切酶

在分子内的特异序列位置切割 DNA 的酶（参照"核酸内切酶"）。"限制"一词来源于这样的事实：这些酶最初在细菌中发现，它们的功能是通过切断病毒的 DNA 而限制（restrict）病毒进入细菌。这些蛋白质有时就被简单地称为"限制酶"。

反向遗传学

一种遗传学/分子生物学研究的形式，它先鉴定出一个基因，之后探索与这一基因相关的功能/表型。

反转录

从 RNA 生产 DNA 的过程。这是一个与转录相反的过程。

反转录酶

行使反转录功能的酶。

核糖核酸酶

切割核糖核酸（RNA）的酶。

Ribonucleic acid (RNA)
A nucleic acid made by polymerization of ribonucleotides.
Ribonucleotides
Nucleotides containing the sugar ribose.
Ribosome
Large macromolecular complexes comprised of RNA (ribonucleic acid) and protein.
RNA interference (RNAi)
A process that uses siRNA or miRNA to induce degradation of a target mRNA. The process 'intereferes' with the normal stability of mRNA.
RNA polymerase
Enzyme that polymerizes RNA strands.
RNA polymerase core
The smallest set of RNA polymerase subunits required for transcription. Unable to initiate transcription correctly.
RNA polymerase holoenzyme
The whole set of RNA polymerase subunits, able to initiate transcription at the promoter.
rRNA
RNA that is directly used to make ribosomes. Is not translated.
Rudder
Part of the RNA polymerase structure that ensures separation of DNA strands in the transcription bubble. In normal English, a rudder is that flat, thin part of a boat that is used to steer. The RNA polymerase rudder somewhat resembles a boat rudder in shape.
RVDs
Repeat variable di-residues, two adjacent amino acid residues in TALE structural module that recognize specific DNA bases.
SDS
Sodium dodecyl sulfate. A detergent used to denature proteins and cover them in uniform charge before gel electrophoresis. This allows proteins to be separated solely by length.
Secondary structure
The simple, generalized folded structures that make up a protein.
Self-splicing
The process of some mRNAs to splice out introns without help from outside factors like snRNPs.
Semi-conservative replication
A style of DNA replication in which produces a DNA with one strand from the parent, and one newly synthesized strand. Prefix 'semi-' = 'half' or 'some'. In

核糖核酸（RNA）
通过聚合核糖核苷酸产生的核酸。
核糖核苷酸
含有核糖的核苷酸。
核糖体
由 RNA（ribonucleic acid）和蛋白质组成的大型大分子复合体。
RNA 干涉（RNAi）
一种应用 siRNA 或 miRNA 诱导目标 mRNA 降解的过程。该过程"干涉"了 mRNA 的正常稳定性。
RNA 聚合酶
聚合 RNA 链的酶。
RNA 聚合酶核心
对转录而言需要的 RNA 聚合酶亚基的最少组合。不能正确地启动转录。
RNA 聚合酶全酶
RNA 聚合酶亚基的全套组合，能够在启动子的位置启动转录。
核糖体 RNA
直接用来产生核糖体（ribosome）的 RNA。它不被翻译。
方向舵
RNA 聚合酶结构的一部分，用来确保 DNA 双链在复制泡中分开。在日常英语中，rudder 是小船上用来掌握方向的扁平、薄的部件（即船舵）。RNA 聚合酶方向舵的形状看起来有点像小船上的舵。
重复序列可变的双氨基酸残基
重复序列可变的双氨基酸残基，在 TALE 结构模块中的两个相邻氨基酸残基，可识别特定的 DNA 碱基。
十二烷基硫酸钠
十二烷基硫酸钠。一种在凝胶电泳前用来使蛋白质发生变性并用统一的电荷覆盖蛋白质的去污剂。这使得蛋白质可以仅通过长度来进行分离。
二级结构
组成蛋白质的简单、通用的折叠结构。
自我剪接
一些 mRNA 不需要外来因子如 snRNPs 的帮助而能将其中的内含子剪接掉的过程。
半保留复制
一种 DNA 的复制方式，产生的 DNA 中一条链来自母本，另一条链是新合成的。前缀"semi-"＝"一半"或"部分"。在日常英语中，conservative＝

normal English, conservative = keep the old. Semi-conservative replication keeps some of the old.

Semi-discontinuous replication

A style of replication in which one strand is replicated continuously and the other is replicated discontinuously. Prefix 'semi-' = 'half' or 'some'. Prefix 'dis-' = 'not'. In semi-discontinuous replication, half of the DNA is not replicated continuously.

sgRNA

Single-guide RNA that forms a complex with Cas nuclease to exhibit target DNA recognition and cleavage activity.

β sheet

A secondary structure in proteins, relatively flat and formed hydrogen bonding between two parallel or antiparallel stretches of polypeptide.

Shine-Dalgarno sequence

A consensus sequence in *E. coli* that marks which AUG sequences should be used as start codons. Named for the two scientists who discovered the sequence, Shine and Dalgarno.

Silencers

Regulatory DNA elements that bind to repressors and cause repression of transcription.

Silent mutations

Point mutations that change a base in a codon, but do not change the amino acid coded for by the codon. Therefore, there is no change in the protein produced, and the mutation is functionally unnoticeable, or silent.

Single-strand DNA binding proteins (SSBs)

Proteins that bind to single-strands of DNA at a replication fork, protecting the strands and preventing them from rebinding to each other.

Sliding clamp

A subunit of RNA polymerase Ⅲ that allows it to synthesize RNA with high processivity. In normal English, a 'clamp' = something that holds something else. The sliding clamp holds the DNA, but also slides along the DNA as RNA polymerase moves.

Small-interfering RNA (siRNA)

Small double stranded RNAs that lead to RNA interference.

snRNPs

<u>S</u>mall <u>n</u>uclear <u>r</u>ibonucleoproteins. These are macromolecules composed of RNA (ribonucleic acid) and protein that are found in the nucleus. They are required for splicing of most introns.

保持旧的。半保留复制即保持了一部分旧的分子。

半不连续复制

一种复制类型，其中一条链连续地进行复制，另一条链不连续地进行复制。前缀 "semi-" = "一半" 或 "部分"。前缀 "dis-" = "不"。在半不连续复制中，DNA 的<u>一半</u>是<u>不</u>连续地复制出来的。

单链引导 RNA

单链引导 RNA 与 Cas 核酸酶形成复合体后展示出目标 DNA 识别和切割活性。

β 折叠

蛋白质的一种二级结构，相对平坦，在两条平行的或反向平行的肽段之间形成氢键。

SD 序列

大肠杆菌中标明哪个 AUG 应该被用作起始密码子的共有序列。根据发现这一序列的两个科学家（Shine 和 Dalgarno）的名字命名。

沉默子

结合到阻遏蛋白上引起转录阻遏的具有调控作用的 DNA 元件。

沉默突变

在密码子中改变了一个碱基但没有改变密码子所编码的氨基酸的点突变。因此，生产出的蛋白质没有发生改变，该突变在功能上注意不到，或者说是沉默的。

单链 DNA 结合蛋白（SSB）

在复制叉处与单链 DNA 结合的蛋白质，它们保护单链防止它们互相重新结合。

滑行夹

RNA 聚合酶Ⅲ的一个亚基，能使 RNA 聚合酶Ⅲ保持高的持续合成能力。在日常英语中，a "clamp" = 将……抓住的东西。滑行夹套住 DNA，并在 RNA 聚合酶移动的时候沿着 DNA 链滑行。

小干涉 RNA（siRNA）

能产生 RNA 干涉作用的双链 RNA 小分子。

核内小核糖核蛋白

核内小核糖核蛋白。是由 RNA（核糖核酸）和蛋白质在细胞核中组成的大分子，在大多数内含子的剪接中都需要。

Southern blotting

Technique to identify specific DNA molecules after gel electrophoresis. Scientist who developed the technique was named Southern.

Specific transcription factors

Transcription factors that only act to regulate transcription of specific genes.

Splice sites

Sequences that mark the beginning and ends of introns and exons.

Spliceosome

The collection of factors, especially snRNPs, that help with the splicing of introns.

Spontaneous mutations

Mutations that occur as the result of natural processes in the cell, such as deamination caused by reaction of bases with water.

Start codon

The codon that marks the start of translation.

Sticky ends

The single stranded, over-hanging ends produced by restriction enzymes. These ends are complementary to the other ends produced by cleavage. As a result, the two ends can stick together easily.

Stop codons

Codons that mark the end of translation.

Subunit

An individual polypeptide that forms part of larger protein. Prefix 'sub-' = 'below'. A subunit is a unit of a protein whose importance is below that of the whole protein.

σ subunit

Component of prokaryotic RNA polymerase holoenzyme. Required for recognition of promoters.

Sugar-phosphate backbone

The repeating structure of ribose (a sugar) and phosphodiester bonds in DNA molecule. In normal English, a 'backbone' = a long, central component that gives support to a structure. The sugar-phosphate backbone lines the double helix and supports the bases inside the helix.

Supercoil

A coiling of the chromosome formed to relieve tension within the double helix. Chromosomes normally have some coiling, but this is beyond the normal level of coiling. The prefix 'super-' = 'beyond' or 'above'.

TALENs

Transcription activator-like effector nucleases, an arti-

Southern 印迹法

在凝胶电泳后用于鉴定特异 DNA 分子的技术。发展这一技术的科学家姓名是 Southern。

特异性转录因子

只在调控特殊基因的转录中起作用的转录因子。

剪接位点

表明内含子和外显子开始和结束位置的序列。

剪接体

各种因子的集合体，尤其是 snRNPs，帮助将内含子剪接掉。

自发突变

细胞中因自然过程而产生的突变，例如碱基与水反应所引起的脱氨基。

起始密码子

标明翻译起始位置的密码子。

黏性末端

由限制酶产生的单链突出末端。这些末端与切割产生的相同末端互补。结果，这样的两个末端可以容易地结合在一起。

终止密码子

标明翻译结束位置的密码子。

亚基

一个单独的多肽，是形成更大的蛋白质的一个部分。前缀 "sub-" = "在……之下"。一个亚基是蛋白质的一个部分，其重要性在整个蛋白质的重要性之下。

σ 亚基

原核生物 RNA 聚合酶全酶的组成成分。在启动子识别中需要。

糖-磷酸骨架

DNA 分子中核糖（一种糖）与磷酸二酯键的重复结构。在日常英语中，"backbone" = 支撑某一结构的长的中心部件。糖-磷酸骨架构成了双螺旋并支撑螺旋内部的碱基。

超螺旋

为了释放双螺旋中的张力而形成的染色体卷曲形式。正常情况下染色体会有一定的卷曲，但超螺旋是超出正常水平的卷曲。前缀 "super-" = "超过" 或 "高于"。

转录激活因子样效应物核酸酶

转录激活因子样效应物核酸酶，一种人工设计的

ficially designed gene editing system using TALE structural module as DNA recognition domain and Fok Ⅰ nuclease as DNA cleavage domain.

TATA box
A eukaryotic promoter element with the consensus sequence TATAAA.

TATA-binding protein (TBP)
A protein subunit of TFⅡD that often binds to the TATA box.

Tautomers
Molecules that can interconvert between two different structural forms.

TBP-associated factors Ⅱ (TAFⅡ)
Subunits of TFⅡD that associated with TBP (TATA-binding protein), another subunit of TFⅡD.

Telomerase
An enzyme responsible for extending the parent strand of DNA at the telomere during replication so that telomeres can be fully replicated.

Telomere
The ends of a eukaryotic chromosome, made of heterochromatin.

Tertiary structure
The structure of individual polypeptides.

TFⅡB recognition element (BRE)
Promoter element in eukaryotes that binds to the TFⅡB general transcription factor.

TFⅡD
A general transcriptional factor for RNA polymerase Ⅱ that binds to multiple promoter elements and specific transcription factors.

Three prime end
The end of a DNA strand that terminates with the three prime (3′) carbon (which has an —OH group attached) of the ribose in a deoxribnoucleotide. 'Prime' means nothing here; it simply denotes that the label 'three' comes from arbitrary numbering of the ribose carbons.

Transcription
The process by which RNA is made from DNA. In normal English, to 'transcribe' = to copy.

Transcription bubble
Separation of single-strands in DNA helix creates a somewhat circular opening where transcription can take place.

Transferrin
Protein that brings, or transfers, iron into the cell. 'Ferr' comes from the Latin word for iron.

TATA 框
共有序列为 TATAAA 的真核启动子元件。

TATA 结合蛋白（TBP）
通常与 TATA 框结合的 TFⅡD 蛋白亚基。

互变异构体
可以在两种不同结构形式之间转换的分子。

TBP 相关因子Ⅱ（TAFⅡ）
与 TBP（TATA 结合蛋白）结合在一起的 TFⅡD 亚基，TBP 是 TFⅡD 的另一个亚基。

端粒酶
在复制过程中负责延伸端粒处 DNA 亲本链的酶，可使端粒复制完整。

端粒
真核生物染色体的末端，由异染色质组成。

三级结构
单条多肽的结构。

TFⅡB 识别元件（BRE）
真核生物中与通用转录因子 TFⅡB 结合的启动子元件。

TFⅡD
一种能与多个启动子元件和特异性转录因子结合的 RNA 聚合酶Ⅱ通用转录因子。

3′-末端
DNA 链的一个末端，以脱氧核糖核苷酸中核糖的三一撇（3′）碳（上面连接着—OH）结尾。"一撇"在这儿没有含义，它只是简单地表示"三"这一记号来自对核糖上碳人为编号。

转录
从 DNA 生产 RNA 的过程。在日常英语中，"transcribe" = 抄写。

转录泡
DNA 螺旋的单链分离产生类似环形的开口，是将要发生转录的地方。

运铁蛋白
将铁带进或转移（transfers）进细胞中的蛋白质。"Ferr"来自"铁"的拉丁文。

Transformation

A technique in which DNA is directly transferred into a cell.

Translocation

The movement of a ribosome relative to the mRNA, to allow a new codon to be recognized and a new tRNA to bind. Prefix 'trans-' = 'across' or 'change'. In a translocation, the ribosome changes location on the mRNA.

Transposase

A protein coded for by many transposons that allows transposition into a new DNA site.

Transposons

DNA elements that can change positions, or locations, in the genome. Prefix 'trans-' = 'across' or 'change'.

tRNA

Short for 'transfer' RNA. Is used to transfer amino acids to the ribosome so that they can be incorporated into growing proteins.

***trp* operon**

Operon containing genes coding for the synthesis of tryptophan.

trp repressor

Repressor protein that regulates transcription of the *trp* operon.

Type Ⅰ topoisomerases

Topoisomerases that undo supercoiling by introducing single-stranded cuts in the DNA.

Type Ⅱ topoisomerases

Topoisomerases that undo supercoiling by introducing double-stranded cuts in the DNA.

Ultraviolet radiation (UV)

Electromagnetic radiation that has a slightly higher frequency than visible light. The highest frequency visible light is violet. The prefix 'ultra-' = 'very' or 'even higher'. Ultraviolet light has a frequency even higher than that of violet light.

Untranslated regions (UTRs)

Regions on each end of an mRNA that are transcribed but are not translated.

Upstream region

The region of a gene above the start site of transcription. In normal English, 'upstream' = in the opposite direction of flow. The upstream region of a gene is above the start site of transcription, in the opposite region that RNA transcription synthesizes.

Van der Waals forces

Weak attractions caused by shifts in the electron clouds

转化

一种将 DNA 直接转移到细胞中去的技术。

移位

核糖体相对于 mRNA 所做的移动，目的是识别新密码子并允许新 tRNA 与它结合。前缀 "trans-" = "跨过" 或 "改变"。在移位中，核糖体改变它在 mRNA 上的位置。

转座酶

许多转座子编码的能使它们转座到新 DNA 位点的蛋白质。

转座子

基因组中能改变位置（positions）或位点的 DNA 元件。前缀 "trans-" = "跨过" 或 "改变"。

转运 RNA

"转运" RNA 的缩写。用来将氨基酸转移到核糖体上，这样它们才能被整合到合成出的蛋白质中。

trp 操纵子

含有编码色氨酸合成酶基因的操纵子。

trp 阻遏蛋白

调控 *trp* 操纵子转录的阻遏蛋白。

Ⅰ 类拓扑异构酶

通过在 DNA 中产生单链切口而消除超螺旋的拓扑异构酶。

Ⅱ 类拓扑异构酶

通过在 DNA 中产生双链切口而消除超螺旋的拓扑异构酶。

紫外辐射（UV）

频率比可见光稍微高一点的电磁辐射。最高频率的可见光是紫光。前缀 "ultra-" = "很" 或 "甚至更高"。紫外线具有比紫光更高的频率。

非转译区（UTR）

位于 mRNA 两头的、被转录出但不被翻译的区域。

上游区域

位于转录起始位点上游的基因区域。在日常英语中，"upstream" = 液体流动的相反方向。基因的上游区域位于转录起始位点之前，与 RNA 转录合成的方向相反。

范德瓦耳斯力

由围绕原子的电子云发生变化产生的弱引力。根

around atoms. Named for the scientists who discovered them.

Virus load
The measure of virus quantity in an infected cell or in a test sample.

Western blotting
A technique to identify specific proteins during gel electrophoresis. (*see* Northern blotting). The word Western here is also a play on words.

Wobble
The ability of the third base in an anticodon to recognize more than one kind of base in the complementary position in the codon. In normal English, something that 'wobbles' = moves loosely, in irregular patterns. The third base in the anticodon binds loosely and irregularly to other bases.

Xeroderma pigmentosum (XP)
A disease caused by a defect in nucleotide excision repair. The root 'derma' = skin. These patients have some skin pigmentation problems.

X-ray crystallography
A technique that determines the three-dimensional structures of proteins by shining X-rays on crystals of the protein.

Yeast two-hybrid
A technique to test for binding between two proteins. The technique is used in yeast and involves two fusion (hybrid) proteins that bind to each other to activate transcription of a gene.

ZFNs
Zinc finger nucleases, also known as zinc finger protein nucleases, an artificially designed gene editing system using zinc finger structural module as DNA recognition domain and FokⅠ nuclease as DNA cleavage domain.

Zinc fingers
DNA binding domains that are elongated, with a shape that is somewhat like a finger. The shape of the domain is coordinated by one or more zinc atoms.

病毒载量
被感染细胞中或试样中病毒的数量。

Western 印迹法
在凝胶电泳过程中用来鉴定特异蛋白的技术。（请参照 Northern 印迹法）。单词 Western 也是一个双关语。

摇摆
指反密码子第三位碱基具备识别密码子互补位置上不止一个碱基的能力。在日常英语中，something that "wobbles" = 以松弛的、不规则的方式移动。反密码子第三位碱基与其他碱基以松弛的、不规则的方式结合。

着色性干皮病（XP）
由核苷酸切除修复功能缺陷造成的疾病。词根"derma" = 皮肤。患者存在皮肤色素（pigmentation）沉着方面的问题。

X 射线晶体衍射
用 X 射线照射蛋白质晶体来测定蛋白质三维结构的技术。

酵母双杂交
检查两种蛋白质之间是否能够发生结合的技术。此技术在酵母中使用，涉及两种融合（杂交）蛋白，两种蛋白质互相结合能激活基因的转录。

锌指核酸酶
锌指核酸酶，也称为锌指蛋白核酸酶，一种人工设计的基因编辑系统，使用锌指结构模块作为 DNA 识别功能域，使用 FokⅠ核酸酶作为 DNA 切割功能域。

锌指结构
伸展的、形状有些像手指的 DNA 结合域。这一形状由一个或多个锌原子协调。

References 参考文献

[1] Bruce A, Johnson A, Lewis J, et al. Molecular biology of the cell. New York: Garland Science, 2002.
[2] Malacinski G M, Freifelder D. Essentials of molecular biology. Boston: Jones and Bartlett, 2003.
[3] Weaver R F. Molecular biology. New York, 3rd edition. McGraw Hill, 2005.
[4] Berk A J. Activation of RNA polymerase II transcription. Current Opinion in Cell Biology, 1999, 11: 330-335.
[5] Erkmann J A, Ulrike K. Nuclear export of mRNA: from the site of transcription to the cytoplasm. Experimental Cell Research, 2004, 296: 12-20.
[6] Gallo A, Thomson E, Brindle J, et al. Micro-processing events in mRNAs identified by DHPLC analysis. Nucleic Acids Research, 2002, 30: 3945-3953.
[7] Gebauer F, Hentze M W. Molecular mechanisms of translational control. Nature Reviews Molecular Cell Biology, 2004, 5: 827-835.
[8] Grindley N D F, Whiteson K L, Rice P A. Mechanisms of site-specific recombination. Annual Reviews in Biochemistry, 2006, 75: 567-605.
[9] Iizuka M, Smith M. Functional consequences of histone modifications. Current Opinion in Genetics and Development, 2003, 13: 154-160.
[10] Jiricny J. The multifaceted mismatch-repair system. Nature Reviews Molecular Cell Biology, 2006, 7: 336-346.
[11] Keegan L P, Gallo A, O'Connell M A. The many roles of an RNA editor. Nature Reviews Genetics, 2001, 2: 869-878.
[12] Krogh B O, Symington L S. Recombination proteins in yeast. Annual Reviews in Genetics, 2004, 38: 233-271.
[13] Lieber M R, Ma Y, Pannicke U, et al. Mechanisms and regulation of human non-homologous DNA end-joining. Nature Reviews Molecular Cell Biology, 2003, 4: 712-720.
[14] Machida Y J, Hamlin J L, Dutta A. Right place, right time, and only once. Replication initiation in metazoans. Cell, 2005, 123: 13-24.
[15] Meyer S, Temme C, Wahle E. Messenger RNA turnover in eukaryotes: pathways and enzymes. Ceritical Reviews in Biochemistry and Molecular Biology, 2004, 39: 197-216.
[16] Orphanides G, Reinberg D. A unified theory of gene expression. Cell, 2002, 108: 439-451.
[17] Roberts S G. Mechanisms of action of transcription activation and repression domains. Cellular and Molecular Life Sciences, 2000, 57: 1149-1160.
[18] Sontheimer E J, Carthew R W. Silence from within: Endogenous siRNAs and miRNAS. Cell, 2005, 122: 9-12.
[19] St Johnston D. Moving messages: The intracellular localization of mRNAs. Nature Reviews Molecular Cell Biology, 2005, 6: 363-375.
[20] Vega L R, Mateyak M K, Zakian V A. Getting to the end: Telomerase access in yeast and humans. Nature Reviews Molecular Cell Biology, 2003, 4: 948-959.
[21] Wilusz C J, Wilusz J. Bringing the role of mRNA decay in the control of gene expression into focus. Trends in Genetics, 2004, 20: 401-207.